Time and Time Again

Supplements to

The Study of Time

VOLUME 1

Time and Time Again

Reports from a Boundary of the Universe

By

J. T. Fraser

BRILL

LEIDEN • BOSTON
2007

This book is printed on acid-free paper.

On the Cover: Caspar David Friedrich, *Wanderer Overlooking a Sea of Fog*, (1818).
It is reproduced by kind permission of the Hamburger Kunsthalle.

Hamburger Kunsthalle / Stiftung offentlichen Rechts, Glockengiessenwall / 20095 Hamburg / Telefon ++49 (0) 40-428-131-200.
E-mail info@hamburger-kunsthalle.de / Internet www.hamburger-kunsthalle.de.
FOTOBESTELLUNG: SEIT JULI 2002 NIMMT DAS BILDARCHIV PREUSSISCHER KULTURBESITZ EXCCKUSIV DIE VERWERTUNG VON NUTZUNGRECHTEN AN FOTOS DER HAMBURGER KUNSTHALLE WAHR.

A Cataloging-in-Publication record for this book is available from the Library of Congress.

ISSN 1873-7463
ISBN 978-90-04-15485-8

PRINTED IN THE NETHERLANDS

For Jane
my wife and beloved companion

CONTENTS

ACKNOWLEDGEMENTS

The Essays of J. T. Fraser in this volume are reprinted by permission and with appreciation of the author and the publisher from the following sources.

"The Whir and the Bell" unpublished

1. "Complexity and its Measure," in J. T. Fraser, *Time, Conflict, and Human Values*, pp. 235–242 & 290–292. Urbana and Chicago: University of Illinois Press, 1999.
2. "Out of Plato's Cave: The Natural History of Time," *The Kenyon Review*, NS2, Winter, (1980), pp. 143–62.
3. "The Extended Umwelt Principle: von Uexküll and the Nature of Time." Jakob von Uexküll: a Paradigm for Biology and Semiotics" Kalevi Kull, ed. *Semiotica*, (134–1/4, 2001), pp. 263–74.
4. "Mathematics and Time," in *KronoScope*, Vol. 3. No. 2. (2003) pp. 153–167.
5. "The Secular Mystery of the First Day," in Walter Schweidler, ed. *Zeit: Anfang und Ende*, Sankt Augustine: Academia Verlag, 2004. pp. 431–445.
6. "From Chaos to Conflict," in *Time, Order, Chaos—The Study of Time IX*, J. T. Fraser, Marlene Soulsby and Alexander Argyros, eds. Madison, CT: International Universities Press, 1998, pp. 2–17.
7. "Clockworks Beyond Themselves," introductory essay to the catalog of the *Clockwork* exhibit at MIT's List Visual Art Center, 1989.
8. "Space-time in the Study of Time," *KronoScope*, Vol. 5, No. 2, (2006) pp. 151–176
9. "Time and the Origin of Life," in J. T. Fraser and M. P. Soulsby, eds. *Dimensions of Time and Life: The Study of Time VIII*, Madison, CT: International Universities Press, 1995, pp. 3–27.
10. "Temporal Levels: a Fundamental Synthesis," in *J. Social and Biological Structures*. v. 1. (1978) pp. 339–355.
11. "Temporal Levels and Reality Testing," in *Int. J. Psycho-Analysis*, Vol. 62 No. 3 (1981) pp. 3–26.
12. "Time, Infinity, and the World in Enlightenment Thought," in *Time, Literature, and the Arts: Essays in Honor of Samuel L. Macey*. T. R. Cleary, ed. Victoria, B.C. University of Victoria, 1994, pp. 192–211.

13. "Human Freedom," in *KronoScope*, Vol. 2, No. 2, (2002) pp. 223–45.
14. "Time Felt, Time Understood," *KronoScope*, v. 3 No. 1. (2003) pp. 15–26.
15. "Change, Permanence, and Human Values," In J. T. Fraser and Lewis Rowell, eds. *Time and Process*, Madison, CT: International Universities Press, 1993, pp. 3–24.
16. "Truth as a Recognition of Permanence: an Interdisciplinary Critique," in *Zeit und Wahrheit*. Heinrich Pfusterschmid-Hardtenstein, ed. Vienna: Ibera Verlag, 1995, pp. 120–31.
17. "Homer, Borges, and the Pied Piper." Unpublished
18. "The Problems of Exporting Faust," in *Time. Science and Society in China and the West*, J. T. Fraser, N. Lawrence and F. C. Haber, eds. Amherst: University of Massachusetts Press, 1986, pp. 1–20.
19. "Time, Globalization, and the Nascent Identity of Mankind," in *Time & Society*, v. 9, No. 2/3 (2000), pp. 293–302.
20. "Hamlet's Castle in Cyberspace," in M. P. Soulsby and J. T. Fraser, eds. *Time: Perspectives at the Millennium*, Westport, CT: Bergin & Garvey, 2001, pp. 1–16.
21. "Reflections upon an Evolving Mirror," in *KronoScope* 4–2 (2004).
22. "Expanding the Universe." Unpublished

ABOUT THE COVER

The painting on the cover is Caspar David Friedrich's *Wanderer Overlooking a Sea of Fog*, (1818). It is reproduced by kind permission of the Hamburger Kunsthalle.

For art historians, it is an example of German romanticism, created by one of its masters who was also a forerunner of the American Hudson River School landscape painters. For the artist himself it was a reinsertion of his transcendental beliefs into the majesty of Nature. For the twenty-first century viewer it is visionary realism. It is an allegory for Man, alone at the boundary of the cosmos, toe-to-toe with the Almighty, asking questions and demanding answers. Such as about the nature of time.

THE WHIR AND THE BELL

Old-fashioned clocks used to whir briefly before they struck the hour. The whir was the sound of self-winding, of the clock getting ready to sound the bell. For the clock watcher the whir and the bell were reminders of the passage of time. Although the subject of this book is time and not clocks, even if clocks do appear now and then as important walk-ons, the whir before the bell is a rich enough image to serve as the title of an introduction to a collection of essays in the integrated, interdisciplinary study of time.

The purpose of this whir—of this introduction—is to sketch the plan of the book, to suggest the stations of the tour the reader will be taking among ideas concerning the nature of time in the physical, organic, mental and social worlds. Each of the chapters focuses on a theme important for the integrated study of time. Each, except Chapter 17, consists of a brief introduction and a substantial, published essay. The themes are as follows.

1. Why we may regard ourselves as participant-observers of the universe, beholding it from a position along one of its boundaries.
2. Why the most sublime aspects of the world are temporal, the least inspiring ones timeless.
3. The meaning of reality for man, beast and matter.
4. What mathematics reveals about the organization of the cosmos.
5. The coming about or origin of time and the universe.
6. Creativity in nature and man, rooted in the capacity of maintaining conflicts between ordering and disordering.
7. Clocks as metaphysical devices that reach beyond themselves.
8. The relation of time to space-time.
9. The origin of life through natural selection, working among the cyclic processes of the physical world.
10. The origin of man: intraspecific selection for planning, memory, and language.
11. Evolutionarily older and evolutionarily newer assessments of the nature of time simultaneously present in the mind.
12. Nature, measure, society and the idea of God: the Enlightenment.
13. Time and human freedom.
14. The tension between time experienced and time understood.

15. The human awareness of passage: the need for seeking the true, the good and the beautiful.
16. Truth and time.
17. Music and time.
18. Time, science, and the global society.
19. Time and the nascent identity of mankind.
20. Time, cyberspace and human values.
21. Whose past is the prologue to the future?
22. Expanding the universe.

The twenty-two essays, as a unit, resemble a polyphonic musical form called the fugue. Fugues combine distinct melodies into harmonic units that form a single, musical fabric. The repeated melody in the music of this book—in its employment of the integrated study of time—is the hierarchical theory of time, a theory in the natural philosophy of time. It is a melody of ideas, harmonized with the backgrounds of different themes, covering many fields of knowing. The reader-listener will come upon the recurring motives of the hierarchical theory of time replayed—restated—in harmony with different specialties of the sciences and with different departments of the humanities.

So far the whir. Now, to the striking of the bells.

1. THE CHANGE RINGING COSMOS

The noble English art of change ringing consists of the ringing of a set of bells of different pitches, according to sequences that follow stable mathematical patterns. A set of rings, so ordered, is called a change. The maximum number of changes possible with a given set of bells is called a peal. When the changes employ five bells, a peal consists of 120 rings. The number of peals that may be rung with twelve bells is 79,001,600. With 15 bells it is 1,307,674,368,000.

The experience of hearing a long peal leaves the listener with a feeling of having been immersed in ceaseless change. The mind seeks a temporal pattern—a repeated melody, a repeated distinct rhythm—but finds none, unless the peal itself ends and gets repeated. In the case of twelve bells, the time needed to ring a complete peal has been calculated as forty years of ringing, day and night. For most of us, this is too long a period to test for the repetition of a pattern.

Change ringing is an acoustic model for the coexistence of being and becoming in the nature of time: the mathematical rule followed by the bell ringers is stable, unchanging, beinglike. But the peal heard ceaselessly changes. It is becominglike. The universe is calculated to have about 10^{81} particles.[1] If we think of the cosmos as change ringing, using each of its particles as a tuned bell, the demiurgos of Plato—the craftsman who created the world according to unchanging laws—will need a long period to listen for the complete cosmic peal rung by all those bells.

It is to the cosmic change ringing, constrained by the laws of nature, that we now turn. We begin with the "primeval atom," a term introduced in 1827 by the Belgian astronomer, cosmologist and priest, Georges Lemaitre, to describe the primeval universe. Preferentially, I will use the term, "the dot universe."

There are good reasons to believe that, in the beginning, the universe was small enough to have passed through a contemporary atom without disturbing it. We will return to the details of this divine dot in "The Secular Mystery of the First Day." (Chapter 5) Though it was small, its mass had to be the same as that of the contemporary universe, which is estimated as 10^{54} grams. Its temperature is calculated to have been 10^{32} degrees centigrade or higher. The geometry of that dot world did not permit it to have a center or an edge. The popular term, "big bang," signifies the start of the expansion of the dot universe.[2]

The big bang is the origin of space, of time and of lawfulness or ordering. Using our sense of time and the period of our earth's revolution around our sun as our unit of time, all this happened some fourteen billion years ago. Ever since, the cosmos has been bringing forth increasingly more complex forms of ordering and, with them, increasingly more intricate conflicts between ways of ordering and ways of disordering. Life itself, as we shall see in "Time and the origin of Life," (Chapter 9) is a stable equilibrium of conflicts between growth and decay.

Just as it was the case for the dot universe, the geometry of the current universe does not permit it to have a center or an edge. But the laws of nature do provide the universe with well-defined boundaries: nothing may be larger than the cosmos or smaller than about 10^{-33} cm., no meaning may be attached to periods longer than the age of the universe, which is about 10^{17} seconds or to periods shorter than about 10^{-44} seconds. Nothing may be colder than absolute zero temperature or move faster than the speed of light.

Since 1981, Gerald Edelman and members of his Neurosciences Research Institute in La Jolla, California, have carried out extensive research on the intricate dynamics of the human brain and made interesting suggestions about its likely evolution through natural selection in its neural population. What they found supports the idea that the human brain is the most complex system known.[3] Thus, to the variables of length, time, velocity, mass and temperature, was added the variable of complexity. The "answers to many of the fundamental problems of the mind," wrote Edelman, "will come from analyzing the complexity of its [the brain's] organization, which is governed by novel ordering principles."[4] Novel that is, in terms of prior knowledge.

The essay that follows employs algorithmic information theory to develop a measure of complexity and, using it, obtain numerical values for the complexity of physical structures and processes, organic structures and processes and for the complexity of the human brain and minding. By "minding" in this context is meant the ability to create symbolic transformations of experience, to create and maintain self-awareness, the ability to speak a language and to perceive the world in terms of open-ended futures and pasts.

The numbers obtained for the different complexities support the earlier thesis that the human brain is the most complex object of the cosmos, more complex than the cosmos itself. They also demonstrate—importantly for the hierarchical theory of time—the stark, well defined distinctness in the complexities of the integrative levels of matter, living matter and minding.

As humans, we are at immense distances from the physical boundaries of the universe: from the limits of temperature, size, mass, speed, length and periods of time. But, by virtue of having human brains we are at—we constitute as far as we know—the upper limit of complexity in nature.

But it then follows that all articulate reports that derive from minding, such as the chapters of this book, are reports from one of the boundaries of the universe. Having thus located ourselves in the cosmos, we will be ready, beginning in the next chapter, to start thinking about time. But first we will consider the distinct and vastly different complexities of the stable integrative levels of nature.

COMPLEXITY AND ITS MEASURE

This essay turns to certain mathematical principles of information and computation and proposes a universal measure of complexity applicable to any system, whatever the nature of its functions. First, however, I sketch some earlier attempts to define and measure complexity.

In a 1951 paper J. W. S. Pringle noted that both individual development through learning and organic evolution through natural selection lead to increasingly complex patterns of behavior.[5] He chose not to define complexity but did remark that whereas the complexity of nonliving systems remains constant, that of living organisms increases. One may refine this claim by pointing to the hidden issue of time scale. During the eons of cosmic history there arose compounds that one would intuitively label complex compared to elemental gases. But the rate of that complexification depends on passive, chance encounters and hence proceeds slowly, whereas living systems complexify as a consequence of their intentional (teleonomic) behavior and thus much more quickly.

In a frequently quoted 1962 paper on the architecture of complexity, H. A. Simon also bypassed a formal definition of complexity, saying only that it has to do with having a large number of parts that interact in non-simple ways.[6] He suggested (1) that complexity always leads to hierarchical structuring and (2) that process descriptions are more natural for complex systems than are state descriptions. The same two issues were the themes of C. H. Waddington's 1977 *Tools for Thought*.[7]

In a 1969 work John von Neumann proposed to measure complexity by "the crudest possible standards, the number of elementary parts."[8] His interest lay in determining the criteria for self-reproducing automata. In two papers P. T. Saunders and M. W. Ho argued that it is the increase in complexity that gives direction to organic evolution.[9] They also suggested that complexity be seen as a measure of the information content of the instructions required to build a system. They added that measuring the complexity of organic systems by their information content must go beyond the information content of the genome and include that of the epigenetic properties of the system.

J. T. Bonner formulated a measure of the complexity of an individual organism by equating it to the number of different types of cells composing the organism; consistently, he equated the complexity of a community to the number of different species existing in it.[10] He noted that size and complexity are positively correlated and that species with more complex members are more prone to extinction. We will encounter this second correlation again when we consider the complexification of information-driven societies.

Two other possible indices of complexity have been recommended by John Maynard Smith and Eörs Szathmary: the number of protein-coding genes and the richness and variety of morphology and behavior.[11]

In mathematics formal treatment of complexities arose in connection with issues of computability. When is a function too complex to be computed? The *Encyclopedia of Mathematics* defines complex systems as those with a "large number of interconnecting elements in mathematical logic."[12] In the same encyclopedia, hierarchical theory is defined as the classification of mathematical objects according to their complexities.[13] This definition in the domain of logic corresponds to what H. A. Simon asserted from an operational approach, namely, that hierarchical ordering is a policy of complex systems.

For reasons that will become clear later, I now turn to algorithmic information theory. An algorithm is a rule that, if repeatedly applied, leads to the construction or recognition of a new structure or process. A thirteenth-century manuscript described algorithm as "the Craft of Nombryng." That craft may involve successive logical decisions, such as in a calculus or repetitions of the same mechanical motions, as when building a brick wall.

Taking advantage of the method of algorithm, the American mathematician Gregory Chaitin formulated the notion of the algorithmic information content of a number series or set. He suggested it as an index of complexity in terms of diversity. A number standing for the minimal set of instructions necessary to construct a series, a set, or an object from its building blocks is the algorithmic information content of the series, set, or object. "In algorithmic information theory the primary concept is that of the *information content* of an individual object, which is a measure of how difficult it is to specify or describe how to construct or calculate that object. This notion is also known as *information-theoretic complexity*."[14]

Let us test the idea. Consider a series of numbers and examine them in the temporal sequence in which they come to your attention. Then, by a thought process that remains unanalyzed, determine whether the numbers are connected by a stable rule. A rule in this case is any instruction that can be used to generate that sequence of numbers and hence represent it. Here are four examples.

Example 1: "1, 2, 3, 4, 5, 6, 7, 8, 9, 10, 11, 12…." A rule for constructing this series is "Start with $N = 1$ and then keep adding 1 to obtain the next member of the series." The algorithmic information content, used as a measure of the complexity of a series, is the number of elements in the formula that instructs the computer to calculate the members of the series. A suitable such formula is "$N_{next} = N + 1$," starting with $N = 1$.[15]

Example 2: "0, 3, 8, 15, 24, 35, 48, 63, 80, 99, 120, 143. . . ." The instruction that suggests itself is "$N_{next} = N^2—1$," starting with $N = 1$.

Example 3: "3, 1, 4, 1, 5, 9, 2, 6, 5, 3, 5, 8. . . ." Inspection fails to reveal any regularity.

Example 4: "513, 5, 13, 0.0116, –8.4957, 1, 69, 4.83, 10, 0,–256, 65.2. . . ." Here, too, inspection fails to reveal a rule.

Let us assume that each of the four series contains 3 billion members. The first two series could be represented by the short formulas shown. The third and fourth series seem random, lawless, unpredictable, and hence in a sense free. If that is the case, then the shortest way series 3 or 4 could be represented would be to write down all 3 billion digits. The first two series are said to be algorithmically compressible; the other two are not.

Someone might note, however, that the twelve numerals of example 3 are the first dozen numbers of the number π, which has been calculated to 4 billion digits. Though random by inspection, it could still be reduced to operational instructions through a suitable formula. Here is one developed by the Indian mathematician Srinivasa Ramanujan:

$$\frac{1}{\pi} = \frac{\sqrt{8}}{9801} \sum_{n=0}^{\infty} \frac{(4n)!}{(n!)^4} \frac{[1103 + 26390n]}{396^{4n}}$$

Counting the number of symbols, example 1 has a complexity of 5; example 2 has a complexity of 6. The information theoretical complexity of example 3 is more difficult to determine because it depends on the computer program used, but a reasonable figure is 70.[16] The complexity of example 4 is 3 billion: its members are completely unpredictable and hence may not be represented by anything shorter than the whole number series.

We are now ready to use the principle of algorithmic information content to help compress into a simple measure the diversity of objects that make up an integrative level. This task may be accomplished because there is an isomorphism between (1) the compression into algorithms of sets of numbers that first appear to have been randomly selected from the infinite store of numbers and (2) the compression into actual objects of sets of elementary particle-waves, objects that first appear to have randomly come about from the infinite store of potentialities resident in primeval chaos.

The compression of sets of numbers into algorithms is the work of mathematicians. The compression of the variables of physical processes into equations is the work of mathematical physics.[17] The compression of potentialities resident in absolute chaos into actual objects has been the work of inorganic and organic evolution.

If (1) we can measure the complexity of a set of numbers by the length of the shortest algorithms into which the rules of their construction may be compressed, then (2) we may also measure the complexity of an organizational level by the minimum number of distinct building blocks into which nature has compressed the infinite potentialities of primeval chaos when it constructed those integrative levels.

The process described under (1) takes place in the minds of mathematicians. This is mental learning. The process described under (2) is evolution by natural selection. This is physical and biological learning. In both processes the most parsimonious account of diversity is employed as a measure of complexity. Let us, therefore, seek numbers that represent the diversity of the building blocks of each integrative level.

The chaotic, atemporal substratum of the world comprises objects that travel at the speed of light. The set of such objects is small: the photon, the neutrino (once it is observed), and perhaps the graviton. The range 1–5 seems to be that of the diversity of the primeval chaos, though physicists have been working hard to reduce it to 1. The number 10^1 may then be assigned to that world as its index of complexity.

The next higher organizational level, the prototemporal, comprises particle-waves that travel at speeds less than that of light. The behavior of all these objects may be accounted for—at this time—in terms of six different quarks, six different leptons, and the antiparticles of each.[18] But there seem to be other objects associated with these forces of nature that hold the quarks and leptons together. If we include all those which appear in high energy interactions, their total number may be estimated to be 500.[19] This diversity may be represented by the range 10^2–10^3.

The stable constitutive objects of the macroscopic world are the chemical elements and their compounds. The 1998 *CRC Handbook of Chemistry and Physics*[20] lists over 15,000 organic compounds; *Lange's Handbook of Chemistry*, 3,000 inorganic compounds.[21] Diligent chemists could probably identify many others, suggesting a range of 10^4–10^6 as a reasonable one for the diversity of stable objects from which the macroscopic physical world above the quantum level has been constructed.

The classes of different building blocks from which the physical world is constructed are countable with relative ease because they comprise objects that, within their species, are indistinguishable. This is not the case for objects that make up the world of living organisms. Since no two living creatures are exactly alike, to obtain a measure of the algorithmic information content of the integrative level of life, one must count all distinct organisms that ever lived. Instead of using the type of accounting that was suitable for examples 1, 2, and 3, we must use the type of accounting that was necessary for example 4. We must estimate the total number of distinct organisms that ever lived.

For sexually reproducing species the counting of each individual as a different organism is well warranted. But just what constitutes an individual and what constitutes a society become problematic as one descends toward the simpler forms of life. Still, the kind of identity found among physical particles of the same species does not exist in the domain of life: every bacterium is in some ways different from all others of the same group.

In any event, the number of described species, sexually reproducing or otherwise, is estimated at 13,620,000; the number of species alive today is estimated at 18,375,000.[22] The number of species that ever lived is estimated at 100 million.[23]

The total number of organisms larger than insects is insignificant compared to the number of insects, hence it is enough to count insects. Among the 750,000 known insect species, there are some 10,000 species of ants. The number of ants at any instant is estimated as 10^{15} individuals.[24] Ants live for six months to a year[25] and have been around since the Paleozoic era, that is, for the last 260 million years. This leads to a head count of about 10^{23} ants that ever lived, take or leave a few trillion. But ants are vast and rare creatures compared to the microorganisms found inside each ant, guessed to be of the order of 10^9 per ant.[26] It is enough, then, to count those microorganisms. This comes to a total of 10^{29}, take or leave a few quadrillion.

Let us check the reasonableness of this number by starting again along a different tack. The total number of arthropods per hectare (ha) of soil is estimated at 2×10^9. Those above ground add another 50 percent, suggesting a figure of 3×10^9 per hectare.[27] Since the total land area of the earth is 1.5×10^{10} ha,[28] the total number of arthropods, neglecting the variability in density, is of the order of 4.5×10^{19}. Assuming that they have also been around for 260 million years, we get a total arthropod population, alive now or previously, of around 10^{28}.

The total number of microbes per hectare of land is estimated as 7×10^{18}. If we include microbes living in ants, other arthropods, and other animals, a figure of 1.4×10^{19} microbes per hectare is reasonable.[29] For the surface of the earth this is 2×10^{29} microbes at an instant. How does one extend this synchronous figure of today diachronically to the age of microbial life? Perhaps by taking a cue from bacteria. In the life cycle of bacteria the periods of reproduction vary between fifteen minutes and sixteen hours, and bacteria existed since Devonian times, that is, for about 400 million years. I assume an average two-hour reproduction period and also a linear growth in numbers as a first approximation to a better informed weighted figure. This yields 10^{12} generations of bacteria. Assuming the same number of generations for microbes, we obtain 10^{41} as the total number of microbes that ever lived. Since this is twelve orders of magnitude above the total number of arthropods, the number of arthropods may be neglected.

The mass of the oceans is 1.4×10^{24} grams or, to first approximation, that many cubic centimeters.[30] There are an estimated 10^{8}–10^{10} bacterioplanktons per cubic millimeter of water. For the earth's seas this makes for 10^{32}–10^{34} bacterioplanktons. But the productive regions of the oceans are the upper 30 meters; there is hardly any productivity below 100 meters. Although the ocean area represents twice the area of the land, biological productivity on land is nearly three times that of all the oceans.[31] For this reason, I assume that the total number of planktons that ever populated the seas is less than the number of microbes and hence makes a negligible difference to the grand total.

Allowing for a considerable margin of error, the complexity range of the biotemporal world may therefore be given as 10^{40}–10^{50}.

The figures used in the calculation of organic complexity were either generally available and judged reliable (at this writing) or obtained as cited. But the responsibility for daring to carry out such a divine arithmetic is entirely that of the author. Fortunately, as will be seen later, even substantial variation from the 10^{40}–10^{50} figure would not affect the conclusion of the reasoning.

The next higher integrative level, the noetic, is the product of the minding functions of the human brain working through the body and in cooperation with other brains. Placing a number on the complexity of cultures may appear to be a hopeless task. But determining the diversity of states that the brain may occupy may not be. And surely the number of those states must be some kind of an index of the potentialities of human imagination.

Still, to seek the diversity of those states reminds one of the biblical figure of speech, the counting of the stars. Who could do so? Perhaps the Lord, because, says Psalm 147, "He determines the number of stars, he gives to all of them their names." No mortal ever actually counted all the stars one by one, but astronomers have made estimates. There are about 10^{10} galaxies,[32] with 10^{9}–10^{11} stars per galaxy, making for a total of 10^{19}–10^{21} stars. Likewise, one can make an estimate of the number of possible different brain configurations.

The human brain contains an estimated 10^{10}–10^{12} neurons.[33] This happens to be on the order of the number of galaxies. But whereas the galaxies of the universe do not collectively determine a joint cosmic present—they cannot, by the provisions of special relativity theory—the neurons of the brain do define the mental present through their interactions (see sec. 2.2). Assuming that each neuron is either on or off, the upper limit of the theoretically different states possible for the human brain is on the order of $2^{12^{12}}$ or $10^{10^{12}}$.[34]

If each neuron were directly connected to all other neurons (neurons are not so connected), then this numeral would represent the theoretical upper limit of different global brain states. But each neuron is connected through its dendrites (impulse transmitters) and its axons (impulse receivers) to only 10^2–10^4 synapses.[35] This leaves a mere $10^{10{,}000}$ different possible configurations, which may then be thought of as that many different brain states. Brain states, one or more at a time, processed simultaneously or in meaningful order, may form an unlimited number of states of mind with their wisps of thought. Out of this unbounded store arise our fears, hopes, memories, and perceptions. Human values, then, help to select from among their fellow images those fears, hopes, memories, and perceptions that are—or promise to be—useful in maintaining personal and collective identities. The lengths of the shortest algorithms to which the selection rules may be compressed may then represent the measure of complexity of the noetic umwelt. These rules—so many different process descriptions of the brain—are explored and tested through the arts, the letters, the sciences, "and the historical struggles of humankind."[36]

The human brain, being the most complex object in the known universe, cannot be modeled by anything less complex than itself. Any simpler model would lose those properties of the brain that depend on its complexity, such as the ability to construct personal identity, speak a human language, or perceive the world in terms of noetic time.

Let me now summarize the conclusions of this applied speculation based on informed guesses.

Complexity of the Integrative Levels Measured by the Diversity of Their Building Blocks

	Complexity
The chaos of radiative energy	10^1
Subatomic and atomic particle-waves	10^2–10^3
Integrative level of massive matter	10^5–10^6
Organizational level of life	10^{40}–10^{50}
The human brain and its noetic world	$10^{10{,}000}$

Recall the Socratic discussion of what a bee is, that is, the character in respect to which bees do not differ from one another. Similarly, we might seek the character with respect to which members of an integrative level do not differ among themselves but differ from members of other integrative levels.

They share the same range of complexities. If a diligent survey of nature turns up an object whose functions and structures are akin to no more than 10^2–10^3 other objects, then you are dealing with a subatomic or atomic particle. If you can identify in your scheme of ordering 10^{40}–10^{50} other objects whose functions and structures are akin to the one you are considering, than you are observing a living organism.

That the world of human thought is in some hard-to-define way more complex than are biological processes, which in turn, are more complex than physical processes, is a rather old idea. What is new in the conclusions of this appendix, represented by a numerical summary, is the suggestion that the ranges of the complexities of the integrative levels of nature, if defined and measured as suggested, are distinct and widely separated. The conclusion is that the degree of complexity of the human brain (again, if defined and measured as suggested) is some 10,000 orders of magnitude greater than that of any other structure, including the universe at large.

2. FROM TIMELESSNESS TO TIME

Homo erectus appeared 1.5 million years ago, archaic sapient humans 300,000 years ago, fully modern humans 50,000 years ago. The earliest habitations appeared six or seven thousand years ago. A case can be made for the claim that the foundations of Western science are rooted in the Platonic theory of knowledge that sees a sharp division between time and the timeless—when it is combined with Christian theology, that sees the same sharp division. These roots of the family tree of the globalizing mankind of the 21st century may be dated as the turn of the fifth century B.C. When Plato's theory of knowledge is represented by his parable of the cave is combined with his image of a divided line—both explained in the essay that follows—they may serve as a visual-narrative metaphor for the immense human journey.

That journey, Socrates would say. is one from the darkness of the temporal to the light of the timeless. Through the Socratic dialogues Plato bequeathed to Christianity and, through Christianity to Western civilization, the preference of timelessness to time, of the eternal to what is passing. He maintained, for instance, that the true, the good and the beautiful, as we perceive them on earth, are but poor copies of their sources and paradigms: the ideals of eternal values of the true, the good and the beautiful. Twenty centuries after him Spinoza reasserted this view when he wrote that it is only "love toward a thing eternal and infinite [that] feeds the mind wholly with joy, and is itself unmingled with any sadness."[1] As I implied earlier, the Platonic dichotomy between time and the timeless came to be built into the foundations of the sciences as well as into all Western philosophies and religions.

The essay that follows maintains that the evolutionary journey of man may still be described as one from darkness to light but not one from time to the timeless. Rather, it is one from the primitive reality of whatever is timeless, to the immense wealth of whatever is temporal. The least significant aspects—the electric charge on the proton, the speed of light—are eternal, are timeless. The most sublime aspects of the human world—love, beauty, knowledge, our lives—are temporal, are passing.

OUT OF PLATO'S CAVE: THE NATURAL HISTORY OF TIME

The Project

The Platonic theory of knowledge, represented by the metaphor of the cave, is probably the most significant contribution of Antiquity to the Renaissance birth and subsequent growth of Western science. Plato's dichotomy between timeless forms and the temporal world of the senses is implicit in the character of scientific law. However, with the advance of evolutionary biology, psychology, and social science increasingly more doubt has been cast on the validity of any theory of knowledge which sees the world as divided into the temporal and the timeless. Such a division cannot accommodate the post-Darwinian, post-Freudian, and post-Einsteinian understanding of matter, life, and man. In this essay I wish to outline a new theory of time and knowledge, one which is consistent with the evolutionary view of the world.

In Plato's *Timaeus* we find a creation story which depicts time as an impoverished image of the eternally revolving and hence timeless heavens.

> The nature of the ideal being was everlasting, but to bestow this attribute in its fullness upon a creature was impossible. Wherefore [the father and creator] resolved to have a moving image of eternity, and when he set in order the heaven, he made this image eternal but moving according to number, while eternity itself rests in unity, and this image we call time.

Time, however, can no longer be understood as simply related to eternity by contrast. Our age demands a richer metaphor, a new theory which can accommodate our understanding of the world and yet retain a place for the dignity of man.

I argue that time is not a one-way thrust in which all the phenomena of the world equally partake, but a hierarchy of temporalities ranging from human time to the timeless. Each temporality has its distinct qualities and each is associated with one of the stable integrative levels of nature. I will present my case in the form of an epistemic-literary metaphor, as Plato presented his in the narrative of the cave.

Plato's Cave

In the seventh book of *Plato's Republic*, Socrates speaks about a cave where prisoners have been chained to the ground and so restrained from motion that they can see only the wall ahead of them. At some distance behind them a fire is burning. Between the fire and the prisoners, men carry implements that have the silhouettes of the furnishings of the world: of objects and living things. These cast shadows upon the wall. Thus, the visual world of the prisoners consists entirely of shadows which move, merge, separate.

Imagine, says Socrates, that one of these prisoners were freed from his fetters, compelled to stand up, and then dragged up to the mouth of the cave so as to see the sunny world of Greece. His eyes would first be filled with sunlight so that "he would not be able to see even one of those things we call real." In the course of his later adjustment he would first discern shadows, then the images of men and women reflected in water, later the things themselves in the sunlight and, at last, the sun itself. He would then realize that his former world was only a projection of what he now knew to be reality.

Likewise, says Socrates, the soul must ascend through contemplation from the world of the sensible into the world of the intelligible. There it may see, from a distance, the idea of the good, which is the cause of all things, including the right and the beautiful. With the good, the soul will also discern "the author of the light" of the intelligible, "the authentic source of truth and reason."

Plato's theory of knowledge is embodied in this metaphor of the cave. At its foundation lies the division of the world into time and the timeless, with a grey, uncertain region between them. He portrays all things of the intelligible and visible world arranged along a vertical line which may be imagined as connecting heaven to earth. On the top of the line are the eternal, timeless ideas, the unchanging forms. Underneath them are likenesses of these ideas, such as actual geometrical figures drawn on a slate. Further down come animals, plants, and the whole class of objects made by man. These are all temporal. Time itself, identified with what is transient and changing, is but a poor image, an earthly projection in our world of appearances, of the timeless, unchanging, eternal forms of heaven.

Sensible things on earth, in our world of time, remind the soul not only of what it already knows from prior existences, but also of what it cannot know from prior sense experience, namely, timeless truths. The destiny of the soul is to climb from the dark, the sensible, the temporal toward the luminous, the intelligible, in short, the timeless.

The argument of this essay is that Plato's theory of knowledge is backward. It is the timeless which comprises what is dark in the mind, what is primitive, determinate and unfree. It is the temporal wherein intelligibility and light reside; only the temporal is open to creation and freedom. The liberation of the soul is a ceaseless journey from the timeless to the temporal. Plato's divided line, connecting heaven and earth, is upside down, so far as time and the timeless are concerned. We intend to turn it right side up, arguing that the architecture of the world demands this renovation.

A Quiver of Arrows

We have to gather the conceptual tools necessary for a visit to the revised version of the cave. These tools are visual metaphors, sketched here in their simplest forms, stated categorically and without detailed justification. The reasoning that has led to them is based on extended inquiries into the role of time in physics, biology, depth psychology, social history, and the arts and letters.

Let us imagine the picture of an arrow drawn on a sheet of paper and let this image stand for our conscious experience of flying time, of our passage from cradle to grave. I shall call this kind of time *noetic.* A world characterized by it I shall call *nootemporal.* Such a world includes our human environment of language, feeling, and civilization. The head, tail, and shaft of the arrow are well defined: futurity, pastness, and presentness are significantly distinct.

But suppose the head and tail of the arrow were small and ill-defined. The picture then becomes a metaphor for *biotemporality,* of a frame of temporal reference which consists mainly of a broad presentness with futurity and pastness appearing only as hazy and distant edges to the present. I am talking about the temporal world of advanced animals, of infants, and of the instinctual, unconscious functions of the mature mind.

If the arrowhead and tail are totally absent from our image, we are left with the shaft of the arrow, a line. Let this represent an environment wherein, although not everything happens at once, the directional quality of time is, nevertheless, absent. Such a temporality often infuses our dreams. Their manifest contents resemble the second reading of a book: though everything is already known, somehow all events do not emerge instantly. I have called such worlds *eotemporal* for Eos, goddess of dawn.

The universe of stars and galaxies, of massive inanimate bodies in general, is eotemporal. There is nothing in physics that could be used to define a "now." (That comes, as we shall see, with the appearance of life.) But future and past make sense only with respect to a present. Quite consistently, the equations of physics are insensitive to the direction of time. For instance, you have a set of equations which describe the revolution of the planets around the sun. You ask this: "If conditions at $t = 0$ are such and so, where is the planet at $t = +25$ minutes?" You will obtain an answer. Now let time flow backward by making it go negative and ask the question where is the planet at $t = -25$ minutes?" You will obtain the same answer. It is irrelevant whether you assume a flow of time from 5 PM to 6 PM or from 6 PM to 5 PM. What constitutes past and what constitutes future—and what constitutes the present—is something the physicist, in his work of translation from the inorganic to the human world, must supply from conditions that have no representations in the equations of physics.

Let us return to the picture of the shaft of the arrow. This shaft may further disintegrate into slivers of wood, fragments of time. Such *prototemporal* conditions are characteristic of certain pathological states of the mind: events just do not hang together. The same kind of temporality is associated with the environment of elementary particles. In the world of atomic particles it is not possible to say precisely *when* is *then*. Temporal positions can only be stated in statistical, probabilistic terms. And this is not a matter of our ignorance: at this integrative level of nature the world itself is ill-defined.

Once upon a time, on a Scottish heath, two men encouraged three witches to tell the future of individual grains out of a bushel of identical grains, as indistinguishable as two electrons (or two "voters" or two "Labor Day motorists" from the point of view of statistics):

> BANQUO If you can look into the seeds of time
> And say which grain will grow and which will not,
> Speak then to me, ...
> FIRST WITCH Hail!
> SECOND WITCH Hail!
> THIRD WITCH Hail!
> MACBETH Stay you imperfect speakers,
> tell me more: ...

As the witches, so prototemporal universes are imperfect speakers. They are not supernatural mysteries whose laws we have not yet learned but on the contrary, early stages of evolutionary development which cannot yet tell "*when* is *then*" more precisely than in probabilistic, statistical terms.

Back to the arrow, once again. Finally, even the fragments of the shaft may vanish and we are left with a blank paper, the symbol of *atemporal* conditions, better known as chaos, which is space without time. Einstein taught us that for an imaginary observer, riding on a photon, time disappears. Consequently, a universe of pure light is atemporal, for the only thing one might be doing in such a universe is riding on photons. But one may also identify other atemporal conditions, such as in the human, subjective sphere. For instance, two stimuli given to a person within about two milliseconds will appear to him as one single stimulus. That period, from the point of view of the perceptive faculties of the subject, can have no temporal structure; it is atemporal. Analogous examples are known to exist in physiology. Atemporality, like eotemporality, can also be experienced in dreams: events that happened long ago in the dreamer's life usually appear in dreams as figures or things far away in space. The primitive levels of our minds, where dream images are formed, know nothing about the kind of human past that makes up nootemporality. There is nowhere the dream work can place the "long ago" except at an atemporal, purely spatial distance.

When the world is seen as a hierarchy of temporalities, the conclusion is unavoidable that time itself has developed along evolutionary steps. Plato would have found such an idea distasteful, uncouth. So does the routine custom of our language. There is no noncontradictory way in which to state that time evolved in time. But difficulties of prevailing linguistic customs have not stopped poets from singing their songs, because they felt and knew that language has a capacity to extend beyond its formal limits. That is why George Steiner can insist that a genuine linguistics cannot be formal or rigorous, that language is always only partial, and that this uncertainty is both intrinsic and necessary. We shall take advantage of this freedom and discuss the unthinkable.

The Umwelts

Thomas Hardy remarked in *Far from the Madding Crowd* that "the limitation of the capacity is never recognized as a loss by the loser therefrom." Six centuries before him, Thomas Aquinas wrote that "whatever is known is known according to the knower." Identical views were expressed in scientific form in the work of the German biologist Jakob von Uexküll early this century. He drew attention to the fact that an animal's receptors and effectors determine its world of possible stimuli and actions. Thereby, they determine the animal's universe. He called such a species-specific universe the *Umwelt* of the species. What is not in that *Umwelt* must be taken as nonexistent for the members of that species. For instance, ultraviolet patterns on certain butterflies exist for other butterflies but not for vertebrates; vertebrates have no sense organs through which they could read those patterns. What an earthworm cannot know might kill it but it still won't know what hit it.

In modern psychology, umwelt (without the italics and capital) is defined as "the circumscribed portion of the environment which is meaningful and effective for a given animal species and that changes its significance in accordance with the mood operative at the moment." Note that the "environment" of which the animal's umwelt is a portion is our own human umwelt. By means of instruments capable of translating the language of other umwelts into our own, we can uniquely expand our own umwelt. We know of the ultraviolet patterns on butterflies because we make photographic plates sensitive to ultraviolet rays. We know about the time of animals because we experiment. We know about the time of atoms, or of the universe, because we write equations which depict their behavior and those equations tell us about their temporalities. We are thus authorized to talk about the temporal umwelts of matter, beast, and man and assume them to be, so to say, really real. The nootemporal umwelt of man, as I have already argued in the case of dreams, must include not only the reaches of our conscious capacities but also those of our biological and unconscious functions. The lower temporal umwelts are always with and in us; we carry the history of time. They infuse the manifest contents of dreams; they may be identified in conscious and unconscious fantasies, in artistic creations, in the methods and teachings of science; and they are brutally evident in certain psychopathic experiences.

The Cave Revisited

The prisoners of Plato's cave were born in that abyss; they reached their adolescence in complete darkness and have never even seen so much as a patch of light. In *Timaeus* the Demiurge, or Craftsman, orders and arranges the physical world, bringing it into conformity, so far as possible, with the most rational preexisting pattern. In my revised version of the cave he is only a guard who watches over the prisoners.

—In that cave of creation, in the blackness of its depth, for unfathomable reasons of his own, the Demiurge decided to light his pipe. Suddenly, there was a spark. If we may paraphrase the Johannine Gospel, the darkness did not comprehend what had happened, but life did. As the prisoners beheld, on the cave wall, certain ill-defined areas of light, they became filled with fear as well as with immense, inarticulate hope.

Plotinus tells us that time, before it was time, lay "in the Authentic Existent together with the Cosmos itself." There also existed, he continues, an active principle of nature compelled to realize and govern itself

> and it chose to aim at something more than its present: it stirred and the Cosmos stirred with it. And we (the active principle and the Cosmos), stirring to a ceaseless succession, to a next, to the discrimination of identity and the establishment of ever new difference...produced time.

The prisoners conspired, they broke their chains, they rose, they revolted. They felt that they would never be able to rest until they found the source of the spark. The Demiurge, hearing the commotion, lit a candle so that he himself might be able to see the results of his first act. What he saw made him retreat, remaining out of reach of the prisoners-run-wild. In the flame of the candle the rising prisoners began to discern the play of light and dark. But instead of satisfying them, what they saw only increased their fury. A band of brothers, they began ascending along the upward sloping cavern, ignorant of specific goals, driven only by an inner command to reach out.

—The wall of the cave, which is also the mind of modern man, retained the history of nature, in the form of dreams and infantile and subconscious imagery. From the wall—from their own minds—the marching brotherhood read the history of time.

Deeds Without Names

At the instant of the Big Bang the universe began to expand. This does not mean expansion into preexisting space but rather the creation of such conditions as would permit the coming about of novelty; that is, it meant the creation of time. The universe which started to expand was an atemporal chaos of pure light. But in an atemporal umwelt, color can be given no meaning because there is no time in which frequency may be gauged. Therefore, in the beginning was darkness, for even white is a color. Up to our own day the substratum of the universe remains atemporal: there are a hundred million times more photons than all other particles taken together.

Prototemporal umwelts were born out of atemporal chaos when some of the light energy began to take particulate form. In our own day, in the atemporal ocean of pure light, we find vast regions of gas and dust. In such regions space is distinguishable from time, although events and things may be interchanged. An electron, for instance, is either a thing or an event, depending on the way we inquire into its character. The two modes—events and things—have taken on distinct characteristics.

Eotemporality came about with the formation of galaxies. When we observe the sky, we see our own as well as millions of other galaxies. Though each is enormous in human terms, as a class of objects they fill only one-millionth of the volume of the knowable universe. Galaxies are isolated regions of space. Within their umwelts time ceased to be fragmented but it did not yet acquire an arrowhead.

The Birth of Life, of the Now, of Aging, of Death, of Etc

All atoms and molecules are clocks; a crystal is a delightful clockshop. Somewhere in the small hours of the earth, when that spinning ball was only about a billion years old, certain crystals acquired a degree of autonomy. They were able to absorb energy at certain available frequencies and use it to maintain oscillations that were in harmony with the cyclic changes of their local environment. Those which could do so had a better chance to preserve their existence as miniature, integrated clockshops than those which could not do so. Through this developmental step, natural selection replaced chance as order in structure came to be supplemented by order in function. Life was born when these primordial clockshops acquired the capacity to model, with crude accuracy, the cyclic patterns of their environments.

In due course the inorganic chemistry of the crystals was replaced by the DNA-RNA-protein system of modern organisms. Our crystalline ancestors bequeathed to us their structures, their capacity to replicate (both may be identified in the genetic system of life), but not their chemistry.

The clocks of these tiny clockshops had to function in ways that were mutually supportive rather than disruptive. A degree of internal coordination had to evolve to insure viability. If the autonomy of these transitional forms was to be maintained, it was necessary to insure by internal means that some events did, and others did not, occur simultaneously. It is these necessities which constitute a definition of "now" in the presentless umwelt of the physical world. It was thus that life upstaged matter. Biotemporality came about when *praesens tempus* was added to *nunc stans*—present time to the abiding instant. With presentness among the features of the universe, futurity and pastness became potentially meaningful.

The "nowness" of early life was not the hurry-up-instant of an American businessman; it had to be measured in millenia. Even today, it takes two years before the roots of a felled birch tree learn that, in fact, they are dead. The trees in *The Lord of the Rings* knew this kind of "now." They spoke Old

Entish, which, said Treebeard, is a lovely language, "but it takes a very long time to say anything in it, because we do not say anything in it, unless it is worth taking a long time to say, and to listen to."

The biological clocks of modern organisms cover an immensely broad spectrum of frequencies. For instance, the human skin responds to ultraviolet rays that oscillate (tick) ten-million-billion times a second. Most organisms pick up heat that ticks (oscillates) ten-thousand times a second. That is, one cycle lasts for one ten-thousandth of a second. One cycle of a neural signal may last anywhere from three to ten seconds. Probably all living things have diurnal cycles, that is, they oscillate once in every twenty-four hours. Lunar periods of twenty-eight days may be found in many organisms. Many plants and animals show yearly cycles, and some bamboos flower once every seven or eight years.

The clock supply of the early shop was surely rather limited in its spectrum. But the earth already had a large store of cyclicities. Boulders vibrated in response to the frequent earthquakes. Thunders shook the atmosphere in the audio frequency region. Light and electric fields oscillated at many frequencies. Daily rhythms governed temperature variations, changes in the sea's salinity, the intensity of surface illumination. Under the selective pressure of such cycles, the clockshops began to evolve. The widening range of frequencies demanded improved internal coordination among the clocks. Like the uncontrolled spread of chain stores, life began to proliferate.

Biological clocks that had to oscillate at very different frequencies had to have very different forms. The retinal clock, which, in our eyes, responds to electric oscillations (light rays) must be different from the hormonal clocks that control daily rhythms or lunar periods. The widening spectrum of oscillations has forced upon life a division of labor. The upper limit of the spectrum was set by the physical world around the ultraviolet frequencies already mentioned; only gamma rays oscillate faster and, thus far, they have not been found to be important to the living process. Going now the other way, toward the lower limits of the spectrum, we find oscillations which get slower and slower until they reach a limit of linear, nonoscillating change. We know this lower limit as the process of aging. It is a mode of organic change which evolved some time after life itself was born. Aging is not a necessary corollary of life. Early life was purely cyclic, and bacteria still are, as is the DNA. Primitive life forms may die as swiftly and often on average as more advanced forms, but they do not die by aging but by accident. With aging came about the inevitability of nonaccidental death, together with the fateful polarization of futurity and pastness. With the birth of death-through-aging, there evolved a divison of labor between the aging and reproducing tasks of life.

In advanced organisms the body evolved into a throwaway soma entrusted with most of the regular functions of biological existence, while the task of the continuation of life was relegated to such practically unaging structures as the DNA-RNA system that makes up the genes.

Every individual human being, in his or her lifetime, once resided in the depth of the cave. The sense of time in children and even in newborn infants has been investigated, but no one has treated the history of time that precedes the day when the infant first demands periodic satisfaction. The sense of time of the newly born, even though it resembles the time perception of advanced animals, is the end product of a process of growth which begins with the clockshop of the DNA. All of us, at the instant we were conceived, were little more than molecular structures embedded in a small but sympathetic environment. Our umwelts were hardly distinguishable from the world of sticks and stones and dry bones, and chemicals on the shelf.

It has been an awe-inspiring journey for livingkind, from the epoch when environmental and astronomical cycles first came to be mapped onto the mechanisms of those early clocks, barely alive, up to our own days when human infants are born, ready made, with their stunningly varied complements of biological oscillators. Although the developmental process itself evolved and hence changed, the stages of the evolution of time remain on the record and are replayed each time a child grows up in the mystery of the womb. As we learn from T. S. Eliot's "Burnt Norton".

> The dance along the artery
> The circulation of the lymph
> Are figured in the drift of stars.

The evolution of life generally followed a path of complexification. Living things began to build into themselves charts of their external worlds, thereby establishing their umwelts. These charts were useful for predicting the likely behavior of food, friend, and foe. The very broad present of early life narrowed, futurity and pastness became polarized and separated by a hazy present. Just yesterday (on the evolutionary scale), in the most advanced species, anticipation replaced expectancy and became a crucial tool of survival. Even later, long term memory evolved, adding its potentialities to those of simple recall. As organic evolution became faster, the methods available to it (natural selection working on random mutation) proved to be inadequate to produce the adaptive changes demanded by the very success of that evolution. According to the record on the cave wall, it was at this level that in one or more members of a family of organisms, known as hominids, the organization of their inner control system—the one that coordinated all their clocks and is better known as the nervous system—crossed a certain crucial threshold of complexity.

Brain and Mind

The physical world has certain absolute limits. Examples are the lowest possible temperature, the fastest possible speed, the largest possible object, the smallest meaningful unit of distance, the shortest meaningful unit of time. Let us add to these another limit, one that seems to have been reached through the functions and structures of living matter: that of the maximum limit of complexity.

By complexity I mean the number of distinct states that an autonomous system may assume. The lantern I use for camping is such a system: it is either on or off. Its complexity is characterized by the number 2. It has been calculated that at any instant the human brain may be in any of its possible 10^{10^9} states. (This numeral signifies the number 10 followed by a billion zeroes). I believe that the immense number of distinct states that the human brain may assume represents a boundary to the complexity of living matter. In this view, the human brain is more complex than any other known structure, including that of the physical universe. As humans we are not even close to the boundaries of the physical world: our temperature is nowhere near absolute zero, our size does not approach the largest possible size, the speed of our vehicles is well below the speed of light; yet by virtue of possessing human brains, we are right at the boundaries of natural complexity.

Consider now that the physical limits can be reached continuously from garden-variety sizes: things can get colder and colder, faster and faster, or smaller and smaller, until they approach the limits. But the laws of these limiting conditions have not been found that way; these boundaries are so far out that their laws are totally unpredictable from our world of fireplaces, spaceships, and baseballs. They had to be—they were—discovered separately and then glued, as it were, to what was already known about the world. So is it with the laws of the human brain. Its rules are unpredictable from anything else we know about life and matter. The hallmarks of nootemporality cannot be divined from what we know about the temporalities of matter and life, any more than what happens at the speed of light or at absolute zero can be divined from familiar speeds or temperatures.

We cannot even talk about the human brain the way we talk about other structures. For instance, it is truly impossible to describe the state of the human brain at an instant in a way one would describe the state of all

American telephone lines at an instant. A "brain state" description would require descriptive apparatus more complex than the human brain—but we know of nothing more complex. The only description possible is in terms of brain *processes*, not of brain states. That is, we can tell people what we think. To do so is to give a process description of one's brain. Such reports constitute the symbolic transformations of human experience. The signs, signals, and symbols of cultural life are metaphoric reports about our brain processes. The agent we usually hold responsible for giving these reports is the mind. It is, itself, a symbolic entity, a process; it is not a thing.

The immense complexity of the brain makes it possible to store in the mind innumerable charts of the human environment, of the self, of things and events that happened, of events and things that could have happened and which may yet happen. More importantly, our mind is also practically an infinite store of maps of events and things which can never happen. Our nootemporal umwelt is determined by the features of these maps, being, as they are, charts of possible and impossible worlds. Travel reports from among the features of these worlds are best rendered in human language.

Language: Its Freedom and Its Inertia

Glaucon was one of the young men listening as Socrates spoke about the cave. "A very strange image you speak of," said Glaucon, "and strange prisoners." "Like to us," replied Socrates and went on with his teaching. The prisoners, he said later, praised those of their number who best remembered the customary "precedences, sequences and coexistences [of the shadows and were, therefore,] most successful in guessing what was to come." Although Plato does not put it in these terms, we must assume that the prisoners were in full command of their sense of time, for they have demonstrated that they knew how to prepare for future contingencies based on the memory of past experiences.

In the post-Darwinian, revised version of the story of the cave, the prisoners should not be imagined as commanding a nootemporal umwelt until after, in their evolutionary history, they have undergone some radical changes. These changes include certain mutually reinforcing processes: the expansion of delayed expectation to long-term anticipation, the development of the sense of selfhood (including knowledge of the inevitability of death), the extension of recall into long-term memory, and the development of the capacity to separate the emotional from the intellectual in their modes of communication. It is the emergence of human language which cemented together these various processes.

Language, in Steiner's words, is "the main instrument of man's refusal to accept the world as it is." The former prisoners were now able to work out behavioral strategies to deal with imaginary conditions. They could experiment with counterfactual pasts and impossible futures. If Creation is identified with the lighting of the Demiurge's pipe, we may call it Station Alpha along the path of evolution. The birth of language and the coemergence of the nootemporal umwelt may then be Station Beta for Babel. It was only after Station Beta that individual responsibility and commitment could emerge from the matrix of instinctual duty. Anxiety and regret on the one hand, and the feelings of futurity and pastness on the other hand, became mutually sustaining modalities of the mind. Time's arrow had finally acquired a sharp head pointing to the future and even a feather for writing about the past.

—After Babel, some of the emerging cavedwellers became obsessed with the idea of survival after death, some others about purpose in the world and order among themselves. Some of them drew circles on the wall and said that these represent everything there is or can be. Some heard harmonies with their inner and outer ears. Some began to inquire into the nature of the Demiurge, forever in retreat at a safe distance. All of them noticed that the light of the candle which the Demiurge carried had come to blend with a more intense patch of light that filled a small area of the distant darkness. It was toward that area that the cavedwellers continued their march.

In the course of its rapid evolutionary emergence, the mind of man learned to protect its integrity against the demands of life as, earlier, life learned to protect its integrity against the demands of matter. Like two neighboring dictatorships, each afraid of those of its own people who sympathize with the other country, and both agreed upon the creation of a no-man's-land between them into which all nonconformists are to be exiled, so there evolved, between the somatic (bodily) and the nootemporal (mental) integrative levels, a buffer zone. Into this region the soma banished those physiological functions that are necessary to maintain the mind; to it also the mind exiled those mental functions which are too close to the soma. Together these functions comprise what we call the instincts; we characterize this region as the unconscious levels of the mind. We know further that the unconscious retains vestiges of its evolutionary history quite intact, unlike the products of biological evolution where earlier phases are usually absorbed in later ones.

The hierarchy of temporalities we carry within us implies a hierarchy of causations—from the probabilistic, to the deterministic, to final causation, to human freedom, to historic causations. And we are subject to a hierarchy of inertias, which are in turn implicit in the causal modes they subtend.

Let us define "inertia" as the tendency of any change to maintain the sense of its prevailing motion. Consider now that all living things are subject to

Newtonian inertia: the rules of mechanics are the same for rolling stones, cats, and people. This fact is important but not especially interesting because life possesses certain freedoms which inanimate matter does not and, hence, it also has its own conservative trends. These are summed up in sociobiology by the term "phylogenetic inertia": the tendency of living things to maintain their peculiar structures and functions. Human language, being the product of living organisms, is subject to phylogenetic inertia. The laws of living organisms impose certain limitations upon the spoken, written, painted, and sculpted word. These restraints are important but not especially interesting because the noetic world has certain freedoms which mindless life does not and, hence, it also possesses its own peculiar conservative trend. Let us call the latter *linguistic inertia.*

By linguistic inertia I mean that property of language by which it resists change in the direction of the prevailing cultural process. The tremendous inertia of value judgment and accumulated knowledge, deposited in human language and dictating socially preferred modes of thought and behavior, has been manifest throughout political and intellectual history.

—No sooner did the band of former prisoners leave Station Beta, than it broke up into conservative and progressive groups. Curiously, it was not always possible to tell which was which, because they often exchanged their declared principles. Let us say that they split up by temperament rather than by issue. Their budding language already had its conservative aspects, counseling a preferred behavioral pattern. Conduct which could not be named was usually judged as inappropriate: those who favored circles would not take the word "square" upon their lips.

But language, as I already stressed, has a capacity to change its direction of motion, for it offers ways through which the unpredictably new may be named. Abraham was surely subject to phylogenetic inertia; most living things protect their offspring. I would also assume that his thought was subject to linguistic inertia. I doubt that there existed an endearing phrase in ancient Hebrew to describe people who murder their children. Yet he was ready to do just that and offer Isaac for a cause that was hardly more than a name.

I assume that he had a choice and that all humans after Babel often have such choices. The path finally taken may be as dramatic as the Socratic self-sacrifice, as innocuous as preferring a certain style of art, or as dangerous as imagining the death of the king. As a class of actions these, and similar decisions, may be regarded as moral choices. They are made possible by the freedom of the nootemporal umwelt, expressed in language, which permits us to give accounts of possible and impossible worlds in the past, in the future, or in a faraway land. Drawing on memories of imagined pasts and depending on hopes for imagined futures, we then design the steps of daily life.

The well-ordered city of Plato, St Augustine's City of God, the paradisal factory of Marx, or the computerized joy of the futurist are all examples of what Frederick Turner calls Counter-Terra. They are not things in space but processes in noetic time. They embody the labor of altering the sense of motion dictated by linguistic inertia. When the modifications so achieved become significant and start influencing codes of conduct, they are usually interpreted as being in the service of a symbolic cause—which is given a name. The capacity of the mind to create symbolic causes may thus be seen as the most direct way of modifying linguistic inertia. The modifications themselves are instants of *poiesis*, which Turner finds implicit in all of nature. They may be symbolized by the formation of buds on plants. Buds do not expand so as to occupy preexistent forms handed down from heaven. Rather, they become new realities of their own. This was also the case for the men and women who left Babel behind.

—The former prisoners of the cave were still driven by their desire to reach the light. But now their behavior, in addition to physical and biological controls, was also guided by the sense of motion implicit in their language. Once in a while certain of them had to make moral choices which, through the alchemy of language, altered the direction of their collective progress. They inaugurated the enterprise of enlarging their noetic umwelt.

Which Way Is Up?

Where do we find in the contemporary understanding of the world the prefabricated Platonic forms? They are certainly not in language, for what is marvelous about language is its open, undetermined freedom: the possibility of unpredictable modifications of its inertial motion through moral and aesthetic judgment. The forms are not among the devices of contemporary philosophy. With some exceptions (such as Jung) they are not in the kit of psychologists and, with the exception of wild utopians, they are not among the conceptual tools of sociologists. Nor are they in biology: organic evolution is delightfully unpredictable. But the Platonic forms may be found in—and in fact ruthlessly rule, as indeed they should—the exact sciences. Those sciences are the true heirs to the belief in a preexisting, discoverable, timeless order. Let us represent this order by the idea of number.

Consider the stunning appropriateness of number (mathematics and geometry) to the description of physical process. The superb fit, or match, derives from the fact that through the rules of number the mind gives an account of its own lower temporal umwelts, those which it shares with the totality of all atemporal, prototemporal, and eotemporal worlds. It is for the same reason that quantified knowledge is the least controversial, the most universally acceptable, form of truth. It is the primitiveness of the roots of mathematics which guarantees its awesome universality, power, and beauty. It is the same primitiveness that makes mathematical tools increasingly useless for dealing with biological, noetic, and historical causations. As we rise along the integrative levels of nature from matter to life, to man, to society, the world becomes increasingly unpredictable—not because of our ignorance, but intrinsically so. Thus, the strict laws of physics, the less rigorous laws of biology, the statistical laws of sociology, all true in their own ways, numerical in their own ways, were surely all applicable to Leopold and Molly Bloom in James Joyce's *Ulysses*. But they were mostly irrelevant. She had to ask him with her eyes "to ask again yes and then he asked me would I yes... and his heart was going like mad and yes I said I will Yes." Indeed, mind is more imaginative than life, life than inanimate matter. In the vast sweep of inorganic and organic evolution, matter, in its increasingly sophisticated functions, seems to be shaking itself loose of number, even if never totally.

Everything totally quantifiable carries with it the limitations of the atemporal, prototemporal, and eotemporal worlds. The laws (languages, causations) of the lower umwelts cannot accommodate Shakespeare's "bare ruin'd choirs, where late the sweet birds sang." Yet the success of exact science and technology has been taken to demonstrate that truth is best identified with what is quantitatively consistent. A substantial portion of contemporary knowledge fashions itself after the methods that have proved so stunningly useful in physics, and seeks quantifiable, preexistent forms. According to Philolaus of Tarentum, the Pythagoreans believed that "actually everything can be known by number." In this, the methodological sense, we live in a Pythagorean age, having arrived here via Plato's theory of knowledge.

The Platonic image of the divided line represents a metaphysical stance that was probably necessary for the creation of mathematized science and thus also a prerequisite for the industrial revolution. However, together with its priceless gifts, the theory of timeless forms and their temporal imitations also bequeathed to us the negation of the idea of creativity in nature in general, and in the worlds of life, mind, and society in particular. But the very sciences that have arisen from Platonic origins teach us, in their modern forms, that in the physical integrative levels, "boundless and bare, the lone and level sands stretch faraway." It is the most sublime aspects of the world which are temporal and the most primitive ones which are timeless. Our situation in the atemporal substratum of the physical universe might be metaphorically described as that of a volcanic island rising from beneath a sea of dark light; from the crater of the volcano the vapors of higher temporalities issue forth in unpredictable, creative bursts.

We heard Socrates say that the soul's destiny is to ascend from the temporal to the timeless. But, one may ask, when the human soul searches for inspiration, reaches out for the beyond, yearns for truth and beauty, does it really ascend in some sense, or would it be more fitting to picture the process as a journey of descent and return?

I believe that the creative enterprise of the mind consists of a descent into the world of primary passions, unformed thoughts, and paleological causation, followed by the blazing of a new path back to humanhood. Such a journey into the lower temporal umwelts of the mind is, by comparison with noetic time, usually described as "timeless." The descent gives us respite from the burden of anxieties and regrets that populate the nootemporal umwelt, a transient relief from human time. In the depths we reexperience oneness with a hermaphroditic creator; then we climb back along the hierarchy of temporal umwelts into the world of creative human passage with its frightening freedoms to do what we judge right or wrong, to judge what is ugly or beautiful, and to mouth what we know to be lies or declare what we hold to be true.

We are so accustomed to the miracle of reproduction, and the work of woman's womb is so secret, that we forget that every human life arises, individually and alone, from the molecular level. Each new conception is a dive into the somewhat-alive world of giant molecules, followed by new *poiesis*. The growth of the fertilized egg into an infant is the "inspiration" of the body, the infusion of biological structures and functions with their noetic potentialities.

The unique gift of man is to be able to repeat this journey in his mind. We travel back into the shapeless world of primitive time and walk out again in another experiment, a new creation. As Eliot so beautifully put it:

> But the fountain sprang up and the bird sang down
> Redeem the time, redeem the dream,
> The token of word unheard, unspoken.

or again:

> The point of intersection of the timeless
> With time, is an occupation for the saint.
> No occupation either, but something given
> And taken, in a lifetime's death in love,
> Ardour and selflessness and self surrender.

To reach the timeless, said Plato, you must go up. To reach the timeless, we learn from "Burnt Norton," is "something given and taken." The process is not one-way, but reciprocal. Translated into our metaphor: we must descend into Plato's cave and then return, creating new worlds on the way up.

Galatea and Lot's Wife

In some ways it is ridiculous to debate whether we should go up, or down and up; perhaps our destiny is to keep on dancing. The point is, however, that it is the descent-and-creation imagery that matches in metaphor our understanding of nature. It is safe to say that all idealistic thought, not only Plato's, sees the light as being handed down from some eternal or quasi-eternal world. This image represents a view *de profundis*; it is beautiful and appropriate if it means a thrust upward from the lower toward the higher temporalities, rather than a pull toward the timeless. Man, since he became man, seems to have been abiding by the advice of Justice Oliver Wendell Holmes. "If I were dying," he wrote once, "my last word would be, Have faith and pursue the unknown end."

The ideas of the true, the good, and the beautiful have not been copied but created. For that very reason they are immensely precious. There is no heavenly warehouse from which they may be replenished if lost through mismanagement.

All search for new knowledge involves a voyage among the temporal levels, but probably none more than that of the poet and writer who plays with the freest forms. I would like to represent this, the aesthetic adventure, through the images of two women.

The first one is known only by the name of her husband: she is Lot's wife. When her family took to the hills to escape the havoc of Sodom and Gomorrha, she violated the law of her Creator and was changed into a pillar of salt. We may imagine her metamorphosis from a woman of good mind, life, and limb to a mindless living body, to a dead body, to an amorphous rock. Her journey was a collapse from the nootemporal umwelt through the biotemporal and eotemporal worlds to the prototemporal, as she returned to incoherent dust.

The story of the other woman begins with Pygmalion. He fashioned a milk-white image of Aphrodite out of a rock and laid it in his bed. Aphrodite was so moved that she brought the statue to life as Galatea who, eventually, bore a son and a daughter to Pygmalion. Her journey was that of emergence along the evolutionary umwelts from the prototemporal to the noetic. We may imagine the lava cooling, the stone assuming form, acquiring life, eventually achieving selfhood and the capacity to fear, hope, remember, anticipate, hate, and love.

The creative journey of the writer is the paradigm of the way we gain knowledge and, hence, is an image of the destiny of the soul. It involves a round trip into the primitive temporalities of the mind. He must begin by violating the law of some creator or another and thus be damned to go back into the cave. But remaining there is death. He must turn around and come out into a world that will be different because of the journey he has taken. He must begin as Lot's wife and return as Galatea.

At the Mouth of the Cave

—The pace of the band of brothers became faster after they left Station Beta, turning at last into a mad rush. For, at the end of the darkness they saw a surface of brilliant light that grew in size as they approached it. It had hues of blues, greens, and browns. Before they reached the opening the Demiurge had vanished, presumably on his way to Zeus to make his report. When they last saw him he was still holding his candle, but its flame was unnoticeable in the sunlight.

Stepping outside the cave, the former prisoners beheld the immensity of a world in which they were active, creative participants. Since words had failed them, they spoke through their silence. They saw brown earth, green olive trees, and the blue sky. They realized that they were naked and that they were men and women.

They stood on the ledge in front of the cave and held tightly to each other for they knew that they were alone and that even God depended on them. In the distance, on the top of Mount Olympus, where the earthly terminal of Plato's divided line used to be located, they saw resting on an insignificant point, arising from a mere DNA molecule and rising upward, a mighty exclamation mark. They noticed, however, that because of atmospheric conditions, it sometimes looked like a question mark.

Reflections Upon the Fates of Two Women

The fates of the Wife of Lot and of Galatea, the metaphors of their descents and ascents, caught the interest of many readers and the imagination of two poets, who favored the stories with their reflections. "Lot's Wife and

Galatea" is by Alexander Argyros. "Galatea and the Wife of Lot" is by Daniel Corrie The poems are reproduced here with their permissions and with my gratitude.

Galatea and the Wife of Lot
in the furnace wind,
in fire's sulphur,
two cities in God-storm
upending and grinding,
Lot shouting to her too late,
wet gleam of her stare crusting,
face's smoothness graining,
salt corroding, movement crystalizing
until a pillar rears in razoring wind
immobile in his nightmare
he calls *No No*

I love you I love you
words murmuring to
the featureless pillar.
In a room's quietly
persistent tapping,
the shape emerges
beneath the chisel's
tireless searching.
Image searching for
a mirage Pygmalion
glimpsed in the fitful
flickering, somewhere
in sleep, into thought
shifting near, forming
a face's cold contour.
Kissing a shape of lips
softening and warming,
inhaling the alien air
of minutes, melting to
pulsebeats and feeling

–Daniel Corrie

Lot's Wife and Galatea

> Art may praise both women, but I greatly
> prefer a woman named Galatea to a
> salt pillar named "Lot's Wife."
>
> J. T. Fraser

For an instant the two women exchange
the knowledge of a self defined by men.
But sisters share vectors. Not these.
One has no name, no way to range
through time, to dream ahead or hear again
the grinding polyphonies of Sodom. The other tries
to love: Lot's wife, tries empathy,

for a moment, she even tries insanity—
she gazes into her crystallizing eyes,
finding something lovely in their fast
and doll-like anonymity.
But Orphic fasncy, though a cheat, desscries
an asymmetry in time, soft future and hard past,
and Galatea sees a choice,

a horror, more—a wonder of a choice,
to birth and slowly die, sad Margaret's blight.
The smooth elastic of a living thong
then tuckles with a snap so quick
that even demons with Laplace's sight
are quite unable to divine things in the long.
I see the billow in her ivory arms

as blood begins to flow, the mind that farms
the body's slow subservience, until
the incoherent mess folds in itself
and, suddenly become gendarme
to a palimpsest of slower times, wills
a most terrible sacrifice, the molt of self
into God's sweet venery.

Yet, I imagine that at times she softly cries
while in her lover's arms, before she sleeps,
because alive and free, she still feels
the need for the old congeries
of stone and walling air. This, she keeps
quite to herself, wild dreams of descent and reel
back into braids of salt.

–Alexander Argyros

3. REALITY AS EXAMINED APPEARANCES

This chapter sets forth an operational definition of reality that accommodates some aspects of the nature of time, thus far unsuspected.

Dictionary definitions of "reality" are not very enlightening. What they all say, in so many words, is that reality comprises of what is real. Philosophical views, many and varied, are more interesting. But they all seem to share the belief that there is a final reality, one that may be approachable through successive approximations even if it may never be completely known. Contrary to this broadly shared belief, the essay that follows suggests that nothing in nature does or even could correspond to the notion of final reality. It maintains that reality is neither "in the mind" nor is it "out there." Rather, it is a relationship between the knower and the known. It is a family of examined appearances. It is a set of working assumptions that is continuously tested for its usefulness for making predictions about the future, and for explaining the past.

The notion of reality as a relationship between the knower and the known was first suggested, about a century ago, by the German theoretical biologist Jakob von Uexküll. He maintained that the external world is not a store of unambiguous information from which each organism may select, such as a reader may select a word from a dictionary.[1] He proposed that, what we call "reality" is the result of a creative process, of an interaction between an organism and its environment. Reality, for each organism, is the result of the integrated functions of its receptors and effectors. It is these functions that determine its worlds of possible stimuli and actions and hence, the nature and scope of its universe. Members of a species share features of their reality because of the biological or (in man) the psycho-biological uniformity of the members of the species. Von Uexküll called this shared reality the *umwelt* of a species. What is not in the umwelt of the members of a species must be taken as nonexistent for that species.

This holds for us, humans, as well. For instance, ultraviolet patterns on butterflies are real for other butterflies because they can see in the ultraviolet spectrum. But they are not real for vertebrates, because they cannot see in the ultraviolet. Those patterns on butterfly wings entered human reality only when we learned how to expand our umwelt—our species-specific reality—by photographing in the ultraviolet domain, then transposing the ultraviolet images into images in that part of the light spectrum which is visible to humans.

The vetting process applied to appearances, on their way to being admitted as real, has never been simple. Aristotle gave good reasons why the world cannot be infinitely large, Savonarola was burnt at the stake for having insisted that it is infinite. One of the models of the universe in contemporary cosmology described it as finite (its volume may calculated) but unbounded (it has no edge). Such a claim would have been judged by Aristotle, by Savonarola, as well as by Savonarola's murderers as self-contradictory and hence impossible or, to call a spade a spade, outright crazy.

The essay that follows extenders Jakob von Uexküll's umwelt principle. Through that extension, through the understanding of reality as a set examined and tested appearances, it removes Bertrand Russell's concern with "the distinctions that cause most trouble in philosophy—the distinction between 'appearance' and 'reality,' between what things seem to be and what they are."[2]

THE EXTENDED UMWELT PRINCIPLE: UEXKÜLL AND THE NATURE OF TIME

A generalization of Jakob von Uexküll's Umwelt principle permits the formulation of a natural philosophy of time that can accommodate, in a single theory, insights about the nature of time that stem from the different sciences and the humanities.

The Expanding Boundaries of Human Reality

The concept of Umwelt is central to Uexküll's theory of organisms. The German word became naturalized into English and, fortunately so, because its translations—species-specific reality, self-world, phenomenal world, perceptive universe, world horizons—are all awkward. Umwelt is now defined as 'the circumscribed portion of the environment which is meaningful and effective for a given species' (English and English 1958). Note that the environment of which an animal's Umwelt is a 'circumscribed portion', is our human Umwelt. It is our human reality.

The first principle of the Umwelt theory, wrote Uexküll, is that

> all animals, from the simplest to the most complex, are fitted into their unique worlds with equal completeness. A simple world corresponds to a simple animal, a well-articulated world to a complex one. (Uexküll 1957: 11)

It follows that for each of its members the Umwelt of a species appears as a complete world which contains everything that can be known and hence, everything there is.

The Umwelt, wrote Uexküll, is like 'a soap bubble around each creature' (1957: 57). I will consider the content of each 'soap bubble' a report about an instant in the course of organic evolution, an always transient, always incomplete recognition of the world, whether by animals or man.

The Umwelten of animal species change at the rate of evolutionary changes in their biological structures and functions, that is, rather slowly. In contrast, the horizons of human reality may expand rapidly. This is made possible by the mind's capacity for the symbolic transformation and manipulation of experience expressed in, and communicated through language and artifact.

With advances in technology and science it became possible to transpose the contents of certain animal Umwelten, not naturally available to humans, into forms of signs and signals appropriate for people. By this process we learned to incorporate their Umwelten into ours. For instance, patterns on the wings of certain butterflies show up only in ultraviolet light. They are visible to other butterflies but not to humans, because vertebrate eyes are not sensitive to ultraviolet. Those patterns became parts of our Umwelt only after we learned to photograph them in the ultraviolet and display them in the visible spectrum. 'Visible' means visible to humans.

The boundaries of human reality explorable through the aided and unaided senses may be further expanded through the abstract language of mathematics, used to describe the functions (laws) and structures of nature. With the help of mathematical models we can learn about physical and organic processes and systems whose features are outside the domain of any direct human sense experience, such as the domain of genes on a chromosome or the implosion of stars.

But even these expanded horizons do not form the ultimate boundaries of human reality because we can think in terms of time. Our reality includes anticipations prompted by present needs, guided by memories, modified by sense impressions, and modulated by fantasies. This vast domain of knowledge is further influenced by personal idiosyncrasies and by socially acceptable principles concerning reality: the poltergeists of yesterday are the creaking steps of today.

While the tick hangs on the tip of a tree branch, wrote Uexküll, its world shrinks 'into a scanty framework consisting, in essence, of three receptor cues and three effector cues—her *Umwelt*' (1957: 12). In the case of humans, the 'scanty framework' has its analogue in the Umwelten of the different sensory systems. One may think of visual, auditory, tactile, and olfactory worlds.

From such distinct sensory Umwelten, it is a small step to the Umwelten of their exosomatic extensions by scientific instruments and theories. For instance, the Umwelt of a radio antenna is electromagnetic radiation limited to a narrow frequency range; it does not include the light of the moon or the sound of the wind. The Umwelt of the antenna is also a scanty framework, also a soap bubble, one that relates to our reality the same way as does the Umwelt of the tick.

Having subsumed in our Umwelt those of animals, instruments, and theories, we may put questions to nature through any or all of them: we may experiment with ticks, with radio antennas, or with mathematical models.

Umwelten so revealed often fail to display features of time that we normally take for granted. For instance, physics reports about processes that do not respond to the experiential direction of time and about objects that cannot behave in any way other than probabilistically. We also learn of conditions—that of the propagating photon—under which the time span between two events, experienced by us as separated by minutes, hours, or millions of years, will shrink to zero. The two events will be simultaneous.

If extended tests, together with convincing reasoning, suggest that some counterintuitive aspects of time—or of causation—are intrinsic to certain processes, then we must acknowledge that the temporalities or causations of the Umwelten of those processes are different from what, in daily human life, we regard as 'natural'. By the Umwelt principle, we must also admit that such temporalities or causations are not only appropriate but also sufficient—that is complete—for the processes and objects considered.

The extension of Uexküll's Umwelt principle to worlds we know only through experiments and/or instruments and/or mathematical models is the extended or generalized Umwelt principle. Of course, the Umwelten of molecules, galaxies, birds and bees, baboons and babies, as revealed to us, become part and parcel of our own, *noetic Umwelt* or reality. The relationship among the Umwelten is a hierarchically nested one. Our noetic reality includes those of photons and ticks; the Umwelten of photons and ticks do not include the Umwelten of horses or paleolithic artists.

Philosophers have been seeking normative criteria for a definition of reality. But for the construction of a natural philosophy of time it is only necessary—and sufficient—to have a working concept of reality, such as the extended Umwelt principle.

The Hierarchical Theory of Time

The hierarchical theory of time takes advantage of the extended Umwelt principle. It is built on a number of propositions, that is, statements of beliefs.[3] My works published during the last three decades comprise critical examinations of those propositions.[4] In what follows I will state them (in italics) and briefly comment on them, dwelling on their details only in respect to causations and temporalities.

Nature comprises a number of integrative or organizational levels

The oldest, stable integrative level of nature is the chaotic universe of electromagnetic radiation.

Out of that primeval chaos arose objects that had nonzero rest mass and traveled at speeds less than that of light; these particle-waves came to constitute the second stable integrative level of the world.

A billion years later—more or less—massive matter began to freeze out, eventually forming the 10^{10} galaxies of the universe. These islands of matter float in an immensity of almost complete emptiness. The massive matter of the galaxies—chemical elements and their compounds in different abundance—constitute the third stable integrative level of the world.

Upon a small object in one of those galaxies life arose. The life of individual organisms is easily snuffed out but the process of life itself is 3.5 to 4 billion years old. Its age permits us to regard the totality of the organic world as the fourth stable organizational level of the world.

Our species emerged a mere 100,000 years ago; the figure depends on what recognizable features are taken to make us what we are. I see the next higher organizational level of nature in the processes and structures created by the human mind, using its skills for the symbolic transformation of experience and its capacity to appreciate non-present objects and events. Despite the relative youth of our species, this view is justified because the genius and audacity of humans challenge the logic of matter and life, from which they arose.

The major fields of human knowledge display a division of concerns that closely corresponds to the stable organizational levels.[5] Learning about mankind is among the tasks of psychology, history and the social sciences; the study of the life process is the task of biology; the science that deals with the astronomical universe is general relativity; the science of particle-waves is quantum theory; the science of light in ceaseless motion is special relativity theory. Pioneering, embracing, taming, protecting, nurturing, and tending to the concerns of all these forms of knowledge, including the sciences, are the arts, the letters, and the other humanities.

The integrative levels form an evolutionary open system along a scale of increasing complexity

A measure of complexity for the integrative levels has been defined and their numerical values calculated.[6] The strikingly different orders of magnitudes of these numerical values strengthen the validity of the assumption of distinctness among them and supports the corollary assumption of level-specific Umwelten.

Processes characteristic of each of the organizational levels function with different types of causation

A cause is anything that we may interpret as being responsible for change as its effect. Causation is the relationship between causes and their effects. The organizational levels of nature display different, stable forms of causations and hence must be described in different forms of lawfulness. We start with the most sophisticated yet most familiar form of causation, that of the noetic world, and work our way down as it were, toward organic processes, massive matter, and particle-waves.

The level-specific causation of the noetic Umwelt is long-term intentionality in the service of distant, often symbolic goals: the building of a pyramid or planning to secure the equality or inequality of people before the law.

The level-specific causation of the biological Umwelt is short term intentionality in the service of organic needs. For those who must breathe, it is a demand for air, for all organisms, the need for nourishment.

Causation specific to the world of massive matter is deterministic: an unsupported apple always falls. This is the world of Newtonian physics and of General Relativity Theory.

Causation specific to the world of particle-waves is statistical, the level-specific laws are probabilistic. The half life of cobalt 60 is 5.3 years. Although this number itself is stable, it is impossible to tell when exactly the next decay particle will appear or which particle of the aggregate of the parent element it will be. Probability in this Umwelt is not a sign of our ignorance but a fact of nature: probabilistic causation is a step in the evolution of causation between chaos and determinism.

What is the level-specific causation of objects traveling at the speed of light, where everything happens at once? An Umwelt in which everything happens at once can only be described as that of absolute chaos of pure Heracletian becoming.

These are the canonical forms of causation. Let me list them in the sequence they appeared in the course of inorganic and organic evolution. For the primeval chaos causation can have no meaning. Out of it emerged probabilistic causation, then came deterministic causation, organic intentionality, then noetic intentionality.

Because of the nested hierarchical organization of nature, each type of causation subsumes those beneath it. For instance, there can be no organic causation (short-term intentionality) without deterministic, probabilistic, and chaotic components.

Each integrative level determines a qualitatively different temporality

To help explain the counterintuitive idea of qualitatively different temporalities, I shall proceed in two steps. First I will deal with the notion of level-specific presents and with their absence. Then I will turn to level-specific temporalities.

Umwelt-specific Presents and their Absence

For an organism to remain alive it is necessary that the multitudes of its inner clocks be kept cycling according to their intricate demands of mutual dependence. Biochemical events that should happen simultaneously must, those that should not ought not, or else the integrity of the life process will be lost. The instant by instant synchronization in cooperative functioning, governed by stable principles, assures collective viability. That viability is manifest as the organic present of a living system. It is with respect to the organic present that future goals may acquire meaning in terms of present needs, and behavior may be organized with the help of memories in the genetically distilled (evolutionary) and individual (developmental) pasts.

From the integrative level of life, let me step up to that of the noetic world.

For the personal identity of a man or woman to remain continuous, it is necessary that the trillions of neurons in his or her brain maintain their cooperative functioning according to stable principles. These principles are inadequately understood. Whatever they are, if the integration process fails, the mental life of a person comes to harm. The instant by instant integration of the immense neural population of the cortex is manifest as the mental present of a person. It is with respect to the mental present that ideas about future and past may acquire meaning and conduct organized in the service of distant, often abstract goals. And it is in the mental present that the ceaseless reclassification of events into future, past, and present creates the experience described by the metaphor, the flow of time.

From man as an individual, let me take a step to human societies as collections of cooperating persons.

To become and remain a tribe, a society, or a civilization, it is necessary for persons to behave so that whatever ought to happen simultaneously does, and whatever ought not, does not. Just as an individual organism defines its living present through inner coordination, just as neuronal coordination defines the mental present, so groups of living organisms define the social presents of their societies by the exchange of signals and signs. It is with respect to the social present that collective plans and memories may then be organized.

It is not possible to maintain a social present without the mental presents of the people involved or maintain mental presents without functional organic presents. These presents are, therefore, necessarily simultaneous. Together, they form a nested hierarchy of presents in which each present serves as the anchor or reference for its respective future and past.

The physical world, as understood through its laws, in sharp contrast to the higher integrative levels, has no features to which the idea of a present could correspond.[7] The physical universe is nowless. And, since future and past can have meaning only with respect to a now, the flow-of-time metaphor does not apply to the time of the physical world.

Undirected time is consistent with the fact that inanimate objects have no needs to be satisfied and do not display purposeful behavior. The future and past I imagine for a pebble are the future and past of my Umwelt, not that of the pebble. Yet, the physical world is not timeless: but its temporalities are qualitatively different from the experiential time of living organisms. By the extended Umwelt principle, the nowless temporalities of the physical world must be regarded as complete in themselves and appropriate for physical functions, even if from the point of view of our daily experience of time they appear incomplete.

The human experience and idea of time's passage must be brought to physics, it cannot be derived from it. Even the much discussed entropic arrow of time is but arbitrarily assigned to the thermodynamics of closed, rather than to those of open systems. For that reason it is useless for defining a direction of time.[8] The physical universe permits the coming about of temporalities appropriate to living and thinking organisms, but it does not itself demand an interpretation in terms of such higher temporalities.

The absence of directed time from all formal statements of physical change has sometimes been taken as evidence that the foundations of the universe are timeless. This presumed timelessness, contrasted with the human certainty of passage, favors the idea of a Platonic division of the world into whatever is eternal or unchanging and whatever is temporal or passing. But such a division is too coarse to accommodate the different types of causations and qualitatively different types of temporal processes revealed by contemporary understanding of nature. A much richer epistemic framework is needed. The hierarchical theory of time offers such a framework by revealing the structure of what with a single word has been called 'time'.

Umwelt-specific Temporalities

When the time-related teachings of the different sciences are combined with what we know of the human experience of time, and the findings are sorted out, five distinct temporalities may be identified, corresponding to the organizational levels of nature. The description of those temporalities, below, proceeds from the most familiar one, that of the human mind, to the most primitive and least familiar one, that of the electromagnetic world.

Nootemporality is the temporal Umwelt of the mature human mind in its waking state. Its hallmarks are: a distinction among future, past, and present, unlimited temporal horizons; and the mental present with its continuously changing temporal boundaries and cognitive content. The characteristic connectivity among events of the nootemporal world (as discussed earlier) is that of intentionality, directed toward concrete or symbolic goals, serving the continued integrity of the self. Nootemporality may be represented by the picture of a long straight arrow: shaft, head, and tail.

Biotemporality is the temporal Umwelt of living organisms, including man as far as his biological functions go. Its hallmarks are: a distinction among future, past, and present, limited temporal horizons, and the organic present whose temporal boundaries are species-specific. The mental present and the organic present are simultaneities of necessity (necessary, that is, for maintaining mental and biological identities). The characteristic connectivity among events of the biotemporal world (also discussed earlier) is organic intentionality directed toward short term, concrete goals, serving the continuity of the life of an organism. Biotemporality may be represented by the picture of a short arrow.

Eotemporality is the time of the physicist's 't' that is, of the astronomical universe of solid objects gathered into galaxies. The prefix *eo-* has been used to identify the oldest of developing forms such as in *Eohippus*, the earliest ancestor of the horse. Eotemporality is the oldest form of continuous time (compare with prototemporality, below). Eotemporal events are countable and orderable, as are the natural numbers and, as do those numbers, they form a pure succession without a preferred direction. Its characteristic connectivity is deterministic causation. Eotemporality may be represented by the picture of the shaft of an arrow.

Prototemporality is the time of the particle-waves of the atomic and nuclear zoo. The prefix *proto-* has been used to signify 'first formed' or 'parent substance' as in protoplasm or protozoa. It is the most primitive form of time, that of the universe out of the cauldron of the big bang. In a prototemporal Umwelt instants may only be specified statistically. The characteristic connectivity of prototemporal Umwelten is probabilistic causation. Its appropriate visual metaphor is the fragmented shaft of an arrow.

Finally, even the picture of the fragments may vanish and we are left with a blank sheet of paper, a symbol for the *atemporal* world of electromagnetic radiation. Atemporality does not stand for nonexistence but for a world of absolute chaos, a total absence of causation, such as is believed to have existed in the primeval universe at the instant of the big bang, or as it exists in a black hole.[9]

The temporalities I described are the canonical forms of time.

The Umwelt Principle in a Post-Darwinian and Post-Einsteinian World

'Instead of saying', wrote Uexküll 'as heretofore, that without time, there can be no living subject, we shall now have to say that without living subject, there can be no time' (Uexküll 1957: 13).

This claim may now be updated by taking advantage of the latent significance of the Umwelt principle and bringing it into our post-Darwinian and post-Einsteinian world.

That principle asserts that the world must be regarded to be the way we find it to be. That is, it equates the Umwelt with reality, epistemology with ontology. But—an important 'but'—it also allows for the expansion of knowledge, by animals and man. With this background in the philosophy of natural science, the following conclusions may be reached regarding the nature of time.

The physical world is nowless but not timeless: it has its peculiar temporalities, as sketched above.

In that presentless world, life creates those conditions and operational properties of matter that define the organic present and permits the coming about of biotemporality. The human brain creates those conditions and operational properties of living matter that give rise to the mental present and permits the definition of nootemporality. Societies create those conditions and operational properties of living and thinking humans in collectives that define the social present and allow the emergence of sociotemporality.

Instead of Uexküll's 'without living subject, there can be no time' we may now propose that without the life process there could be no biotemporality, nor could there be noo- and sociotemporality, because there could be no live humans to think and form societies.

Received views tend to regard time as a background to reality or equate it with the human experience of passage or define it through its distinctness from the timeless. At variance with these views, the hierarchical theory of time regards time as constitutive of reality, as a symptom or corollary to the complexity of the processes of the different integrative levels. The proposition is that time had its genesis at the birth of the universe, has been evolving along a scale of qualitative changes appropriate to the complexity of the distinct integrative levels of natural processes and remains evolutionarily open-ended. Earlier temporalities are not replaced but subsumed by latter ones.

Speaking about the evolution of time seems to be a contradiction in terms. Yet consider the expanding universe. It does not expand into a preexisting empty space. It is space, with nothing external to it, which expands. The hierarchical theory of time maintains that time does not evolve within a preexisting expanse of time. Rather, it is created as an aspect of the complexification that characterizes evolution.

Uexküll espoused the Kantian idea that categories of understanding determine the perception and conception of sense data and recognized the consistency of his Umwelt principle with that idea. However, he was unsympathetic to the Darwinian idea of organic evolution.

But much has happened during the century that separates the publication in 1909 of Uexküll's *Umwelt und Innenwelt der Tiere* from the early 2000s. It is a mark of Uexküll's scientific gifts that his *Umweltlehre* permits a coherent and self-consistent extension to a world wherein certain inorganic processes are revealed to be totally alien to human experience, and wherein an awareness of the open-ended evolutionary character of nature has become a necessary constituent to all fields of knowledge.

References

Cajori, F. (ed.) (1973). *Sir Isaac Newton's Mathematical Principles of Natural Philosophy and His System of the World*, trans. by A. Motte. Berkeley: University of California Press.

English, Horace B. and English, Ava C. (1958). *A Comprehensive Dictionary of Psychological and Psychoanalytical Terms: A Guide to Usage.* New York: David McKay, (under the word 'Umwelt').

Fraser, J. T. (1978). *Time as Conflict: A Scientific and Humanistic Study.* Basel: Birkhäuser Verlag.

—— (1982). *The Genesis and Evolution of Time: A Critique of Interpretation in Physics.* Amherst: University of Massachusetts Press.

—— (1987). *Time, The Familiar Stranger.* Amherst: University of Massachusetts Press.

—— (1990 [1975]). *Of Time, Passion, and Knowledge*, second edition. Princeton, NJ: Princeton University Press.

—— (1999). *Time, Conflict, and Human Values.* Urbana: University of Illinois Press.
Uexküll, Jakob von (1957). A stroll through the worlds of animals and men. In *Instinctive Behavior: The Development of a Modern Concept*, Claire H. Schiller (ed. and trans.), 5–80. New York: International Universities Press.

4. WHAT KIND OF A UNIVERSE TO EXPECT?

In Chapter Two ("Out of Plato's Cave"): we watched the slaves revolt, break their chains and walk out of their cave. Stepping outside, "the former prisoners beheld the immensity of a world in which they were, active, creative participants." I select to imagine that one of them was a member of a fraternity which, in today's terms, would be dedicated to the study of philosophy, religion, and science. Members of that fraternity were followers of Pythagoras of Samos. Little is known about Pythagoras himself but there is sufficient evidence to support the belief that he was a historical and not merely a legendary figure.[1] For the Pythagoreans, religion and science were not separate modes of knowing but two inseparable aspects in a single way of life. They studied mathematics in a cosmic context. They believed—one may say—that numbers performed a divine dance, through which they exhibited the ultimate, unchanging principles of the universe. They saw the task of philosophers as that of discovering the rules of number and, through those rules, identify the structure of the universe.

The essay that follows maintains that there is, indeed, an isomorphism between the structure of mathematics and that of time, as revealed by the hierarchical theory of time. And, that this isomorphism is not accidental. Rather, it stems from the evolutionary development of the human ability to handle numbers. But, if this be the case, then it follows that if mathematics is found to possess certain undecidable propositions—which it does—then nature must also possess corresponding qualities. The unnamed Pythagorean (former) slave, had he known what we now know, could have examined his understanding of mathematics and, from what he found, should have been able to answer the questions of his fellow revolutionaries as to what kind of a universe they may expect to find, once they were outside the cave.

By its mode of reasoning, this essay pays respect to the Pythagoreans who laid the foundations of mathematized science. A 17th century voice in that tradition was that of the English physician and author, Sir Thomas Browne, who saw the world arranged "according to the Ordainer of order and mystical mathematics of the city of heaven."[2] A more recent voice is that of Eugene Wigner who, in his Nobel lecture, sought the reasons for "The Unreasonable Effectiveness of Mathematics in the Natural Sciences."[3]

MATHEMATICS AND TIME

Abstract

The structure of mathematics, as revealed by the exploration of axiomatic systems, bears striking similarities to the structure of nature, as revealed by the hierarchical theory of time. It is assumed that this isomorphism is not accidental but reflects the evolutionary development of the human capacity of handling numbers. This assumption permits a conjecture. Namely, if mathematics is found to possess certain systematic uncertainties, than nature must also possess corresponding qualities which may be identified. The paper proposes that the theme of the conference, "time and uncertainty," be understood in this broad context.

I would like to demonstrate the existence of certain striking similarities between the structure and properties of mathematics on the one hand and, on the other hand, the structures and processes of nature at large, as revealed by the hierarchical theory of time. Then, using these correspondences, I propose a framework that promises to provide a unified perspective for the rich program we have ahead.

1. Number Born from the Search for Permanence

The Dutch philosopher, Spinoza (1632–1677) remarked that it was only "love toward a thing eternal and infinite [that] feeds the mind with joy, and is itself unmingled with any sadness."[4] The desire for identifying permanence is also present in the view of Immanuel Kant (1724–1804) that the human mind seeks "the coordination of all sensibles according to fixed law."[5] In Kantian philosophy number is a form of *a priori* judgment, knowable by reason alone, independently of experience. Einstein also believed that mathematics is a product of thought, "independent of experience."[6]

It has been convincingly argued that the capacity to count evolved from the skill of subitization that is, from the ability to give an instantaneous determination of the size of a set that contains only a few members. Human infants as young as five months, can subitize.

The capacity to count, I suggest, was favored by natural selection because of the advantages counting has in recognizing continuities. These advantages were then formalized in the idea of number, suggesting a quality of nature that does not seem to be subject to change. The crucial step from subitization to counting was made possible, I believe, by the emergence of the symbol and notion of selfhood and with it, the idea of oneness. Namely, the "I" serves as the paradigm of unity. If this is indeed the case, then it follows that only a creature that can construct and maintain the symbolic continuity known as selfhood, can count. Out of one comes two. For the Pythagoreans and through the sixteenth century "one" was seen as the root of every number but not itself a number, because reckoning started with two. Our animal ancestors subitized their offspring, our human ancestors counted their children.

In sum: the path from subitization to counting, to number, and to the logical rules of number are evolutionary steps that were selected for because of their usefulness for promoting conditions favorable for the survival of humans. In our post-Darwinian age number may then be seen as an expression of the hereditary differentiation of the central nervous system, producing dispositions to think in terms of stable relationships.

This whole, long and impressive historical exercise has been driven by the insecurity conferred upon members of our species by the conflict between our certainty of death and our dreams of conflictless, eternal life.

2. Number in Western Tradition

The identification of the rules of number with eternity has been a part of Western intellectual tradition. Mathematics, the science of numbers has been held to reveal the permanent, timeless order at the foundations of the universe.[7]

During the 6th century B.C. the doctrine that all things and forms are number and that *harmonia* that is, balance and order according to number, is the supreme law of the universe, was central to the teachings of the Pythagoreans. Having found in the rules of number the key to eternal verities, they appealed to those rules for securing for their souls the blessings of everlasting life. The teachings of the Pythagoreans came to maturity in Plato's theory of knowledge. He saw number and geometry as aspects of reality vested in Logos, in the timeless divine intelligence that granted order, reason, and spiritual light to the world. In Plato's theory of anamnesis or recollection, he expressed his belief that the human soul is privy to the unchanging truths of the universe, such as the laws of number.

The Book of Wisdom of the Old Testament, written around 50 B.C. maintains that God created the world out of formless matter, then ordered all things according to measure, number, and weight. Later, the Pythagorean-

Platonic idea of number and geometry was combined with biblical tradition and became a part of the spiritual synthesis worked out by the Scholastics, as they tried to reconcile the eternal laws of God with the temporal and unpredictable fate of man. Around the end of the 4th century St. Augustine took it for granted that the truths of number are certain, universal, and eternal. Twelve centuries after St. Augustine, Kepler echoed Plato's anamnesis. He wrote that, "The idea of quantities has been in God from eternity. Quantities are identical with God; therefore they are present in all minds created in the image of God."[8] He also declared that the Christian knows that the principles of mathematics are coeternal with God.

Enthusiasm for the power of number was shared by the mercantile capitalists of Europe. The skilful use of number brought them prosperity which, in turn, reinforced their faith in the immutable laws of a Christian God. For merchants, no less than for Plato, God had to be a mathematician. In yet another 350 years after Kepler, Einstein wrote that in a man of his type the major interest leaves the personal and strives for the mental grasp of things.[9] What he meant by "mental grasp" is expressed in the reduction of space, time, force, matter, and motion to geometry, in the General Theory of Relativity.

3. Trouble in the Republic of Numbers

Those who seek in mathematics a haven against the fell hand of passage may take heart in the remarkable continuity of certain mathematical forms, such as those of conic sections, from the mud cones of Eratosthenes (3rd c B.C.) to Kepler, (1571–1630) and to our own days. Such impressive continuities notwithstanding, the history of science since Cusanus (1401–1464) looks like a series of coups against certainty and timeless perfection.

Imagine Socrates drawing lines in the sand. "Here is a straight line, here is a perpendicular to it, here is a second straight line perpendicular to that. The first and third lines are said to be parallel. Will they ever meet?"

The answer, judged obvious for millennia, has been, "No, Socrates, they will not."

Around the turn of the 19th century mathematicians began to construct geometries in which the intuitive answer of parallels-never-meet did not hold. Gauss (1777–1855) described one such system as "astral geometry," meaning that its peculiar features become evident only at the immense distances of the cosmos.[10] Astral geometry turned out to be appropriate for 20th century physical cosmology. The currently accepted view is that whether or not parallels meet depends on the geometry of the space in which they are embedded.

The history of ideas about parallel lines may serve as an icon of the profound cultural changes that have taken place since the scientific revolution, to which a reassessment of the position of man in the universe has been central.

Copernicus removed the earth from the center of the universe. For the Greeks, planetary orbits had to be circles because circles are seamless, uniform and hence perfect, as are the heavens. When Kepler recognized that planetary orbits are ellipses, he condemned heaven to imperfection. Darwin removed humans from the pinnacle of creation by revealing how they emerged from lowly creatures, through a process in which they need not be the final products. Einstein's General Theory of Relativity is a super-Copernican manifesto. It declares that not only are we not in the center of the cosmos but that the cosmos cannot have a center. Freud removed us from the company of angels. Then, to add insult to injury, logic and mathematics were removed from the company of Logos, the divine reason, and placed into the intellectual tool kit of humans.

I would like to represent that intellectual tool kit by the axiomatic-deductive system of mathematics, then examine its discontents.

4. Axiomatic-deductive Systems

Since the time of Archimedes and Euclid, (~3rd c B.C.) the variety of objects of interest to mathematicians has increased tremendously. But the working method of mathematics remained the same. It consists of (i) observing the mathematical objects, (ii) identifying in their behavior such properties as appear to be unchanging, (iii) formulating axioms based on these observations and (iv) applying to the axioms certain rules of reasoning, called logic. If an axiom survives logical criticism, then it becomes a *theorem*.

Formulating axioms based on the observation of what we do to mathematical objects and formulating the rules of logic that guide our judgments are mental processes. Their nature and origins are of little or no interest to mathematicians. But they are of interest to the theme of this paper because it is through the formulation of axioms and of the rules of logic that the creative capacities of the human brain enter mathematics. These guidelines to reasoning are historical sediments of judgments about relationships that our species identified as stable. With stability as a test, these guidelines became useful biases of thought.

Early during the last century Bertrand Russell and Alfred North Whitehead set out to demonstrate that all the laws of mathematics are derivable from, and reducible to rules of logic. During this epochal labor, Russell identified certain axioms in mathematical reasoning that permitted the drawing from them logically conflicting statements. Such axioms are called paradoxes. They are conditions of indeterminacy.

Russell also observed that such indeterminacies may sometimes be removed if the paradoxes were examined in a language of higher complexity than the one in which they first appeared. The essence of Russell's observation was stated by John von Neumann as follows.

> In the complicated parts of formal logic it is always one order of magnitude harder to tell what an object can do than to produce the object. The domain of validity of a question is of a higher type than the question itself.... The feature is just this, that you can perform within the logical type that is involved everything that is feasible, but the question whether something is feasible in a type belongs to a higher logical type.[11]

By 1937 Russell was no longer convinced that the principles of logic were independent of experience and hence neither were those of mathematics, derivable from them. In 1958 he wrote that when, earlier, he pursued the logistic thesis of mathematics, he wanted the kind of certainty that people seek in religious faith. He may or may not have known of Sigmund Freud's remark that "Mediocre spirits demand of science a kind of certainty which it cannot give, a sort of religious satisfaction. Only the real, rare true scientific mind can endure doubt, which is attached to all our knowledge."[12]

An example of the kind of certainty that Russell and Freud rejected has been the belief that mathematics is a *totally consistent* and *completable* edifice of the human intellect.

To call a branch of mathematics *complete* is to claim that all theorems in that branch are provable to be valid or invalid, using the language—that is, the mode of reasoning and appropriate signs—of that branch.

To call a branch of mathematics *consistent* is to claim that it does not contain axioms from which logically conflicting statements may be drawn.

In a 1931 paper that became legendary, Kurt Gödel inquired into the nature of axiomatic—deductive method, which is the method upon which mathematical reasoning is based.[13]

He showed that every system of arithmetic necessarily contains propositions which can neither be proved nor disproved within the system. In one fell swoop, the very foundations of mathematics were shown to be intrinsically *incomplete* in the sense "completeness" is defined in this context.

Also, as I mentioned earlier, Russell found that there were axioms in mathematical reasoning that permitted the drawing from them logically contradictory statements. Therefore, mathematics was also *inconsistent* in the sense consistency was defined.

Let me now combine Gödel's findings about incompleteness with Russell's finding of inconsistency and von Neumann summary about languages and express them in one of several possible ways.

It is easier to play a game according to stable rules than to explain the rules to someone. It is easier to speak a language than teach someone to speak it. It is easier to conduct one's self according to the experiential aspects of time than to explain to someone what time is. This last sentence would have sounded familiar to St. Augustine, seventeen centuries ago.

As a parallel example concerning the hierarchy of languages, let us recall the axiom that parallels never meet. Before the boundaries of applicability of that axiom could be determined, it was necessary to recognize a language richer than that of Euclidian geometry. But, since there is no, and can be no algorithm for the creation of unpredictably new languages, we must depend on induction that is, on the creativeness of the human mind.

If mathematics is indeed a distilled, abstract representation of what our minds identify as permanent in nature, then something in nature must correspond to the systematic inconsistency and incompleteness of the axiomatic deductive system, as well as to the inductive steps necessary to create higher order mathematical systems.

5. The Hierarchical Theory of Time

The hierarchical theory of time is a natural philosophy, developed and critically examined in five books and many papers. In its structural features it is a framework for an integrated study of time; in its dynamical features it pertains to the characteristic conflicts that define matter, life, the human mind, and society.[14]

The theory embraces the concept of integrative levels, an idea that has a continuous history from Aristotle to contemporary science and plays an important role in organizational theory. The hierarchical theory enriches and expands the received notions of integrative levels by a definition that permits a numerical measurement of complexities in terms of their of diversities, using algorithmic information theory. The ranges of complexities are then found to be radically different, a fact that increases the credibility of the original proposition concerning the qualitative distinctness among them.

The stable integrative levels of nature are as follows. First, the absolute chaos of pure becoming; this is the world of vacuum with its seething sea of creation and annihilation. Above that level we find the world of elementary object-waves, then the world of massive matter, then the integrative level of life and finally, the world of the functions and collectively created structures of the mind.

These levels are hierarchically nested that is, each subsumes and is restrained by the laws and regularities of the integrative level or levels beneath it. For each level the nature of the whole is more than the nature of the sum of its parts and hence, each is governed by level-specific laws, uses level-specific languages and follows logics that cannot be formulated in the language or languages of lower levels. And, importantly, because each level is of increased complexity, each adds peculiar novelties and new degrees of freedom to the structures and functions from which it arose. The nature of Nature's laws itself changes from integrative level to integrative level.

The theory that I described with brutal brevity employs a definition of reality, proposed by the German biologist, Jacob von Uexküll, almost a century ago. Instead of elaborating it in detail, I would like to take advantage of poetry, because it permits the stating of complex principles by simple metaphors. Here is a stanza from Wallace Stevens' poem, "The Man with the Blue guitar."

> They said, "You have a blue guitar,
> You do not play things as they are."
> The man replied, "Things as they are,
> Are changed upon the blue guitar.[15]

The guitar is reality, the blue guitar is one type of reality, qualitatively different from others. And processes of greatly different complexities work with different types of causations, languages, lawfulness and determine qualitatively different temporalities.

The theory is built on eight propositions, two of which are relevant to my reasoning this morning. One of them recognizes distinct forms of causation, the other distinct forms of conflicts.

Proposition #5. Processes characteristic of each integrative level function with different types of causation

The level-specific causation of the noetic or human *umwelt* is *long-term intentionality* in the service of distant goals. That of the organic *umwelt* is *short-term intentionality* in the service of organic needs. Causation specific to the world of massive matter is *deterministic.* Causation specific to the world of particle-waves is statistical; its level-specific laws are *probabilistic.* In the atemporal world of absolute chaos or pure Heraclitean becoming, no meaning may be assigned to the idea of stable connectedness.

When these, the canonical forms of causations are considered together, they suggest that causal connectedness is an evolving quality of nature. It emerged from the primeval chaos in its most primitive form, which is probabilistic causation. It then evolved through the steps of determinism, short-term intentionality and long term-intentionality.

Because of the nested hierarchical nature,

- All probabilistic processes necessarily include elements of chaos.
- All deterministic processes necessarily include probabilistic and chaotic elements.
- All organic processes that is, all examples of short term intentionality, necessarily include chaotic, probabilistic and deterministic elements.
- All noetic processes that is, all examples of long-term intentionality necessarily include chaotic, probabilistic, deterministic and organic causations.
- Finally, the processes of human history, while displaying collective intent, necessarily include all the lower order forms of causation.

Proposition #8, relevant to my reasoning, claims that the *dynamics of each integrative level comprises specific, unresolvable conflicts*.

The identity and integrity of a society, the identity and integrity of a person, that of a living organism and even of matter itself are seen as sustained by certain conflicting trends.

Thus, a *society* is defined by its conflicts between conduct that supports what is perceived as the common good, and conduct that supports the personal good. If these conflicts cease, there may still be persons around but they will not constitute a society.

A person is defined by his and her conflicts between those processes that tend to create and maintain the identity of that person and the social, biological and physical perturbations that oppose it. If these conflicts cease, there may well remain a living human body but it will have lost its psychological identity.

A living organism is defined by the conflicts between the processes of its growth and those of its decay. If that conflict ceases, we may still have all the inorganic matter that made up an organism, but not a living organism.

Inanimate matter itself is a tangible form of conflicts between entropy increasing and decreasing trends, two trends that define each other. If this opposition vanishes, matter collapses into absolute chaos.

Again: because of the nested hierarchical organization of nature, the conflicts of each level subsume the conflicts of lower levels. For instance, the conflicts that define the human mind subsume the conflicts of living matter as well as the opposing trends that define matter itself.

6. Time, Mathematics, and Nature

I began this talk by claiming that there are striking formal similarities between the structures and properties of mathematics and the structures and processes of nature. With all the foregoing in mind, let me identify some of those similarities.

Incompleteness, in mathematics meant the existence of axioms, properly belonging to a system of mathematics, but not provable in its language.

Incompleteness in nature consists of the impossibility of determining the boundaries of the laws of an integrative level, using only the language of that level.

Here are some illustrations.

Absolute chaos may be so identified only in comparison with a higher order causation, minimally, that of probabilistic laws.

Probabilistic causation or lawfulness may be so identified only in contrast to, and in the language of a universe that admits deterministic causation.

Deterministic causation or lawfulness may be so identified only in contrast to, and in the language of an integrative level that admits short-term intentionality.

Short term or *organic intentionality* may be so identified only in contrast to, and in the language of a universe that admits long-term, noetic intentionality.

In its turn, *noetic intentionality* may be so recognized only in contrast to and in the language of historical causation or collective intentionality.

This interpretation of the nested hierarchy of causations suggests that *nature is chronically incomplete and incompletable.*

Inconsistency in mathematics meant the presence of axioms to which logical contradictions are inherent.

Inconsistency in nature may be recognized in the existence upon each integrative level of certain contradictions or unresolvable conflicts which define that level. By "unresolvable" is meant that if the conflicts vanish, so does the integrity of the organizational level. This interpretation of inconsistency suggests that the foundational dynamics of nature is not Platonic harmony but a nested hierarchy of conflicts.

Mathematics is *axiomatic.* So is nature. Each new structure and its functions constitute a new proposal, a new axiom, subject to the test of retaining its identity through time. This is not a projection of our thoughts upon the cosmos; the situation is the other way round. It is a five billion year feature of the cosmos which we, humans, succeeded in copying. We learned axiomatization from the world at large.

What in mathematics is the need for *inductive reasoning* corresponds, in nature, to the emergence of integrative levels with unpredictable structures and functions. This is what David Park wrote in his *Classical Dynamics and its Quantum Analogues*. "New discoveries usually involve radically new ideas not implicit in what went on before, and deductive connections, when they can be made at all,. must be made backwards."[16] He was in good company. This is also what Einstein meant when he remarked that the evolving thought of physics "can only be attained by free invention."[17] He also wrote that "as far as the laws of mathematics refer to reality they are not certain and as far as they are certain, they do not refer to reality."

Let me stress again a part of my thesis. Namely, that the correspondences between mathematics and nature are the consequences of the evolutionary development of the human capacity of handling numbers. What we recognize as "the laws of number" and what we regard as "reality" are two process descriptions of the human brain that continuously define and refine each other.[18]

Since this open-endedness has necessarily been there from the beginning, the idea comes to mind that human knowledge does not, for it cannot reveal pre-existent, eternal verities. Instead, it raises the incomprehensible to the level of the obvious, then it shows that the new obvious is incomprehensible. Under "knowledge" I include all of its many forms: the arts, the letters, and the sciences.

7. From Uncertainty to Indeterminacy

Now, it is time to pull the various strands together.

The theme of this conference originated in a discussion between Dr. Lestienne and Dr. Harris in a restaurant in Paris, with Heisenberg's principle of uncertainty as an immediate association.

I would like to locate that principle of uncertainty in the understanding of nature proposed by the hierarchical theory of time, then suggest a speculative generalization, one which remains in need of exploration.

Heisenberg's idea may be represented by one of its forms: $\Delta \mathbf{p}\ \Delta \mathbf{x} \geq \mathrm{h}$. The meaning of $\Delta \mathbf{p}\ \Delta \mathbf{x}$ will be discussed by other speakers as will Planck's constant, represented by the letter "**h**". My interest is to note that the product on the left pertains to an ensemble of objects. As all ensembles whose members are indistinguishable, it is necessarily a probabilistic statement, appropriate for the prototemporal world.

In contrast, the right hand side is well-defined, it is a fundamental constant of nature and hence it is appropriate to the deterministic world of the eotemporal integrative level.

The equation recognizes a numerical connection between its left- and right hand sides. It is, therefore a formal, numerical link between two different types of connectedness: between probabilistic and deterministic causations.

The principle represented by this equation is called that of "uncertainty." It reflects a surprise that the well-defined nature of the right hand side cannot be applied to the left hand side. The name, "uncertainty" suggests, that with diligent work our degree of uncertainty may be decreased. But this is not the case.

The original Heisenberg term is "Unbestimmtheit," for which the best translation is not "uncertainty" but "indeterminacy." This indeterminacy is not a shortcoming of our knowledge but an aspect of nature. It is not an epistemic but an ontological condition. The relationship represented by the equation could have been called the "determinacy principle" because, lo and behold, the indeterminate relationship of the left is subsumed in the determinate condition of the right.

Until the twentieth century, our understanding of the physical world has been well represented by the words of Edmund Halley's dedication to Newton, introducing his *Principia*, published in 1686.

> Here ponder too the Laws which God,
> Framing the universe, set not aside
> But made the fixed foundations of his work.

It took the intellectual courage of the Copenhagen school to convince others that fundamental to some of God's laws are the indeterminacies thereof.

If, indeed, the correspondences between mathematics and nature are the consequences of the evolutionary development of the human capacity of handling numbers, and *if* this capacity is an adaptive feature of the mind, and *if* mathematics shows systematic, hierarchical incompleteness—as defined—then the Heisenberg indeterminacy principle must be only one particular relationship between processes that function with qualitatively different causations.

Other relationships of similar nature may not be expressible in neat and simple mathematical forms but all of them should manifest the kind of problems that arise when one is trying to foist a higher order causation upon a lower integrative level which is not sufficiently complex to support it. And, as is the case with Heisenberg's indeterminacy principle, all of them should be found be ontological indeterminacies—that is, aspects of nature—and not epistemic that is, not tokens of human ignorance. What exactly these other regions of indeterminacy are, I do not yet know.

These thoughts give me an opportunity to tie the theme of our conference to the Founder's Lecture at our ninth (1995) conference on *Time, Order, Chaos*.[19] There I remarked that "beneath all natural phenomena lurks [what I called absolute] chaos into which all processes and structures may collapse at any time and out of which, under certain conditions, different permanent structures and processes may arise."

The systematic indeterminacies in nature permit a filtering of absolute chaos by the laws and regularities of the stable integrative levels. This filtering process provides for the appearance of the unpredictably new.

To sum up may talk, I would like to take advantage of the power of metaphor and allegory, so wonderfully used by great writers. Specifically, I want to quote from John Fowles' *Mantissa*.[20] Its two protagonists are the Model and the Sculptor. The Model is holding forth.

> Has it never occurred to your poor little male brain that logic, as you call it, is the mental equivalent of the chastity belt? Where do you think the world would have been if we'd all worn nothing but logic since the beginning? We'd still be creeping around in that sickeningly dull garden.

5. THE BEGINNING OR ORIGIN OF TIME

The former slaves, encouraged by their newly found freedom and by their understanding of the universe, decided to stop identifying themselves as a fraternity and admit women. With the qualitative mysteries of the organic thus added to the quantitative mysteries of matter, they developed interest in beginnings: of the days, of lunations, of the seasons, of the years. Of the origins of their countryside. Of life. Even of the universe.

The belief that the world had a beginning, an epoch which was fundamentally different from all later epochs, could not have come from their or from anyone's observations of the heaven and the earth. In the lifetime of an individual and even in the lifetimes of many generations nature remains cyclic and, on the average, unchanging. Nor could cosmic progress from chaos to order have been divined from the observations of natural calamities. These were likely to have been seen as instants of destruction rather than creation. Ideas about a universal beginning are much more likely to have come from observing the beginnings of individual lives: of plants, animals, men, women.

The world at large was known to the Greeks by the word *cosmos.* It meant order, harmony, arrangement. Greek philosophers also maintained that the cosmos had its own *logos.* In the thought of Heraclitus of Ephesus—some time before 500 B.C.—logos signified the underlying coherence or ordering principle of all things, the innate reason in all things shared by humans. It also meant "word." The translation of *logos* as "word" survives in the Johannine Gospel: "In the beginning was the Word, and the Word was with God, and the Word was God."

Since the mid-seventeenth century the study of the orderly and harmonious structures and functions of the world at large, the mode of reasoning native to the universe, has been called cosmology. A scientific conversation with the logos of the universe is called experimentation. The language spoken in these conversations is mathematics—as demonstrated in the preceding chapter. Since mid-nineteenth century, the study of the genesis or origin of the universe has been called cosmogeny.

The former slaves of the cave and latter day Pythagoreans became the family of humans. Members of this family did not leave their old ideas of the logos and of the genesis of the cosmos behind. Rather, they performed upon them twenty-five centuries of dialectical sculpting. As a result, our current insights into mathematics and into the nature of matter, life and the mind are just as appropriate to the concerns of a high-tech, globalizing humankind as were earlier ideas for their respective cultures and epochs. The essay that follows features the new Craftsman with his new lightning bolts. It also features contemporary philosophy, science and religion. It concludes that the dot universe from which time arose fits treatises of quantum cosmology no less than the mystical vision of William Blake who reminded us that "Men forget that / All deities reside in the human breast."

THE SECULAR MYSTERY OF THE FIRST DAY

Abstract

It has been a task of the sciences to consider natural phenomena that appear inexplicable and, through the use of collectively agreed upon modes of reasoning and testing, make them explicable without necessary reference to the sacred or the divine. It is a lesson learned from the history of science, however, that sooner or later, the new understanding will reveal phenomena that are inexplicable in previously accepted terms of reasoning and testing. For those temperamentally so inclined, the newly identified conditions may then appear in need of reference to the sacred and the divine. The essay illustrates this open-ended dialectic in the family of problems concerning a beginning or origin of time.

The history of the universe, as seen in physical cosmology, demonstrates the emergence of increasingly more complex modes of causation. Tracing the same history in reverse reveals corresponding losses in the degrees of freedom available for natural phenomena, and attendant impoverishments in the quality of time necessary and sufficient to interpret those phenomena. As the reverse survey approaches the initial singularity, it draws certain conclusions about the origin of time.

The Birth of an Idea

The belief that the universe had a beginning could not have come from observing nature at large because, never during the life of our species, was the world anything but cyclic. Day and night followed day and night, season followed season. Such a belief is more likely to have come from observing the births and deaths of humans, animals and plants.

Identifying the epoch when our ancestors first possessed a sense of time and hence could have thought of a creation of the world, must remain a conjecture. Such as, for instance, that burying the dead with food and ornaments, about 100,000 years ago, demonstrates a belief in postmortem existence and hence, a capacity to respond to imagined, future challenges and not only to current needs. Between those ages and the formulation of the first narrative cosmogonies, such as the Babylonian Enuma Elish dating to the second millennium B.C., some four-thousand generations of people passed. During these eons, so I speculate, the observed facts of births and deaths gave rise to feelings which, if translated into ideas and expressed in contemporary terms, could be called creativity and destructiveness. The articulation and amplification of these ideas through language, magic, religion and the arts helped in generalizing them into notions of the birth and death of the big everything.[1]

The earliest stories of cosmic births are set in the countryside where the narratives originated and report about conflicts among gods, between people and their gods, and among people. The later history of cosmogonies and cosmologies shows an expansion of spatial and temporal horizons as well as changes in the preferred modes of perceiving and explaining the nature of the world. Whereas in narrative cosmologies humans were part and parcel of the universe at large, the physical cosmologies of our days are mathematical abstractions in which humans are not explicitly present. For that reason, if we wish to relate our concerns as mortals to a contemporary understanding of the history of the universe, a link between the mathematical formalism of physical cosmology and our awareness of passage must be established, such as through an integrated study of time.

The creation of time, as a part of or a corollary to the story of the cosmos, entered intellectual history along a fascinating and tortuous way. For the purposes of this essay it is sufficient to join that history with Augustine of Hippo. Early during the fifth century he gave good reasons for maintaining that "the world and time had one beginning and one did not anticipate the other."[2] By "time" he could not have meant the nature of time as it enters Special Relativity Theory, where it displays an unsuspected elasticity and where it vanishes for objects traveling at the speed of light. Nor could he have meant the nature of time in the quantum world where instants are not well-defined points but washed out, probabilistic likelihoods. Nor could he have been thinking about the insides of black holes where distinctions between time and space disappear, where they are churned into an incoherent foam.

He must have meant the human experience of passage, our awareness of an ever-changing present between a past that includes our births and a future that includes our deaths. I shall call this kind of time noetic time or *nootemporality*. Let me, therefore, rewrite Augustine's words. "The world and noetic time had one beginning and one did not anticipate the other."

Preliminaries

The reasoning of this essay is based on the hierarchical theory of time, a theorem in the natural philosophy of time. In its structural features it is a framework for an integrated study of time; in its dynamical features it pertains to the characteristic conflicts that define matter, life, the human mind, and society. The theory is built on the following eight propositions:

(1) Nature comprises a number of integrative levels
(2) which form a hierarchically nested and evolutionarily open system
(3) along a scale of increasing complexity.
(4) Processes characteristic of each of these levels
(5) function with different modes of causation
(6) and must be described in different languages.
(7) Each level determines a qualitatively different temporality, and
(8) each adds new, unresolvable conflicts to those of the level or levels beneath it.

For detailed elaboration and critical assessment of these propositions reference has to be made to extensive, prior publications.[3] Here it is necessary only to introduce the theory's definition of reality. It is an extended version of the *Umwelt* principle formulated a century ago by the biologist Jakob von Uexküll. In his original definition *Umwelt* meant "the circumscribed portion of the environment meaningful and effective for a given species."[4] The word became naturalized into English and fortunately so, because its translations—species-specific reality, self-world, perceptive universe or world horizons—are all awkward. Umwelt now signifies what may be reasonably assumed to be the reality—the totality of the world—for the members of a species.

The extension of the principle consists of this. Just as we can discover and explore the nature of time in the umwelts—realities—of infants, apes, dogs, ticks and fish, using experiments and theory, so we may explore the nature of time in the umwelts of solid matter, particles and photons, using appropriate experiments and theories. And if some of the temporalities so revealed appear incomplete compared with noetic time—so be it.[5] The extended umwelt principle thus makes possible the formulation of a natural philosophy of time that can accommodate, in a single theory, insights about the nature of time that come from physics, biology, psychology and from the immense store of the humanities.

Consider next that between the creation of the world-and-time and our epoch some fourteen billion years passed. During this immense journey, with the increasing functional and structural complexification of matter, new laws of nature emerge. They correspond to the increasingly more sophisticated forms of causal connectedness possible among events. These groups of laws form a nested hierarchy. That is, each group can describe its own and lower-order processes using its level-specific language but the lower order processes do not demand the higher order language for their intelligibility.

The hierarchical theory of time notes the increasing functional and structural complexification of matter and its laws and adduces reasons in support of the idea that time itself had its genesis in the early universe, that it has been evolving and that it remains qualitatively open-ended. This unorthodox claim will seem less unorthodox if it is remembered that the expanding universe does not expand within preexisting space but consists of expanding space.[6] In a similar historical process, time does not evolve in preexisting time. It is the quality of temporality that changes in distinct, well-defined steps.

Ascending from soon after the Beginning

This section identifies the major steps in the evolution of causation during the history of the cosmos. Causation is the manner in which events may be connected by the laws of nature and/or by the intent of animals or man.

For the purpose of physical cosmology, cosmogenesis is divided into periods for which different laws of nature apply. Quantum cosmology is appropriate for the cosmos between about 10^{-43} sec. to about 10^{-11} sec. after the theoretical zero point of time. Particle cosmogony deals with history between 10^{-11} sec. and about one hundredth of a second after the theoretical zero time. Standard cosmology applies thereafter.[7] Temporal positions along cosmic history are identified with the help of imagined presents (nows) with respect to which futures and pasts may have meaning. Time units employed are those of mother earth: years, days, hours, minutes and seconds. This is done to accommodate, within our noetic umwelt, realities knowable only through abstract reasoning and instrumental findings and often alien to our senses.

We take our leave from a very early world, from where and when the universe consisted of uncombined particle-waves. A visit to the earlier world will be arranged later. The laws of particle-waves—as all laws of the physical world—are iron-fisted. They always apply, there are no exceptions. But their specifications are probabilistic, statistical. Einstein never felt at ease with the probabilistic character of that world. He would have preferred to think of the universe in terms of precisely definable positions in space and instants in time. He remarked that God does not play dice.

But the statistical laws of that primeval umwelt do not hide God's and/or nature's plans behind probabilistic screens. Its laws are statistical because they pertain to the most primitive level of nature which, as we shall see below, is just out of chaos. In it, precise locations in space and precise instants in time do not yet exist. By a simile, the babbling of an infant does not hide his thoughts. He does not yet have well-formed thoughts to hide. During that period of cosmogenesis time and space were—figuratively speaking—babbling infants. Particle-waves of the same species are indistinguishable and hence predictability, vested in probabilistic laws, applies only to average behavior, to statistical aggregates. This ontological view is a part of what is known as the Copenhagen interpretation of quantum theory.

During the 500 million years that followed the epoch of particle cosmology, those particle-waves coalesced into atoms, atoms clustered into molecules, molecules into solid matter, solid matter into stars. By the end of a billion years the first galaxies are formed. By this process of gathering—during this process—deterministic connectedness is born. Determinism means a single-valued relationship between causes and effects. It forms the basis of Newton's and Einstein's laws of the motion of ponderable masses. Deterministic laws did not replace the earlier modes of the governance of the physical world but joined them. Henceforth, the universe functioned under the simultaneous control of laws that govern through probabilistic causation and other laws that govern through deterministic causation.

After a total of about nine and a half billion years the sun and the earth came about. In another billion years the simplest forms of life appeared. It introduced a novel form of connectedness among events. To appreciate this new mode of causation, it is important to understand the necessary, central feature of life.

Life processes are defined by—life is identically equivalent to—the ability of a system to synchronize its physical and chemical processes so that they help maintain rather than destroy their collective functions, that they secure the dynamic integrity of the whole. One may describe living organisms as coordinated clock shops, the clocks being oscillating molecules and assemblies of molecules. Reproduction consists of passing on the ability of creating and maintaining that synchronization.

The instant by instant synchronization that is necessary for collective viability defines presentness in—inserts presentness into—a physical world in which presentness is otherwise undefinable.[8] Once viably established, presentness serves as the necessary anchor or reference point in time with respect to which future, past and the flow of time may acquire meaning.

An alternate way of recognizing the significance of the "now" for life appeals to thermodynamics. Namely, the appearance of life amounts to the introduction of a thermodynamically open system into the prior, thermodynamically closed system. The much discussed decay arrow of the Second

Law of Thermodynamics is definable only in the presence of a growth arrow. To recognize the decay arrow of inanimate matter it is necessary to have a growth arrow of living systems.[9]

The new form of causation, a corollary to the necessity of maintaining instant by instant coordination, is *organic intentionality.* Organic intentionality, directed toward concrete, short-term goals, is a hallmark of all life forms and is absent in inanimate matter. It did not replace probabilistic and deterministic lawfulness, but joined them. Organisms are governed by probabilistic and deterministic lawfulness as well as by organic intentionality, anchored to—referring to—the organic present.

Three billion years after biogenesis, humans emerged. To the short-term intentionality of living organisms, mental capacities added a yet more sophisticated mode of causation. It is long-term intentionality in the service of symbolic as well as concrete goals. Members of our species are thus under the governance of long-term (mental) intentionality, short term (organic) intentionality, as well as deterministic and probabilistic lawfulness.

These several modes of connectedness, as a family, make up the canonical forms of causation.

Descending to the Beginning

During our climb along cosmic history we observed the emergence of increasingly more sophisticated modes of causations. We are now ready to turn around and descend back to the origin of time. This is a useful about-face because the return trip offers perspectives not easily evident on the way up. Specifically, we will note successive losses in the degrees of freedom available to natural phenomena, together with attendant impoverishments in the quality of time that is necessary and sufficient to accommodate those phenomena. The freedoms lost along the descent are, of course, those that were acquired along nature's inorganic and organic evolutionary passage.

We start in the temporal umwelt of the mature mind of modern humans. Anthropologists differ in the details of the stages of human evolution and also about the time spans involved, but they agree on the general shape of that history.[10]

As already said, the world of the human mind is nootemporal, it distinguishes among future, past, and present and possesses unlimited horizons of futurity and pastness. Noetic futures and pasts are anchored to the mental present, created and maintained by the instant to instant coordination of the trillions of neurons in the human brain.[11]

When, perhaps four million years ago, one of my hominid ancestors roamed Ethiopia, she could employ a hazy awareness of a narrow horizon of future and past, all of which were superimposed upon a biologically programmed behavior. She is unlikely to have experienced conflicts between the certainty of her death and a dream of survival after death. Therefore, those forms of human creativity and destructiveness that are driven by such conflicts, did not exist in her. Also absent were all forms of human freedom of action that require long term intentionality, employing personal and collective memories embodied in language, art, and artifact.

As we approach five or six million years into the past, we watch humans and chimpanzees merge into their common ancestry, on their long journey back to Old World Monkeys.

Although individual, long-term memory thus vanishes, a form of long-term memory remains. It consists of behavioral preferences that were encrypted, by selection, in the genetic endowments of the species. These grant the behavior of different species different advantages, but they do not extend the boundaries of the temporal reality of individuals. The most advanced form of temporality that survives (as we gaze backwards) is *biotemporality* with its organic present and organic intentionality.

It is appropriate here to take a philosophical glance. Namely, the history of non-human life on earth does permit interpretation in terms of imagined, long-term goal-directedness in nature but, since organic evolution is explainable in terms of short-term organic intentionality and natural selection, teleology in earthly history becomes, by the principle of parsimony an unnecessary assumption.

In another 3.5 billion years we pass the epoch of biogenesis. Life disappears, together with definable nowness, leaving only inanimate matter to populate the planet. The lines of Shelley's "Ozymandias" describe that world in poetic imagery.

> Nothing beside remains. Round the decay
> Of that colossal wreck, boundless and bare,
> The lone and level sands stretch far away…

The idea of temporality without a "now" is a strange one, in spite of the fact that physicists have been living with it for a century, and even though nowless time is well known from daily experience, where it is mistakenly described as timelessness.[12]

With intentionality and nowness gone, distinctions between futurity and pastness also vanishes and with it, a direction of time. "All the successful equations of physics are symmetrical in time," wrote Roger Penrose. "The future and the past seem physically to be on completely equal footing,"[13] As Michael Heller showed, the physical world is time orientable but is not time oriented.[14] In that world events may be connected only by deterministic and probabilistic causation. Its level-specific time, *eotemporality*, may be conceptualized as one of pure succession without direction, in analogy with the set of natural numbers.

After another 9–9.5 billion years back all massive structures that is, all galaxies vanish. We are left in a universe of gases and radiation. The world of that universe is *prototemporal*: in it, instant may only be defined statistically, only by likelihood.

It is at the time of the creation of galaxies along the evolutionary path (the disappearance of galaxies along the devolutionary path) that the morning stars could have sang together celebrating, says the Book of Job, the laying of the foundations of the world. The sons of God who, also according to Job, celebrated the event with shouting for joy were, then, still some fourteen billion years in the future.

Another half a million or million years back the universe reaches a temperature of around 10^{32} degrees centigrade. We arrive to the instant whence we took our leave when we began to survey the universe along its evolutionary path, at 10^{-11} sec. from the theoretical zero point of time. Beyond 10^{-11} sec. it is quantum cosmology that takes survey of all the world, down to about 10^{-43} second.

Quantum cosmology is the application of quantum theory to the entire gravitational universe, now rather small and dense. At this writing a theory of quantum gravity is in the process of being born.[15] The difficulties of constructing such a theory reside in the fact that Quantum Theory is the lawfulness of the world of particle-waves in their prototemporal umwelt, whereas General Relativity Theory is the lawfulness of the world of massive matter in its eotemporal umwelt.

When we are 10^{-43} seconds from the theoretical singularity of time zero, we come upon "a totally impenetrable wall, the Planck barrier."[16] Beyond that "wall" that is, in the primeval universe, space and time have no distinct existence. They, together with all the forces of nature, are scrambled into a furious quantum foam. In that foam no meaning maybe assigned to causation, to any notion associated with lawfulness in time. It is a world best described as one of absolute chaos, to distinguish it from the formal chaos of chaos theory.[17] It is an atemporal world. *Atemporality* does not stand for the theological, philosophical or poetic idea of timelessness. It stands for conditions of pure becoming.

Atemporality, prototemporality, eotemporality, biotemporality and nootemporality are the simplest and clearest forms of temporalities and for this reason, they have been called the canonical forms of time.[18] If all that has been claimed for the structure of time is correct, it then follows that time itself is an evolving quality of the universe. In the words of Alexander Argyros, time in the hierarchical theory of time "is no longer understood as a background for reality, nor simply as the human experience of flux, but as an evolving palimpsest of emergent temporal levels constitutive of reality."[19]

"Not with a Bang but a Whimper"

Not only in narrative cosmologies but also in the formal reasoning of philosophy, the idea of a beginning of time had a beginning.[20] The universe of Aristotle and Parmenides are bounded in space but not in time. In Christianity, by the fifth century, the question is well formed and dealt with by Augustine, as cited. In medieval Judaism and Islam the question of whether time is or is not coextensive with a created world was widely considered.[21] In these and in other examples of which I am aware, whenever questions of a beginning or origin of time are discussed, there is an assumption of a sharp contrast between time and the timeless. "Time" always stands for nootemporality. But, what is meant by timelessness?

The term has several distinct meanings. In theology it is an attribute of God, sometimes interpreted as endless duration, sometimes as being outside time. It may be associated with eternal life and with the validity of religious dogmas. In classical philosophy, timelessness is an attribute of the Heraclitean Logos and of the unmoved prime mover. It is sometimes invoked in connection with the assumed, eternal validity of the rules of numbers. In psychology it signifies a feeling associated with the transient dissolution of identity.[22]

The ordinary notion of timelessness carries a heavy cultural load. It tends to conjure up dreams of complete fulfilledness, of a world where one always wanted to be, whether it is one of perfect peace or glorious battle, depending on traditional preferences. In such a dream world there is no decay or illness, no suffering, no bodily needs, only indestructible, healthy life. There is also final and true justice and accomplished love. But, in whatever guise and in whatever context the idea of timelessness appears, its roots may always be traced to the human need for refuge from the threat of passage in general and from the knowledge of the inevitability of an end to the self, in particular. A perception of harsh contrast between time and the timeless, follows.

What does contemporary cosmology reveal or at least suggest about the nature of time, as it arose from the atemporal absolute chaos? Did it "hit the ground running?" This idiom originated during the mid-nineteenth century

California gold rush when tens of thousands of fortune seekers stowed away in freight trains for free rides to the west. To avoid discovery at the next station, they had to jump off the train before it stopped. The safest way to do so was to hit the ground, running. Applying the idiom to time: did time begin with "there was evening and there was morning, one day" of the Book of Genesis? If it did, then Bishop Ussher's calculations in 1686, concluding that the date of Creation was October 23, 4004 B.C. might have been wrong in its numbers but not in its principles.[23]

The calendrical location of the First Day, based on the biblical account of Creation is not acceptable to contemporary cosmogony, yet it remains the universal and seldom questioned model of the beginning of time. G. J. Whitrow summed up this traditional understanding when he wrote that "at all levels [of nature's evolutionary organization, from radiation to particles to life, man and society] time is essentially the same, although certain aspects of it become increasingly significant the more complex the nature of the particular object or system studied."[24] In other words, time is a Platonic "moving image of eternity" whose features remained unaltered since it was created.[25] But this model of time is at variance with contemporary understanding of the history of the universe. Specifically, as we progressed in our reverse journey toward the Big Bang, we observed changes in the modes of causation, in the character of nature's laws, and in the quality of temporalities. The temporalities became increasingly poorer, as judged by the degrees of freedom which they allowed. We witnessed the loss of long-term intentionality, then short-term intentionality, then nowness, then deterministic connectedness and finally even probabilistic connectedness. Beyond prototemporality we encountered a universe in which distances, temporalities, mass, energy and all of nature's forces were scrambled in a quantum foam, making for a world that I described as atemporal.

We were in the world of quantum gravity, in the initial cosmos which we bypassed at the beginning of this essay by starting with particle cosmology. That universe, as I stressed, was one of atemporal, absolute chaos, of pure becoming, one that had infinite potentialities precisely because of its total lawlessness. Because of its homogeneous lawlessness, physicists describe such a world as possessing perfect symmetry.[26]

For the human intellect and emotion, the atemporal cosmos qualifies as Hell and disqualifies as Heaven. Its atemporality, as I mentioned, is not to be mistaken for the paradisiacal timelessness of poetic dreams. Also, the temporality that arose from that non-paradisiacal universe was not nootemporality with its "evening and... morning, one day," a kind of time that could support Dante's *L'Amor che muove il Sole e l'alatra stele* but prototemporality, that could not.

In "The Hollow Men," T. S. Eliot laments as well as dares the mental and moral penury of passing life and of the inevitable end of the self in death: "This is the way the world ends / Not with a bang but a whimper."[27] The ending of time is not a subject of this essay but the second line cited does fit what the natural philosophy of time reveals about the beginning of time. Namely, that time began with a whimper. That of prototemporality. Furthermore, the relation between the primeval atemporality and the nootemporality of our species is not one of mutual contrast and exclusiveness—as have been ideas of timelessness pitted against those of time—but one of a nested hierarchy. Noetic time subsumes all the canonical forms of time: it subsumes long-term intentionality because we have minds and organic internationality because a person with a mind is, necessarily, also alive. It also includes deterministic, probabilistic and chaotic (atemporal) components.

From Becoming to Being

What does the emergence of prototemporality from the atemporal world suggest about the subsequent evolution of time? For an answer, I turn to a millennia old debate.

"Of the many controversies carried on among early Greek philosophers, that of *becoming or change* versus *being or permanence* was one of the most prominent," wrote the philosopher, Cornelius Benjamin.[28] That controversy may be represented by the opposing views of Heraclitus of Ephesus and Parmenides. Heraclitus (540–480 B.C.) perceived the world in terms of ruthless conflict and change. The Heraclitean world was a totality of processes rather than things, a world of becoming. Parmenides, a contemporary of Heraclitus, focused on what he regarded as the true reality of the world. For him change and time were only chimerical impressions of the senses. The true world was one of being.

A few centuries later these differing stances came to be embedded in the history of natural philosophy that, wrote G. J. Whitrow,

> is characterized by the interplay of two opposing points of view which may be conveniently associated with the names of Archimedes and of Aristotle, those intellectual giants of antiquity whose writings were of decisive importance for the late medieval and renaissance founders of modem science.[29]

The thought of Archimedes may be classed with those who see the essence of the universe in being, Aristotle with those who see that essence in becoming.

In the development of medieval and to a considerable extent modem philosophy and, certainly, in the practice of science, being has been given prominence while becoming assumed an inferior role. Being, embodied in the laws of nature, are taken to be ontologically prior to becoming, represented by the boundary conditions to be applied to those laws. Consistently, the cosmic origin of time itself is broadly equated with the emergence of the unpredictable (of becoming) from the eternal, timeless permanence or being.[30] But the natural philosophy of time suggests that this formulation is backward.

Namely, the probabilistic laws of the quantum world are the simplest forms of being or permanence. They are the first-borns of pure becoming. It was being that arose from pure becoming, not the other way round. The origin of time is identically equivalent to the emergence of the most primitive form of orderability. It is the first break in the symmetry of primeval chaos, the first step in the evolution of other orderabilities or beings, involving increasingly more intricate forms of causations. From that distant event to our days, beinglike and becominglike conditions remained identifiable and define each other in all the canonical forms of time.[31]

In view of the reputable age of this debate, it is interesting to speculate about its sources. From the store of the natural sciences, reasons may be adduced in support of the idea that in living organisms and in the minds of humans there are instinctual drives toward separating the permanent (predictable, beinglike) from the unpredictable (becominglike) aspects of time in the environment and in the self.[32] One may even view the human experience of time's passage as a ceaseless, incomplete and incompletable integration of whatever is perceived as predictable with whatever is perceived as unpredictable.

In the Beginning

We have arrived at the beginning of the world and learned that time did not "hit the road running," that it was not born with a bang but with a whimper. The reasoning I employed were those of physical cosmology, a science which, by collectively agreed upon methods of reasoning and testing may thus succeed in making explicable what was earlier inexplicable—the origin of time—without necessary reference to the sacred and the divine. But the very same reasoning also leads to a description of the initial singularity, to a specification of its properties that are alien not only to our senses but are rather bewildering to our imagination. Those specifications are unmeasurable but not uncalculable. They pertain to a universe in which all known laws of nature break down. That cosmos was named the "primeval atom" by the Belgian physicist, astronomer and ordained Catholic priest, Georges Lemaître, formulator of the Big Bang theory of cosmogony.

Its mass had to be the same as that of the present one, which is estimated to be 10^{54} grams.[33] Its size was forty orders of magnitude smaller than the size of a current proton. This makes it small enough to be imagined passing through a current atom, without as much as being noticed.[34] Its temperature was 10^{32} degrees centigrade or higher.[35] The world whose physical measures these are, is one that would be beheld by an observer—in her noetic umwelt—when she was 10^{-43} sec. from the theoretical zero point of time.

In the primeval, atemporal universe time, space, electromagnetism and gravity were all merged into a furious foam. It had no features of the cosmos as we know it. It follows that it could not even be called an object, that it exists as ideas in our minds, though we are free to assign to it a past reality because our noetic world is temporally open-ended. The universe from which time arose thus fits treatises on quantum cosmology no less than the mystical visions of William Blake, who reminded us that "Thus men forgot that/All deities reside in the human breast."[36] While we are getting ready for the next adventure in exploring cosmogenesis, we might as well describe the First Day as a secular mystery.

Quod erat demonstrandum.

6. CONSTRAINING CHAOS

Twenty-five centuries after Pythagoras of Samos, a few members of the family of men and women became convinced that the world did not begin with a neat and orderly build up but with an overwhelming, chaotic whisper, followed by a painfully slow evolution. Also, that the power of that whisper—the Big Bang—was so immense that it could have made the dead rise if there had been any dead around, but there were none. Even the simplest forms of life were still 9,500 million years in the future.

We already encountered the dot universe (known to cosmologists as the "initial singularity") and noted that that it was one of pure becoming or absolute chaos. It was a microscopic cosmos to which none of the known laws of nature apply. But, since we humans are here, we may appeal to our species-specific reality, our noetic umwelt, including our noetic sense of time, apply it to a world for which it is not really valid and ask, what happened after the initial singularity big-banged?

Human infants are delivered. Lambs and fawns are dropped. Chicks are hatched. The universe may be said to have emerged. By "emergence" is meant the appearance of any of the uncountably many forms of permanent constraints upon the substratum of the absolute chaos or pure becoming, constraints that may be identified in the ordered structures and processes of the universe. "Emergence" thus joins other examples of creativity in nature, whether described as cases of delivery, dropping, hatching or creation.

The essay, to which this is an introduction, employs chaos theory to help it identify the dynamics of creativity in nature, of which the origin of time, "not with a bang but a whimper"—to quote from the preceding chapter—is an example.

The essay that follows notes that the first forms of permanence, or being, that emerged from absolute chaos were particle-waves. But neither those early objects/conditions of permanence, taken in themselves, nor pure change or becoming taken in itself, could serve as the source of creativeness in nature because they define each other. They acquire meaning only with respect to each other. The roots of creativeness must, therefore, be sought in their coexistence, in the dynamics of their conflicts. The essay concludes that creativeness in nature, including in man, is vested in constraining chaos through conflicts between ordering and disordering processes, between forms of being and forms of becoming.

The history of nature comprises the resolutions of increasingly more intricate conflicts. The trend is that of increasing complexity—the theme of Chapter One. The suggestion comes to mind that the origin of time should be identified with the earliest conflict ("earliest" as judged by our noetic sense of time) between the most primitive forms of being and becoming, between the most primitive processes of permanence and change.

FROM CHAOS TO CONFLICT*

Abstract

This paper examines deterministic unpredictability from the point of view of the interdisciplinary study of time. The first part reports about the existence of regions of unpredictable numerical solutions to certain deterministic functions, revealed by recent work on the behavior of nonlinear systems. The second part locates deterministic unpredictability in the developmental history of time and causation. The third part traces the origins of predictability to the emergence of primitive forms of permanence from the primeval chaos of pure change. This perspective permits a reconceptualization of creativity in nature and man as a capacity to generate and maintain conflicts between ordering and disordering processes. Inorganic and organic evolution, no less than mental and cultural development may then be viewed as the complexification of conflicts between increasingly more intricate forms of permanence and the ever-present chaos of pure change. The paper concludes by noting that the creation and maintenance of conflicts between ordering and disordering processes, which has only lately achieved formal representation in exact science, has been the time-honored task of the arts and letters.

The story of Creation, as recounted by Hesiod during the 7th century B.C., covers a long stretch of events in a few sentences. First of all, he wrote, did Chaos come into being, then broad-bosomed earth and misty Tartaros, and Eros, fairest of immortal gods. Next, earth brought forth mountains, those lovely haunts of the divine Nymphs.[1] My paper also addresses, among other matters, the issue of Creation but instead of the Nymphs, it concludes with Dr. Frankenstein.

The Rediscovery of Chaos

Let us assume that the current understanding of the primeval world is correct. Namely, that the elementary particles acquired their masses through interaction with a hypothesized, all-pervading field, called the Higgs field. If so, then there were no objects in the pre-Creation universe that moved at speeds less than that of light and hence no objects for which the passage of time could have had any meaning. I will describe that world as one of an atemporal, *absolute chaos*, a state of pure becoming to which none of our time-related ideas may apply.

Absolute chaos is to be distinguished from the *formal chaos* of nonlinear dynamics. All future states of formally chaotic systems are calculable and hence predictable in principle, but never without error. Nor are they predictable without error from inspection because they never repeat their pasts, exactly. Yet, they do display repeating patterns; that is, they possess permanent features in their long-term, global behavior. The copresence in formal chaos of unpredictability and predictability accounts for its significance for the study of time, where it conjures up the ideas of the Eleatics.

Theoretical studies of nonlinear behavior have been neglected, mainly because they are difficult to handle mathematically. The situation changed when computers became available. They began to serve as tools that helped reveal the fine structure in the evolution of mathematical functions, the same way microscopes reveal the fine details of spatial structures.

In the *Republic* (X, 597b), Plato speaks about three kinds of beds. One is the eternal idea of bedness. The other is an actual bed, a copy of the eternal bedness. Finally, an artist may paint a picture of the bed which is, therefore, only the copy of a copy of the ideal bed. "The painter... the cabinetmaker, and God," he wrote, "preside over three kinds of beds."

Nonlinear dynamics has its own three beds: the empirical, the mathematical, and the geometrical. Just as people are well advised to know whether they share their bed with God or the cabinetmaker or the painter, so in dealing with nonlinear systems one must be clear whether one deals with equations, with geometrical figures, or with the phenomenal world. Let us start with geometry and mathematics and then turn to the phenomenal world.

Central to the theoretical study of nonlinear behavior are two conceptual tools: the phase or state space, and the attractor.

A phase space is a bookkeeping device. It is a multidimensional space with as many mutually orthogonal coordinates as are the degrees of freedom in the process considered. It is not possible, nor is it necessary to imagine what a coordinate system of more than three dimensions would look like; it is enough to understand that a single point in phase space represents the full description of a system at an instant. Also, that as the system keeps on doing whatever it is supposed to be doing, that point will generate a line, called a trajectory.

A master trajectory toward which nearby trajectories of a system evolve is called an attractor. It is an organizing principle stated in mathematical language. The geometry of an attractor is a report about the behavior of the system. If it is a point, the system is stationary; if a closed curve, the system cycles, repeating its past; if it never reenters itself, the system never repeats its past. If its structure is infinitely intricate, the system's behavior is going to be infinitely intricate and may be formally chaotic. These geometrical and mathematical properties of attractors are independent of whether or not there are or may conceivably be any physical, biological, cognitive, or social processes of which they may serve as trajectories.

If it be true that in the trillion-dimensional state space of humanity all roads lead to Rome, then Rome is the attractor for the trajectories of the pilgrims' progress. This way of speaking is, of course, an analogy; I described a model of dynamic changes, using geometrical and mathematical metaphors. But attractors are themselves models. They are metaphors for processes. One often sees bicycles whose motion could be described by an attractor in phase space, but one never sees those attractors traced out around the bicycles in a multidimensional space.

Let me leave bicycles and their phase spaces and consider, instead, certain simple mathematical functions which have well-behaved solutions for some regions of their variables, while for some other regions of the same variables their solutions are unpredictable by inspection and not calculable without error. Furthermore, the transition from the predictable to the unpredictable regions of the functions follows stable routes. These routes have such names as *intermittences, crises, quasi-periodicities, border skirmishes*, and *period doubling*. From among them, the period-doubling route to chaos has become an icon of chaotic behavior. I will use it to represent the family of all routes to chaos.

In its most general form, period doubling is an adventure in the life of a quadratic equation, such as,

$$x_{n+1} = ax_n(1 - x_n)$$

where a is a control parameter and x is to be iterated for $n = 0, 1, 2, \ldots$. The equation arose from the mathematical modeling of the annual fluctuations of insect populations and belongs in the family of growth curves whose mathematical forms are called logistic equations.[2] Iteration is a method of computation where the output of an equation is repeatedly fed back as an input to the same equation. It is a computer-friendly process, which is the reason why humans have become adapted to it.

Figures 1.1 and 1.2 are loci of the logistic equation just given.

For the logistic equation, unpredictability means that the values of the function to the right of the onset of chaos (Figure 1.2) may be calculated, but cannot be guessed from inspection, nor can it be calculated without error, because of the extreme sensitivity of the function to the value of the control parameter.

And yet, the jungle of points that make up the chaotic region display certain recurring patterns; they repeat their general gestalts. In Figure 1.3, observe the clear areas, called windows of periodicity, where the jungle

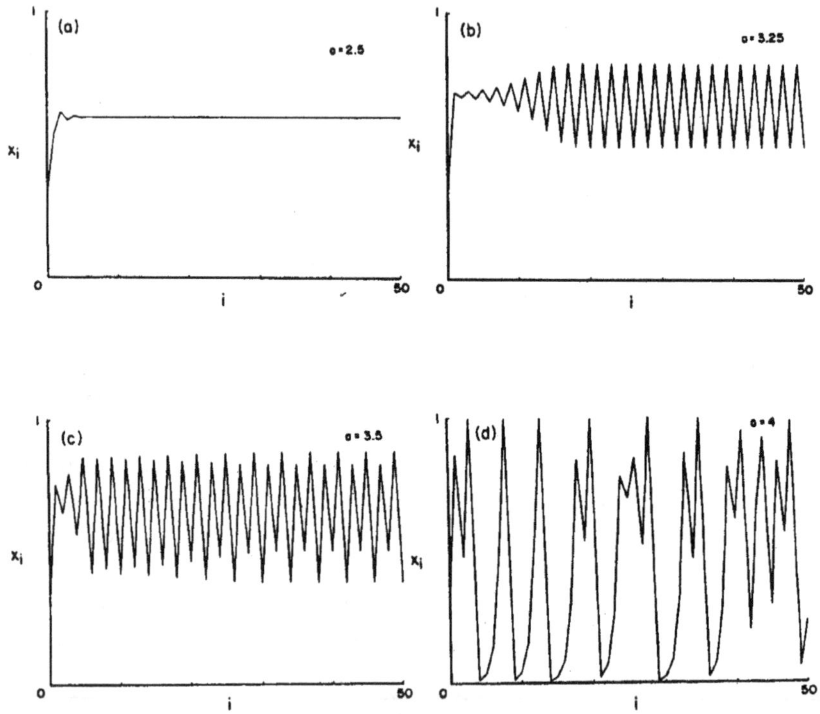

Fig. 1.1. Four time series, showing the development of the logistic equation,

$$x_{n+1} = ax_n(1 - x_n)$$

for four values of *a*. (Reproduced from Glass & Mackey [1988], pp. 30–31, courtesy of the publisher). The variable *i* along the abscissa, running from 0 to 50, is the number of times *n* has been iterated. The ordinate is the value of function, between 0 and 1. For $a = 2.5$ the solution reaches a steady state of a single value. For $a = 3.25$ it oscillates between two values; for $a = 3.5$ it oscillates among four values; for $a = 4$ the solution becomes unpredictable or chaotic.

clears up and yields a well-defined number of distinct, single-valued solutions. There are an infinite number of such windows. Each of them, along its right edge, relapses into formal chaos via the period doubling path; the route to formal chaos contains an infinite depth of self-similar elements. It is the existence of such global regularities which demand that a distinction be made between absolute chaos and the formal chaos of chaos theory. The repeating regularity makes for an infinite depth of self-similarity. This may involve exact replication, it may be statistical or random; it may even consist

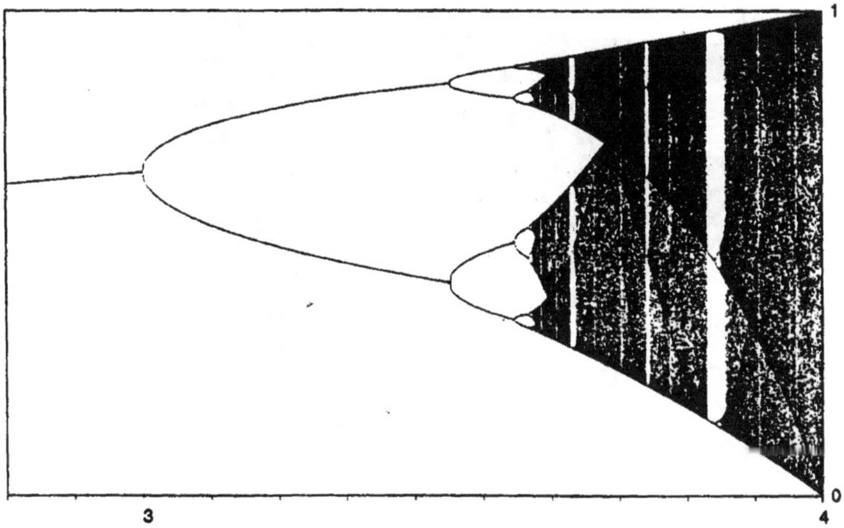

Fig. 1.2. This is called the final-state diagram. It collects all the final states, understood in the manner explained in Figure 1.1. The abscissa is the value of what in Figure 1.1 is the variable *a*. The ordinate, again, is the value of the function, for appropriate figures of *a*. (Reproduced from Peitgen, Jürgens, and Saupe [1992, p. 589], courtesy of the publisher).

When we see two branches, that means that the system is alternating between two different states, two modes of organization. Its period is said to be two. The periods double as powers of *2* until at $a = 3.5699\ldots$ called the Feigenbaum point, it becomes chaotic.

of continuously emerging new patterns mixed with continuously reemerging old patterns. An infinite depth of self-similarity means that, no matter how much the details are magnified, self-similarity remains.

Note that self-similarity signifies the presence of a pattern of behavior or structure which retains its identity in a world of pure becoming; it represents the birth of permanence from pure change.

Remembering Plato's three beds, we may now turn from geometry and mathematics to the phenomenal world. An infinite depth of self-similarity suggests that if there were processes of which such trajectories were the attractors, then those processes would be governed by certain stable rules, intimately mixed with unpredictable changes. Here again, one is reminded of the ideas of Heraclitos and Parmenides. There are geometrical objects which correspond to such requirements, and they have their own provenance. It may be traced back to what early this century the mathematician George

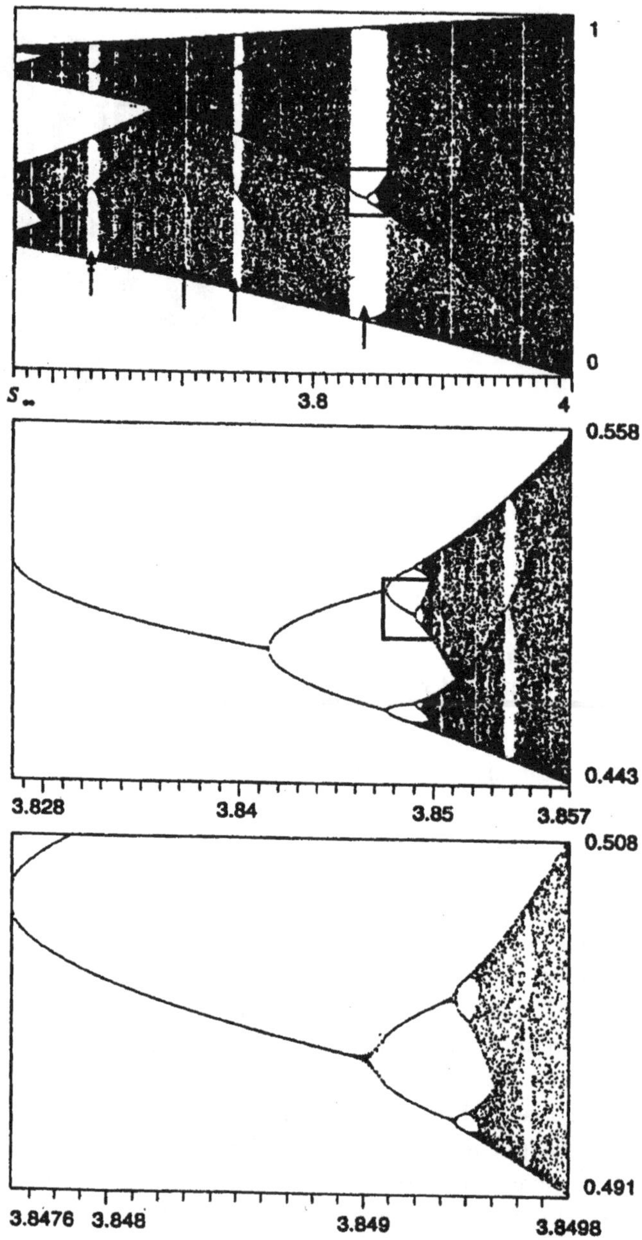

Fig. 1.3. Self-similarity in the final-state diagram. (Reproduced from Peitgen, Jürgens, and Saupe [1992, p. 637], courtesy of the publisher). The arrows point to windows of order in the chaotic domain. The heavy squares enclose the regions which are shown, enlarged, in the next diagram.

David Birkhoff described as "remarkable curves."[3] He judged them remarkable because the spaces into which they had to be inscribed were not of integer but of fractional dimensions or, in later terms, fractals.[4] Like planetary orbits, they have two aspects: a geometrical and a dynamic. In their geometry they are fractals; in their dynamics they are formally chaotic.

How do people determine whether a set of empirical data is random noise or formal chaos or a mixture of both? They look for patterns familiar from the study of nonlinear systems, then search for an equation which, for some ranges of its control parameter, yields that pattern of unpredictability. The part of the noise so identified is then declared to be chaotic, whereas whatever is left is just ordinary, undistinguished noise. Formal chaos may in fact be defined as random behavior which can be shown to have arisen from nonrandom behavior via a stable mathematical relation.

Let me now assume, from the witness of many tests, that formal chaos may indeed be identified in physical, biological, cognitive, and social processes. The suggestion then emerges that beneath all natural phenomena lurks chaos into which all processes and structures may collapse at any time and out of which, under certain conditions, different permanent structures and processes may arise.

That there is no facet of human experience without elements of apparent chaos is hardly a novelty. But it is now official because it has found expression in science, the favored language of our age. However that may be, one should not sneer at the rediscovery of chaos. Everyone in Newton's time already knew that detached apples will fall, but they did not know exactly how and why.

The question which remains, is this. Does the mathematical legitimation of such processes permit us to enlarge our knowledge of predictability in ways which would not have been possible earlier? The answer is, yes.

The Evolution of Predictability

In the second part of my paper I propose to locate deterministic chaos in the evolutionary history of causation and time. To do so, I will enlist the assistance of a class of principles in the natural philosophy of time, developed and elaborated during the last twenty-five years, under the name, the hierarchical theory of time.[5]

A sharp distinction between what is temporal and what is timeless may be found all across Western philosophy. The same distinction is fundamental to the great religions and, as a metaphysical stance, it was a necessary ingredient in the creation of early modern science. The hierarchical theory of time was formulated in response to my belief that, in spite of its success as an assumption, a bifurcation of the world into time and the timeless is incompatible with what has been learned about the nature of time through contemporary scientific knowledge and humanistic insight.

Received views either regard time as a background to reality, or equate it with the human experience of passage, or define it through its distinctness from the timeless. At variance with these views, the hierarchical theory of time regards time as constitutive of reality, specifically, as a symptom or corollary to the complexity of nature.

It is a generally accepted hypothesis of contemporary science that the history of the universe involves the emergence of increasingly more complex structures and processes, in the continued presence of surviving earlier structures and processes. If time were indeed a correlate of complexity, as I propose it is, then it would follow that time itself has evolved with the increasing complexification of nature. That this in fact is the case, is one of the claims of the theory.

It has been said that the devil is in the details. Therefore, let me turn to the devil.

In contemporary cosmology, the history of the world is understood to have begun with the coming about of particle waves of nonzero restmass from a hot, primeval chaos. As the universe cooled, many of these objects jelled into ponderable matter and collected in galaxies. Upon a speck of such matter living organisms arose; from among them came our species with its ability to transform experiences and feelings into symbols and employ these symbols in the building of civilizations.

Corresponding to these evolutionary steps, the hierarchical theory of time recognizes five stable, hierarchically nested integrative levels of nature. By hierarchically nested is meant that each integrative level subsumes the functions and structures of the one or ones beneath it, and each adds to the potentialities of its predecessors certain new degrees of freedom. Beginning with the most complex one, those organizational levels are the mental processes and their universe of symbolic representations; the processes of life; the physical processes of the astronomical universe of massive matter; the processes of elementary objects; and the chaos of radiative energy.[6]

Complexity is defined as an index of the riches of an integrative level. It is measured by applying Gregory Chaitin's definition of the algorithmic complexity of a number series to the minimum number of distinct structures from which that integrative level is constructed.[7]

When the time-related teachings of the sciences and the humanities are systematically surveyed and arranged to correspond to the hierarchy of the stable organizational levels, five distinct concepts of time and five different types of causations are recognized.

Nootemporality is the temporal reality of the mature human mind. Its hallmarks are continuity and a clear distinction between future and past, that is, between categories of time that acquire meaning in reference to the mental present. The level-specific causation of the nootemporal world is *intentionality in the service of symbolic causes.*

Biotemporality is the temporal reality of living organisms, including humans, so far as their biological functions are concerned. Its hallmarks are continuity as well as distinctions between future and past; that is, between categories of time which acquire meaning by reference to the organic present. Compared to the mental present, the temporal horizons of the organic present are limited. The level-specific causation of the biotemporal world is *intentionality in the service of biological needs.*

Eotemporality—named after Eos, the goddess of dawn—is the time of the universe of massive matter, it is the time of the physicist's *t*. Instead of trying to eliminate it by saying that it hides a direction of time, it should be taken at its word as it were, and understood as standing for a continuous and nowless time, to which ideas of a direction of time cannot be applied. The level-specific causation of the eotemporal world is *determinism.* Predictions and retrodictions here are of equal authority.

The most ancient form of time is prototemporality—for proto-, the first in a series—the time native to the world of elementary objects. It is a undirected as well as discontinuous form of time. Its level-specific causation is *probability*; here predictions and retrodictions can only be stochastic.

Finally, the world of electromagnetic radiation is atemporal. Atemporality does not signify nothingness but a condition for which none of our ordinary notions of time has any meaning. It is a state of absolute chaos, a condition believed to reign in a black hole. For an atemporal world *no mode of causation may be given meaning* for it is one of pure becoming and hence of complete unpredictability.

These are the *canonical forms* of time and causation.[8]

To grant ontological status to these epistemic conditions, the theory introduces a definition of reality, called the extended umwelt principle.[9] It asserts that what we know of the world is the way the world must be assumed to be at least—in Newton's words—"till such time as other phenomena occur by which [those beliefs] may either be made more accurate, or liable to exceptions" (Newton, 1687, p. 400).

The extended umwelt principle authorizes the assumption that each temporality is complete for the organizational level where it is identified, even if incomplete when compared with the human experience of time. Trying to imagine them as human experiences is inhibited by the difficulties of regressive sharing. It is not easy to imagine the temporal world of an infant; imagining the temporal world of a snail or a hummingbird is even more problematic. And we seem to have poor capacities to imagine ourselves into the temporal umwelts of viruses or objects traveling at the speed of light. One can proceed only by trusting the validity of reasoning.

Having identified five different concepts of time and equated each with the reality of time appropriate for the respective integrative levels, the theory makes one of those leaps of imagination that are not at all arbitrary, yet may be shown as having been justified only ex post facto because it is not possible to arrive at them through deductions from propositions admitted as correct by earlier reasoning.

That leap of imagination is the postulate that the canonical forms of time and causation are not preexisting forms of reality which become evident as nature complexifies, but that they emerge as new kinds of temporal qualities. In this view, time ceases to be a background to events and is understood, instead, as constituting an evolving aspect of reality and a correlate of complexity.

In trying to assimilate the idea that time itself has evolved, it is useful to remember that the expanding universe does not expand into preexisting space; it is space itself which expands. Likewise, the evolution of time is not a progress into preexisting forms of time but the creative emergence of increasingly more complex temporalities.

Corresponding to the canonical forms of time, the hierarchical theory of time recognizes the five types of causations I have mentioned, and sees them as having arisen from the primeval and ever-present absolute chaos. They are (1) probability, (2) determinism, (3) organic intentionality, (4) noetic intentionality, and (5) social intentionality.

Consider now the final-state diagram of Figure 1.2. It is routinely assumed that, so far as the diagram is concerned, time flows from left to right; the process depicted is one of period doubling, not period halving. What happens if one is unconventional enough to imagine that time flows from right to left?

First there is a condition of total formal unpredictability. Then there emerge elements of self-similarity. The notion of self-similarity, as I mentioned, amounts to the existence of patterns which endure, which maintain their identities in a background of pure change. The bubbles of self-similarities seen, as one progresses from right to left, are the mathematical traces of the earliest forms of permanence arising from a world of pure change.

Time itself was born when being arose from pure becoming, when permanence and the possibility of making predictions arose from pure unpredictability. I will return to this issue in the third part of the paper.

Consider next the hierarchy of causations. Because of the nested hierarchical organization of nature, the modes of prediction corresponding to the different organizational levels can never be found in pure forms; each is an amalgam of qualitatively different modes of predictions. Let me list the different types: chaotic, probabilistic, deterministic, organic intentionality, noetic intentionality, and social intentionality.

It is not possible to make probabilistic predictions without allowing for chaotic contributions.

It is not possible to make deterministic predictions without allowing for probabilistic and chaotic contributions.

It is not possible to make predictions about the behavior of a living organism without allowing for deterministic and probabilistic causation, as well as for totally unpredictable, chaotic contributions.

It is not possible to make predictions about human conduct in the service of concrete or symbolic causes without allowing for biological intentionality, as well as for deterministic, probabilistic, and chaotic contributions.

It is not possible to make predictions about the conduct of a human group, acting in the service of concrete or symbolic causes, without allowing for noetic and biological intentionalities, as well as for deterministic, probabilistic and chaotic contributions.

The contribution from chaos takes different forms upon the different integrative levels because it is shaped by the boundary conditions that inorganic evolution, organic evolution, and cultural development provide. Upon each integrative level the existing governing principles restrict the regions of absolute unpredictability to the uncontrolled niches in their lattice of rules. The laws of nature serve as filters to the infinite wealth of absolute chaos.

As a consequence, chaotic events in quantum physics, unpredictable from the probabilistic laws of the particle world, are different from chaotic, unpredictable events in astronomy, unpredictable mutations in mice, the unpredictable conduct of a person, or the unpredictable behavior of a nation. What the study of complex nonlinear systems reveals is a tunneling or feedthrough of absolute chaos into the deterministic, eotemporal world. This is but a single example, the emergence of chaos in the higher integrative levels of the physical world. Tunneling of pure becoming may also be observed in bio-, noo-, and sociotemporal processes.

Similar tunneling may also originate from integrative levels above the atemporal. Probability, for instance, native to the prototemporal world, feeds through into the deterministic, organic, cognitive, and social worlds where it takes different forms. Determinism, native to the world of massive matter, reappears in the organic, mental, and social worlds, again, taking different forms.

It is interesting how the use of computers led to the discovery that certain simple mathematical relations which have been known for centuries, represent formal expressions of the coexistence of chaos and determinism. But, it is no more remarkable than the fact that conic sections, discovered during the third century B.C. by Apollonius of Perga, were found appropriate for the description of planetary orbits nineteen centuries later, when the instruments for observing the behavior of the planets were ready to assist Kepler in his search for stable laws of nature. What all these examples illustrate, in the classic phrase of Eugene Wigner (1967), is the "unreasonable effectiveness of mathematics in the natural sciences" (p. 222f.).

A summary of the second part of my paper should say that deterministic chaos is a recognition, in mathematical form, of the feedthrough of the primeval state of total unpredictability into the evolutionarily more recent levels of nature.

The Genesis and Evolution of Conflicts

The third part of my paper suggests a particular humanistic perspective upon the ideas of formal chaos.

In the absence of universally acceptable transcendental contexts in terms of which human values could be formulated, the world at the end of the twentieth century appears extraordinarily chaotic. If there is a science whose name suggests that it might reveal order beneath that chaos, people want to know it. There is, therefore, an understandably broad mass appeal to such names as *chaos theory*, *attractor*, and *strange attractor.* And the hype of the popular press, together with the cold attractiveness of fractal forgeries, oblige.

Fortunately, the scientific study of chaotic behavior has also spawned a great deal of critical and highly disciplined humanistic work which in its intensity, though not in its extent, is reminiscent of the effects of the Newtonian theory of gravitation upon the literature and intellectual life of the eighteenth century.

What may be the reasons for the liveliness of these responses among humanists? For an answer, I will again enlist the assistance of the hierarchical theory of time. But at this occasion I want to appeal not to its structural but to its dynamical perspectives. Those perspectives interpret the processes of the different integrative levels as constituting an open-ended hierarchy of unresolvable, creative conflicts.

By conflict is meant the coexistence of two opposing trends, regularities, or groups of laws in terms of which those processes may become intelligible.

Unresolvable means that if such conflicts cease, so does the integrity of the process. *Creative* means that under certain conditions the conflicts may give rise to the processes of a higher integrative level—itself subject to the principles of unresolvable, creative conflicts. Upon each organizational level the opposing trends of growth and decay, the rise of order from and the relapse of order into chaos define each other.

Society, for instance, is said to be a group of people organized for a common purpose. This is correct, but it is only half of the story. The whole story is a ceaseless conflict between social growth and social decay. If that conflict disappears there may still be people around, but society will have collapsed.

The minding process—the use of one's mind—certainly involves self-organization in the brain. But again, this is only half of the story. The whole story is one of a ceaseless conflict between processes of self-organization and those of disorganization. When that conflict ceases there may still be a living body but not a person.

Life is also usually thought of as one of self-organization and growth. But the whole story is that of a conflict between growth and decay, coordinated from instant to instant in the organic present. If the conflict ceases the matter of the body will remain, but it will not be alive.

Unlike in the social, mental, and life processes where the growth and decay contributions are impossible to disentangle, processes of the physical world may sometimes be thought of as consisting of growth or of decay. Still, solid matter not only arose from radiation but can return to it promptly, such as upon the dropping of an atomic bomb.

Ninety years ago the poet Rupert Brooke wrote about the dream of a fish: "Somewhere behind Space and Time/is wetter water, slimier slime." In the ritualized dream called natural philosophy, absolute chaos is the slimier slime of the primeval state of the universe as well as of its ever-present foundation. If so, then the simultaneous ordering and disordering processes of the world lead our contemplation back to the epoch when the Ancient of Days created the world. The Biblical wisdom, "for dust thou art and unto dust shalt thou return" must then be revised: "for chaos thou art and unto chaos shalt thou return."

The reasoning suggests a need to reconceptualize cosmic creation. Philosophical, religious, and scientific cosmogonies identify the origin of the world with the emergence of order from chaos. This is a true but truncated perspective. A complete view of Creation must insist that what emerged from the primeval, absolute chaos was not order but conflict between orderability and disorderability, and that the history of the universe has been one of the generation of increasingly more sophisticated, new forms of conflicts.

Among the simplest types of conflicts are the physical ones, such as those on Shakespeare's seashore, where soil and water struggle, "increasing store with loss and loss with store." Among the most sophisticated ones is the noetic conflict, informed of such dilemmas as whether or not to "take arms against a sea of trouble and by opposing, end them." But common to all conflicts along the integrative levels of nature is that they are between different forms of being on the one hand, and the chaos of pure becoming, on the other hand.

The logistic diagram of Figure 1.2, scanned from right to left, as I remarked earlier, suggests that what was born at Creation was the possibility of self-similarity or being. Everyone seems to be laboring to answer the question how becoming may arise from being or, in Zeno's paradox of the flying arrow, how motion may arise from rest. But there is nothing in nature to which these questions could correspond. What we observe is the coming about of different forms of being from the mother lode of chaos, the coming about of different forms of rest from the mother lode of motion. It is not becoming which was born of being but the other way round: it is being, permanence, and continuity that emerged from the primeval state of pure impermanence and discontinuity. Once born, being and becoming began their journey of unresolvable, creative conflicts between corresponding types of ordering and disordering processes. The human experience of time, I believe, is a recognition of these conflicts in ways appropriate for the human body, mind, and society.

The reconceptualization of Creation I am recommending does not add new data to what is known about the origin of the universe nor does it prohibit anyone from seeing order arising from chaos; it is only a reminder that there is no up without down, forward without backward, ordering without disordering. It tells us that the origin of time ought to be identified with the coming about of being from becoming. The evolutionary making of a diamond from chaos amounts to the creation of permanence. More precisely, the creation of processes, manifest to us as structures, which may represent our idea of permanence because a diamond outlasts the lifetime of a man or woman, by far. Some of the consequences of this reconceptualization I have discussed elsewhere; right now, I want to return to the logistic equations and the humanities.

The significance of the logistic equation examined earlier is that it can accommodate deterministic order and formal chaos in the selfsame expression. This is another example of Wigner's unreasonable effectiveness of mathematics in natural science, already mentioned. To use a strange mathematical phrase, the technology of computing removed the degeneracies from the logistic equation; that is, it helped reveal its hidden potentialities. What was found was a common formal home for the opposition between ordering and disordering.

Leaving the Pythagorean path of numbers for the Protegorean idea of man the measure of all things, I would say that the humanistic interest in nonlinear science comes from the recognition that an exact science has finally discovered, in the mathematical traces of the phenomenal world, the coexistence of predictability and unpredictability. Creation celebrates the coming about of being or permanence, and with it predictability. Unpredictability or becoming has been there, all along.

Recognizing the power and presence of conflicts may be new in mathematics but it is what the arts and letters have been doing all along. Artists and writers have not been changing chaos into order but have been creating and perpetuating conflicts. Whether intending to put the world afire, or speak of thoughts wise or foolish, or trace feelings distant or intimate, artists and writers deal with the unresolvable, creative conflicts of man, animal, and matter. Through that work they not simply report on, but actively partake in the cosmic process of creation and destruction.

Great pieces of art contain in their sensitivities those oppositions between order and chaos which certain nonlinear equations reveal in their mute and primitive umwelts. The period doubling route to chaos is an Apollonian recognition of a Dionysian reality, "Double, double toil and trouble / Fire burn and cauldron bubble."

To join the efforts of those who have been introducing the human agency into the deliberations of formal chaos, and also to sum up my chapter, I propose to leave the innocent Brazilian butterfly to stir up a storm in Texas. Instead, I turn to a not so innocent Dr. Frankenstein, on his way somewhere in the Swiss mountains, to meet the human storm of his imagination, idealism, and labor (M. W. Shelley, 1818). "The path, as you ascend higher, is intersected by ravines of snow, down which stones continuously roll from above; one of them is particularly dangerous, as the slightest sound, such as even speaking in a loud voice, produces a concussion of air sufficient to draw destruction upon the head of the speaker" (p. 93).

The Frankenstein effect, as I would like to call it, may serve as a logo to humanistic perspectives of the study of nonlinear systems. It is a reminder of the fateful power of the human voice, even if whispered, in establishing conditions to which the human predicament is extraordinarily sensitive.

References

Fraser, J. T. (1975), *Of Time, Passion, and Knowledge*, 2nd ed. Princeton, NJ: Princeton University Press, 1990.

—— (1978), *Time as Conflict: A Scientific and Humanistic Study*. Basel: Birkhäuser.

—— (1982), *The Genesis and Evolution of Time*. Amherst, MA: University of Massachusetts Press.

—— (1987), *Time, the Familiar Stranger*. Amherst, MA: University of Massachusetts Press & Redmont, WA: Tempus Books.

Glass, L., & Mackey, M. C. (1988), *From Clocks to Chaos: The Rhythms of Life*. Princeton, NJ: Princeton University Press.

Hilborn, R. C. (1994), *Chaos and Nonlinear Dynamics*. New York: Oxford University Press.

Kellert, S. H. (1993), *In the Wake of Chaos: Unpredictable Order in Dynamic Systems*. Chicago: University of Chicago Press.

Kirk, G. S. & Raven, J. E. (1957), *Presocratic Philosophers*. Cambridge: Cambridge University Press.

Newton, I. (1687), *Sir Isaac Newton's Principles of Natural Philosophy and His System of the World*, tr. Andrew Motte, rev. Florian Cajoli. Berkeley: University of California Press, 1962.

Peitgen, H.-O., Jürgens, H., & Saupe, D. (1992), *Chaos and Fractals: New Frontiers of Science*. New York: Springer Verlag.

Prigogine, I., & Stengers, I. (1984), *Order Out of Chaos*. Toronto: Bantam Books.

Republic X, 597b. In: *The Collected Dialogues of Plato*, ed. Edith Hamilton, & Huntington Cairns. Princeton: Princeton University Press, 1961.

Shelley, M. W. (1818), *Frankenstein*. New York: Penguin Books, 1983.

Stewart, I. (1989), *Does God Play Dice? The Mathematics of Chaos*. Cambridge: Blackwell.

Wigner, E. (1967), The unreasonable effectiveness of mathematics in the natural sciences. In: *Symmetries and Reflections*. Bloomington: Indiana University Press, 1967.

7. THOSE METAPHYSICAL DEVICES

In the taxonomic organization of the species our full home address is as follows. Kingdom Animalia (animals), phylum Chordata (with backbones), subphylum Vertebrata (segmented spinal cord), class Mammalia (suckles the infant), subclass Eutheria (gestates its young with the help of a placenta), order Primates (has five-digited extremities and a collarbone), suborder Anthropoidea (has stereoscopic vision and a large brain), family Hominidae, genus Homo, species Sapiens. *Homo sapiens*. Man the wise. But this is only one of our alter egos, the one that has been searching for meaning and order in the open-shelved library of our passions and modes of knowledge.

The other alter ego, *Homo faber*, is man the craftsman. He and she have been making useful objects and arrange the details of activities. He and she organize our lives with the assistance of natural processes that they judge stable and therefore, predictable. Such regular, predictable processes are the functional bases of clocks and calendars.

In the earlier chapters we met revolting slaves, believing mathematicians and brilliant philosophers. The essay that follows is about devices made by ingenious tinkerers, by people who constructed them for social, religious, metaphysical and scientific uses. All our fellow creatures possess biological clocks or, more precisely, are made of such clocks, as we shall see, by-and-by. But only homo faber *makes* clocks.

All clocks employ one or another of the regular processes of the universe or are constructed to be such a process, themselves. Their "regularity" and "stability" are judgments by the clockmaker. Clocks are also, and often, objects of art. They are ticking, chiming, blinking, smoldering reminders of our passage from cradle to grave and hence of love, passing, joy, sadness and beauty. They connect each clockwatcher to the change ringing cosmos, to his and her local environment and to the lives and thoughts of their fellow humans. They are the most general machines. It is not surprising, therefore, that they readily lend themselves to serve as metaphors across the vast domain of the arts, the letters and the sciences. In their structures and functions, they reach much beyond themselves.

CLOCKWORKS BEYOND THEMSELVES

My backyard is a permanently changing exhibit of clocks and calendars. One of those clocks comes each morning at 9:25—as measured on another clock—and sits on the top of the birdfeeder. After a while it gets frustrated by not being able to reach the seeds and with a shake of its bushy tail, goes elsewhere. Our calendars come male and female on the same tree. They signal the arrival of June by mating, because in the spring a pinecone's fancy lightly turns to thoughts of love.

Complementing this earthly exhibit we also have a heavenly clockwork whose shiny indicators follow their regular rounds from east to west. During the day the sun does the same. In its yearly journey, though, it moves from west to east with respect to the constellations, telling us to celebrate the calendrical rites of time's passage.

The objects in the *Clockwork* exhibition share with squirrels the capacity to serve as clocks but unlike them, they also carry statements about human values and concerns. More than any other machine, clocks and calendars are scientific, social, artistic and philosophical devices as well. One cannot say much about them without an appeal to our understanding of nature and to the needs of man, the clockmaker. This essay illustrates the extent to which philosophical traditions, unquestioned beliefs, unnoticed social conventions and pragmatic common sense enter the apparently simple act of telling time by the clock.

The broad tapestry of issues is surveyed under four headings, each of them stated in the form of a question. They help us focus our inquiry but they do not pretend to be independent of each other. On the contrary, they are interwoven aspects of what is meant by time measurement, as seen from the perspective of the interdisciplinary study of time.

I want to begin with a summary that consists of the questions, with the answer given following each of them. Sometimes the adventure is not in any mystery of where one is supposed to arrive, but in finding a path that leads there.

Summary:

What do clocks reveal about time?
Nothing.

How does a clock measure time?
It does not. We need at least two clocks or clocklike processes to measure time.
How can we tell whether a clock is right?
By comparing its readings with those of other clocks, then judging all of them together in terms of what we think they ought to be showing.
Why is a clock a metaphysical machine? Because its works and displays reflect the unquestioned beliefs of its makers concerning the nature of time.

What do Clocks Tell us About Time?

By "time" one usually means the idea of passage as well as the experience of future becoming past with respect to an ever-changing present. The question of the title may, therefore, be rephrased. What do clocks tell us about the passing of time?

Let someone reverse the conventional sequence of numbers around the periphery of a clock or mix up their sequence, then let us watch the hand crawl from *12* to *11* to *10* or, say, from *12* to *4* to *9*. No one will say that time began to flow backwards or else it is out of joint (at least on account of the clockface). People would say that someone designed an unconventional dial. For we bring to the reading of clocks our awareness of time's passage. The experience of a present, wedged between expectations and memories, between future and past with respect to the now, cannot be read off the clock.

But then, in their role as timekeepers, what do clocks do?

They count events which we judge to be uniformly repeatable and assign some kind of identification to them. They may be pictures, such as numerals, or else directions with respect to a reference direction. They may be bars of music as in the Westminster Quarters of Big Ben in London, based on four notes of the aria, "I know my Redeemer liveth" from Handel's *Messiah*. The events counted may be the number of candles burnt (as was traditional in Japanese Geisha houses) or else the changing scents of incense in certain old Chinese clocks. (Figure 1) Some clocks count drops of water, others grains of sand. A bacterial culture may also serve as a clock if the hours are laid out along the radius of the circle that is being filled by the expanding colony. My car battery could be used as a calendar if someone calibrated the height of the accumulated sludge.

The point is that any process that consists of countable events may be employed as a clock. But finally, what makes a clock a clock is *our* conviction, based on a web of philosophical, aesthetic and scientific reasons that its events follow each other uniformly, according to intelligible principles. No clock can tell us what to mean by the passage of time or how to distinguish between past and future with respect to the present.

Fig. 1. A Chinese incense clock from the 14th century, bearing witness—as do all clocks—to the values, skills and ways of life of their makers. The pattern is that of a groove carved into a hardwood disk. Incense made from a variety of aromatic powders was placed into the groove and lit, probably in the center. Burning like a slow fuse along its path, it identified the different hours by different scents. From J. T. Fraser, *Time, the Familiar Stranger*, Redmond, WA: Tempus Books, 1988.

How does a Clock Measure Time?

An act of measurement is an act of comparison between like dimensions: weight with weight, electric charge with electric charge, dollar with dollar, length with length. My yardstick may be laid down along our driveway 85 times: the driveway is 85 yards long. I could use a 100 yard tape measure—awkward perhaps, but it helps explain length measurement—lay it down along the road and note two coincidences. One end of the road was (let us say) at the numeral *97*, the other end at the numeral *12*. The road would still be 85 yards long. Let me formalize the spatial coincidences I had to identify: *road's end at 97* and *road's* (other) *end at 12*.

Yardsticks and tape measures only *are* whatever they are, but clocks always *do* something; we say they measure time. Yet, all that a clock can do is to supply and count units of process, a service which in itself is no more a time measurement than owning a straight edge is a length measurement. An act of time measurement must involve a comparison of the counted reference process with another process. What is the other process when I look at my watch or sundial and solemnly declare that it's eight o'clock?

I would like to illustrate what we do whenever we measure time by visiting a shepherd, his lover, his dog and his sundial (with directions on it carefully numbered in the 24 hour system). Here is the short-short story.

When the shadow on the dial points to *18* the shepherd's lover appears, carrying a dish of baklava. When the shadow points to *8* the dog barks.

Let us recall the length measurement, including the two coincidences observed (*road's end at 97* and *road's end at 12*). When measuring time we could also talk about coincidences but it is more telling to speak of simultaneities. What are the two sets of simultaneities that we may use to measure time in the pastoral story? They are "lover arrives at 18" and "dog barks at 8" or, in an abbreviated form, *lover at 18* and *bark at 8*.

But, knowing these simultaneities does not yet amount to a time measurement. In the case of the road, the direction of the road did not matter. In time measurements, the direction of time always does. We must know which of the two simultaneities happened first and whether or not the sun had risen more than once between them.

Fortunately, both people remembered that she arrived in the afternoon, that just before sunrise he had told her that she was sweeter than the baklava and that the dog barked when she left. Whatever else might have happened is irrelevant. Important is that a time measurement has been made: they spent 14 hours together.

What were the two processes involved? One was the motion of the shadow, the earthly trace of a heavenly process which we judge regular, continuous and intelligible. The other was the events of the shepherd's life, woven into a continuous, intelligible pattern. Had the shepherd lived on Mars, his lover on Venus and the dog on the Dog Star (better known as Sirius), there would not have been a single, homogeneous story (process) and no time measurement could have been made.

What are the two processes and two sets of simultaneities in the time measurement, "Lincoln was born in 1809?" One process is the flow of Western history with its events woven into a continuous, intelligible pattern through time. Let us call it the history scale. The other is the flow of astronomical events, woven into another continuous and intelligible pattern on the astronomy scale. We selected certain events which repeated themselves, such as the sun crossing the celestial equator on its way north, and began to count them. On the history scale we assigned the numeral *1* to the sixth such event after the birth of Christ and noted that the birth of Lincoln (also on the history scale) coincided with event called *1809* (on the astronomy scale).

The remark about assigning the numeral *1* to the sixth year after Christ's birth gives us a glimpse into the unobvious intricacies of chronology. Dionysius Exiguus (Dennis the Little) was a Scythian monk who lived in Rome. At the request of Pope St. John I he computed the year of Christ's birth. Using one of the chronologies current in his days, he identified the year as 753 A.U.C. *(Ab Urbe Condita* "from the founding of the city [of Rome]"). The year so identified became the epoch (reference year) of the Christian calendar. By his computation Dennis the Little made himself live during the first half of the sixth century. But for some reason, his calculations were a little off. Modern scholarship places the birth of Christ in 6 B.C.

If we leave the very complex worlds of history and life and enter, through physics, the much simpler world of non-living matter, the explicit identification of processes and simultaneities involved in a time measurement becomes obligatory and, in general, clear-cut. Conceptual difficulties arise only when we deal with those integrative levels of nature where our garden-variety ideas of time do not apply. Difficulties or not, time measurement becomes the very means through which scientists search for order. A law of nature, for instance, is a stable principle that tells us how to time the goings-on of one process by means of the goings-on of another process, and vice versa.

The reader may now amuse him- or herself by identifying the processes compared and the two sets of simultaneities involved in the following results of time measurements: "The switching time of our latest device is one-hundredth of a nanosecond" and "The universe was created fifteen billion years ago."

Summing it up: Time measurements always involve at least two processes that unfold in a continuous and intelligible manner, and two sets of simultaneities of the *lover at 18* and *bark at 8* type.

How do We Know that the Clock is Right?

For this essay, a clock will be said to be accurate if the length of a second, a minute, an hour, a year—any unit of time—as measured by that clock is exactly what we had agreed to call a second, a minute, an hour or a year. A clock would be absolutely right if its time units, put end to end, would flow as uniformly as would (an assumed) cosmic time which, one could further assume, flows uniformly whether or not we have a gadget with which to measure it.

If I doubt the accuracy of my clock, I can readjust it with the help of another one which I judge to be more accurate. That second one, in its turn, may have to be adjusted with the help of a third. How would I know that I have arrived or at least I am approaching ideal accuracy and therefore, show the right time? Here is a plan of attack.

Let us manufacture 100,000 of the best clocks, made to the same exacting standards. Let them run simultaneously and let them vote by their ticks: the majority ought to be judged correct or at least offer the best approximation we could achieve. But, what if the clocks have a systematic error which, let us say, makes them identically sluggish after lunch? No democratic vote would ever reveal that to be the case, yet their time signals would differ from those of other clocks. Which ones should we believe? Let us expand our horizons and use the sun as our clock: it is surely a reliable one. But the sun does not move uniformly, as may be seen by plotting the position of the shadow on the shepherd's dial against the signals of the sandglass that is turned each time it finishes its cycle. The sun regularly moves ahead and falls behind the time indicated by the dancing sandglass or by the corps de ballet of many sandglasses. If based on a precise theory of the sun's motion we could write an equation to help us calculate, instant by instant, by how much it runs ahead or falls behind an ideal, uniform scale of time, then the solar clock could be corrected. We could always tell the right time or at least a good approximation thereof. This is what we have to do.

First we must formulate an idea : that of a uniformly flowing time. Next, we have to learn the laws that govern the motion of the sun and finally, solve the equations to help us discover by how much we must correct the sun clock, from instant to instant, to give us the indications of the ideal clock (which we failed to be able to construct from actual clocks). These calculations can be performed but—a very important "but"—the need for them makes time measurement an entirely new game. It ceases to be the reading of the position of a heavenly hand on the Platonic dial. It becomes a function of what we *think* the sun ought to be doing. But what we think the sun ought to be doing depends on our understanding of the laws of nature—and that changes each time there are new managers in the front office: Plato & Aristotle, Galileo & Kepler, Newton & Einstein and others to come.

How about the planets? The story is the same. No observational data can serve as a time signal unless first corrected by our understanding of the laws that govern the motion of the planets.

How about the fixed stars that form, at their immense distances, the ultimate framework of the astronomical universe as seen from earth? They are not really fixed but one can cement together an imaginary framework using primary radio sources and assume it to be fixed—until further observations and theories are developed.

Having come this far, our measurements will be precise enough to make us become concerned with the motion of the earth. The earth rotates about its axis and revolves around the sun; the orbit itself revolves around the sun and the plane of that orbit precesses (moves like a spinning top whose axis is not vertical). The axis of the earth also precesses as well as nutates (weaves in and out of the circle of precession) and has other predictable and unpredictable motions. Also, the axis of the earth moves with respect to the surface of the earth. If readings upon our sophisticated clock hand (the framework of fixed stars) differ from those of other clocks (sun, moon, planets, earthly clocks) which clock or clocks are we to believe? Let us try atomic clocks. Two of the best ones, laid side by side, will need three million years to get out of phase by one second. This is an impressive achievement, made more so considering that we know how to measure the difference without having to wait that long. But which of the two identical clocks should be believe? Besides, atomic clocks cannot give us the time of day—that reading must come from astronomy. But alas, the length of a second defined for and obtained from atomic and astronomical readings shift with respect to each other.

Let us give up this wild goose chase and reach our conclusion. There is nothing in nature to which our idea of a uniformly flowing time would correspond, nothing that our increasingly more precise time measurements could be said to approach. (This conclusion is not to be confused with the motional and gravitational variations of time in relativity theory.) Instead, we have a myriad of processes usable as clocks. We select some of them, appeal to our understanding of what they ought to be doing and correct their readings accordingly. Finally, we employ their ticks, tocks, vibrations and winks to help organize our individual and collective lives in pursuit of our plans and hopes.

We may paraphrase a well-known remark of the naval officer Stephen Decatur during the War of 1812. "May our clock always be in the right; but our clock, right or wrong." All clocks are always wrong, compared with the imagined uniform flow of time; all we can discuss is our guesses of by—approximately—how much.

Why is a Clock a Metaphysical Machine?

A metaphysical belief is an assertion about the nature of ultimate reality, one that appears to be a self-evident truth to people of an age and at a place. The history of timekeepers demonstrates that metaphysical beliefs have been built into the ways clocks function.

It all began, I suppose, when a hairy, intelligent creature noted that a tree's shadow changed its direction during the day, and its length through the year, with reliable regularity. The crucial point is that he—she, they—judged those changes more regular and reliable than the cycles of their own needs. Therefore, to continue my speculation, they used the sun's motion to time their life. What reasons might have been given for this trust in the sun? Surely, for millenia, the question itself had no way of arising. It was all—as we would say it today—a fact of life. Later, through the advanced religions and the philosophical genius of ancient Greece—and beyond—the answers would point to a deity or deities. Since the rise of natural science the received answer has been: the laws of nature. That the laws of nature are absolutely reliable and permanent, though subject to revision as our understanding grows, is a metaphysical belief.

In the regularity of the moving heavens Plato perceived a representation of eternity—by "eternity" he meant a changeless, ultimate unity—and contrasted it with time. God, he wrote, created the rotating celestial sphere to serve as an image of eternity that moved "by number," that is, according to numerical law. That image of eternity, he added, we call time.

Consider a clock dial, numbered along its periphery from *1* to *12* in the sense we call clockwise. Remove the long hand and place into the clockwork two gears so as to make the small hand make one clockwise revolution in 24 instead of 12 hours. Renumber the dial and direct *24* to the north: the small hand would then follow the motion of the sun or rather, that of an evenly moving sun. You are holding "a moving image of eternity," a representation of the Platonic metaphysics of time.

Thus far the issue of the display. How about the inside of the clock?

The oldest known ancestor of geared clockworks is the Antikhytera mechanism, a complex gearwork salvaged by pearl fishers in 1901 near the Greek island of that name. A hand-driven device with extraordinarily clever gearing, it showed the length of the synodic month (new moon to new moon), the sidereal month (one revolution of the moon with respect to the fixed stars), the length of the year, the location of the sun along the ecliptic and the location of four naked-eye planets. This remarkable clockwork-before-clocks was a model of the machinery of the heavens, a first century B.C. planetarium. It demonstrated its maker's belief that the motion of the stars was governed by mathematical laws, represented by his gear ratios. No one knows in what words he would have stated his philosophy but I cannot imagine it to have been anything beyond the assertion: that is how it is.

Nothing comparable to the Antikhytera mechanism may be found in the history of clockworks until the appearance of the great astronomical clocks sixteen centuries later. Those astronomical clocks internalized in their mechanical features the hosts of intricacies that their makers believed had governed stellar, planetary, solar and lunar motions. Though they were precise ecclesiastical calendars and showed the time of day, their prime purpose was to demonstrate the lawfulness of the world and through that demonstration praise the ordainer of order, the God of the Christian universe. (Figure 2) Plato's student, Aristotle, defined time as the "number of motion in respect to 'before' and 'after'." Consider a digital watch: its clockwork counts regularly repeated events of one sort or other, assigns numbers to them, then displays the "number of motion." You are holding a representation of one half of the Aristotelian metaphysics of time. The other half of Aristotle's definition of time ("in respect to 'before' and 'after'") must be supplied by the clockwatcher's sense of time.

I said earlier that the usual clock dial and the gearwork moving the hands represent the Platonic metaphysics of time. It would have been more felicitous to say that they served as metaphors for that metaphysics. Likewise, a digital display may serve as a metaphor for the Aristotelian metaphysics of time.

Digital watches are also metaphors for the prevailing attitude to history. Time ceased to be a path for the soul to reach eternal life; the motion of the planets became quite irrelevant to human affairs as did, to a great extent, the motion of the sun. When it gets dark we turn on the light, never mind the sun. We even erect buildings without windows. Also, the density and swiftness of communication networks across the earth created a set of new conditions I have been describing as those of the time-compact globe. These new features of social life include the narrowing of collective attention to problems of a limited present, with little genuine readiness to meet long-term contingencies anticipated with the aid of long term memories. Although the Platonic clock dial did not tell us what we are to mean by future and past with respect to the present, it did at least offer a visual metaphor for the passage of time. The digital watch is a visual metaphor for a reality that is limited to a narrow present.

With these remarks in our memory bank or wallet (as in "Time hath, my lord, a wallet in his back/Wherein he puts alms for oblivion") let us turn to time.

Temporal experience, more than any other aspect of existence, is all-pervasive, intimate, and immediate. Time, life, and death meet in an experiential unity which has been recognized in all great religions and has been the staple of all great philosophies. Time is also a constituent of all human knowledge, experience and mode of expression, it is intimately connected with the functions of the mind and appears to be a fundamental feature of the universe. Explicitly or implicitly, time has occupied the center of man's intellectual and emotive concerns.

A consequence of this central role of time in human affairs is that the question, "What is time?" is emotionally loaded; answers to it tend to be dogmatic. For physics, it is generally an undefined *t*, for biology the cyclic and aging orders

Fig. 2. The Strasbourg astronomical clock, first finished in 1574, rebuilt in 1842. Though it is remarkable for its precision, its primary purpose was to serve as a model of the universe that works according to mathematical order under the tutelage of the Christian deity.

The clock shows solar, solar mean, and sidereal times, the times of the sunrise and sunset, the motion of the moon and of course the days, months, and years. It takes into account leap years, including their fine-tuning: from among century years only those divisible by 400 are leap years. It shows the dates of Easter, the movable feasts, fixed feasts and the names of saints celebrated each day. It gives advance notices of lunar and solar eclipses, shows the position of the year in the 28 year cycle (after which the days of the months return to the same days of the week) and indicates the precession of the equinoxes. A seven day succession of allegorical figures represent the days of the week. In another stage a genie strikes a bell every first quarter hour; the second, third and fourth quarters are struck by a young man, a man and an old man. After the last stroke of 12 noon the twelve apostles appear, pass beneath the figure of Christ, each turns toward Him, bows and receives His blessing.

The clock does much more than what I have listed, leaving some surprises for the visitor to discover.

From J. T. Fraser, *Time, the Familiar Stranger*, Redmond, WA: Tempus Books, 1988.

of life, for geology the eons of the earth's evolution, for psychology a mode of perception, for sociology a convention of the group, for Christianity an arena in which God's will may be manifest, for the lover or the dying an enemy or friend. The list may be extended.

When we speak of or contemplate a clock—unlike when we contemplate a standard voltage cell, a kilogram weight or a yardstick—our memories, hopes and fears, both conscious and unconscious, become easily mobilized. A clock is a master process and hence a reminder of passing. And the awareness of passing and of the inevitability of death are at the roots of the tension which is responsible for the immense creativity and destructiveness of our species. It is not surprising, therefore, that clocks readily lend themselves to serve as metaphors across the complete spectrum of the arts, letters and the sciences. The present exhibition may be properly described as that of visual metaphors upon the idea of time, applied to clockworks.

8. HOW TO USE A CLOCK

Clocks—those metaphysical instruments that go beyond themselves—help coordinate the activities of our daily lives. They help us learn about the habits of our fellow creatures with whom we share the earth. They make possible the keeping of records about the motions of the planets and stars. They are even useful when we are away from home: think of the long and fascinating history of determining longitudes at sea.[1]

One may imagine an advertisement in the paper: HAVE GOOD CLOCK, WILL NAVIGATE. After all, a good clock is a good clock. But, during the twentieth century, advances in our understanding of the nature of time revealed some totally unexpected problems in comparing clock readings. Such as: relating the rate of time's passage, as measured by a clock here, to the rate of time's passage measured by an identical clock at a distance, must employ a family of theorems that have counterintuitive details. Measuring time at a distance brings up questions that travelers of old would not have thought of asking. For an introduction to this intricate matter, see the relevant sections in Whitrow's magisterial *The Natural Philosophy of Time*.[2]

Starting, as humans did, with moving shadows, burning incense and swinging pendulums, how did we get to this strange new world of clock use? Let us begin with our old friends, the disciples of Pythagoras. For the members of that ancient brotherhood, as already mentioned, what today we would identify as religion and science were not different domains of thought, but a single fabric of a way of life. That coherence remained alive throughout the history of Western thought. Immanuel Kant expressed it this way: "Two things fill the mind with ever new and increasing wonder and awe, the more often and the more seriously reflection concentrates upon them: the starry heaven above me and the moral law within me."[3] This tie between man and the cosmos paved the way for the employment of time measurements for a better understanding of the universe. A system of such an understanding, grossly misnamed relativity theory, is an amalgamation of Western philosophy with the skilled use of geometry, an imaginative command of natural science and a desire to seek whatever seems permanent in nature, whatever seems undying in this world. The essay that follows considers the instructions of the theory of relativity about correlating clock readings so as to make the laws of physics remain formally independent of where, when and under what conditions they are applied. The theory makes the laws of nature appear eternal and hence, makes our old enemies—passage and death—appear insignificant.

But also—importantly and interestingly—the reasoning of the essay shows that the vast intellectual edifice of relativity theory, the dance of its geometer god, does not reveal anything about the sources of time's passage. Relativity theory tells us how to correlate clock readings but the passage of time, as an experience and an idea, must be brought to those calculations by the living, and thinking homo faber.

SPACE-TIME IN THE STUDY OF TIME[1]
AN EXERCISE IN CRITICAL INTERDISCIPLINARITY

Abstract

Space-time is a four-dimensional abstract space, an analytical tool used in carrying out the program of relativity theory. That program requires that the laws of physics be stated in mathematical forms that remain unaffected—that are not changed—by the motion of the coordinate system in which they are applied. The nature of time, as understood in the context of space-time, differs from the common idea of time in that it makes no provisions for a "now"—for a present—and for the passing of time. Ideas about a present instant and about the passage of time have to be imported into physics from domains of knowledge outside physics. With the help of evolutionary epistemology based on evolutionary ontology, this paper locates the time of relativity's space-time in the integrated, interdisciplinary study of time.

Introduction

An integrated study of time, built on an interdisciplinary foundation, faces certain difficulties because of the necessary compartmentalization of knowledge. They include the problems of specialized vocabularies, used by the many and different fields upon which the study must draw.[2] Those vocabularies often employ words that signify ideas which are different from what the same words mean to the uninitiated. Unless the specialized meanings are carefully defined, confusion follows.

Consider, for instance, two words in common use: "space" and "time." When, in the physical sciences, they are joined in the notion of "space-time," the hyphenated expression acquires a special meaning. Eric Kincanon already warned against the uncritical use of physical theorems for gaining an understanding of time.[3] By exploring the nature of time in the space-time of relativity theory, this essay extends Kincanon's warning to the uncritical use of the vocabulary of physical theories in making generalizations about "time that surveys all the world" and not only the physical universe.[4]

1. The space-time of relativity theory

Interest in the natures of space and of time spans the history of Western thought, but the concept of space-time is of recent origin.[5] It was first used early in the 20th century in the context of Einstein's theory of relativity and Minkowski's refinement of the mathematics of that theory.

The theory of relativity, of which space-time is an analytic tool, is built on an assertion that was intellectually as brave as, if less life-threatening than, Galileo's *Eppur si muove*. Einstein asserted that "the phenomena of electrodynamics as well as of mechanics possess no properties corresponding to the idea of absolute rest."[6] Absolute rest, if it did exist, would have been a cosmic, experimentally identifiable zero speed with respect to which absolute speed could acquire meaning. If there was no absolute rest, no meaning could be attached to absolute motion. Only relative motion and relative velocities have meaning. The Special or Restricted Theory of Relativity, so named eleven years after it was published, replaced the unidentifiable absolute rest of Newton's natural philosophy by an experimentally identifiable absolute motion. By absolute motion is meant a cosmic reference velocity, that of light, which may be identified anywhere and has the same value irrespective of the reference frame, everywhere.

The recognition that only relative motion has meaning is essential to Einstein's reasoning. But the name "relativity theory" is misleading because the theory does not concern itself with what is relational but with what is absolute; it does not focus on what is changeable but on what is permanent. It demands that "the laws by which the states of physical systems undergo change [should not be] affected, whether these changes of state be referred to the one or the other of two systems of coordinates in uniform translatory motion."[7] None of Einstein's papers have the phrase "relativity theory" in them until 1911, long after others began to refer to it that way.[8] In 1910 the mathematician Felix Klein suggested the name *Invariantentheorie*, the theory of invariance.[9] This is an unassailably correct name for a theory whose essence is that "only those equations are admissible as an expression of natural law which do not change their form when the coordinates are changed by means of a Lorentz transformation...."[10] By "natural law" Einstein seems to have meant the laws of physics.

Searching for invariance is, in itself, of interest to the study of time because it illustrates the human need for seeking examples of permanence. Such examples serve as reassurances against the fell hand of passage, as witnesses to the unimportance of time and by inference, to the unimportance of death.[11] As the mathematician Edmund Halley said in his ode that introduces Newton's *Principia*,

> ... Here ponder too the Laws which God,
> Framing the universe, set not aside
> But made the fixed foundations of his work.[12]

It is in seeking some of those fixed foundations that Einstein's theory of 1905 employs the idea of space-time. Specifically: the geometry of space-time makes possible a form of record keeping of distances and times in such a manner as to keep the equations of physics invariant—unchanged in their forms—for all observers in linear translation. The name "Special" or "Restricted Theory of Relativity" refers to the limitation of the validity of the theory to observers (coordinate systems) in linear translation. The General Theory of Relativity (1916) extends the validity of the theory to classes of observers in nonlinear translation.

In three consecutive steps, I will now sketch the nature of time as it appears in the space-time of Special Relativity Theory. (i) A description of the coordinate system employed in the geometry of space-time will reveal the sense in which space-time may be said to be a continuum. (ii) A sketch of the method used for measuring intervals in that continuum will help us appreciate the cosmic evolution of causation and of time.[13] What will have been learned in (i) and (ii) will then be employed in (iii) to help place the nature of time, appropriate to the space-time of relativity theory, into the perspective of an integrated, interdisciplinary study of time.

(i) *What is meant by space-time being a continuum?*

The space-time of relativity theory is an abstract space. Here "space," deviating from its common meaning, is understood as a pointset of any number of dimensions: 1, 2, 3, 4, 5 or 1492. For simplicity, let each point be identified with an ordered set of numbers, one number for each dimension of the space. In the space-time of *Invariantentheorie*, alias relativity theory, each point represents a location in three-space as well as an instant in the human experience of time. Therefore, each point in the four-space of space-time represents an event. The distance between any two events is called a *space-time interval.*

Each of the four axes of relativity's space-time are taken to be orthogonal (perpendicular) to the other three. It is not easy nor is it necessary to imagine what such a coordinate contraption may look like. It is only necessary to master its geometry.

The three coordinates of experiential three-space are calibrated, as one would expect, in units of length. Not so obviously, the fourth axis, called the time axis, is also calibrated in units of length. Specifically, in the distances traveled by light in the selected units of time. Thus, the post with a sign, "1 second" also has another sign, "2.997,91 $\times$ 10^{10} cm" because that is the distance that light travels in one second. With the four axes properly calibrated, the world stage is set. However, to calibrate the time axis (in units of length) it is necessary to know the world in terms of the passage of time that is, in terms of future changing into past, in the present. I hope to show that such knowledge cannot be derived from relativity theory itself but must be introduced into it from natural processes that are not within the purview of physical science.

But first, we must attend to some other details of bookkeeping in the space-time continuum. Whereas the numerals along the three space axes are real numbers, the numerals along the time axis are imaginary numbers; they are multiples of $\sqrt{-1}$, the unit of imaginary numbers.[14] The use of imaginary units along the time axis makes it possible for that axis to be orthogonal to all the three space axes and hence serve as the fourth dimension of space-time, independent of the other three dimensions. Minkowski called equating 2.997,91 $\times 10^{10}$ cm—mentioned above—with $\sqrt{-1}$ sec. a "mystic formula."[15] Yet, what impresses this writer the most—and would tempt him to call it mysterious—is not the decision to employ imaginary numbers along the time axis, or the many types of number within mathematics, such as real, imaginary, complex, natural, rational, irrational, transcendental, cardinal, ordinal, transfinite etc. but the fact that numbers and their relationships are so universally and unfailingly appropriate for the description of natural processes.[16]

In conclusion: the "continuum" of space-time does not mean that, within space-time, the extra cubic inches of your body may be changed to extra years of your life, nor that the ticks of your clock may be changed into watermelons. It means that in the handling of space-time intervals, to be discussed below, one does not deal with periods of time or with different distances but, consistently, exclusively and only, with space-time intervals.

(ii) *Space-time intervals and the evolution of causation and of time*

Time is already in space-time, there is no time external to space-time. It follows that a space-time interval is not a line along which an observer could see a body move and, if he wishes, even measure its speed. A space-time interval is a trace that lies timelessly in a four-dimensional mathematical space.

How is one to measure the magnitude of a space-time interval? By extending the Pythagorean theorem to a four dimensional mathematical space and, with its help, assigning a number to the interval. Here are the details.

The distance between two points in a Cartesian plane may be calculated by the Pythagorean theorem. If dl is an element of that distance and if dx and dy are the orthogonal projections of dl upon the x and y axes, then $dl^2 = dx^2 +$

dy^2. In a three-dimensional space, if dx, dy and dz are the orthogonal projections of dl upon the x, y and z axes then $dl^2 = dx^2 + dy^2 + dz^2$.

Next, on to four-space. If the time taken for a particle to travel a short distance is dt—as measured by a clock traveling with it—then the projection of dt upon the time axis (calibrated in units of length) will be $cdt\sqrt{-1}$, as explained. I also mentioned that the numbers along the time axis are imaginary. In the space-time convention, therefore, the distance will be cdt.

The space-time interval between two neighboring events is conventionally represented by ds, an element of such an interval. Let a particle—or for that matter, a planet—move between two close-by events. The orthogonal projections of ds upon the four axes will be dx, dy, dz and $cdt\sqrt{-1}$ We obtain $ds^2 = dx^2 + dy^2 + dz^2 - c^2dt^2$

The total space-time interval—the geometrical sum of all the elementary distances ds between the two events whose separation in space-time we want to measure—is the new absolute: all observers in linear translation will assign to it the same number.

The trace, lying timelessly in space-time, represents an interval along which, depending whether the value of ds^2 is negative, zero or positive (three conditions to which I shall later return) information transfer and with it, causal connectedness has, or may or may not propagate.[17] But what kind of causal connectedness has actually propagated or may propagate, cannot be learned from the trace alone. For an answer we must turn to the evolutionary history of the cosmos and decide upon the epoch for which the trace is taken to be valid. Here is that history, reversed in time, along the story of evolution.

The level-specific causation of living-and-thinking matter—of us humans—is long-term intentionality. Our reality includes an open future and past, referred to the mental present where the future changes into past. It is this form of causation that makes human civilizations possible.

The level-specific causation of living matter—causation characteristic of and unique to processes studied in biology—is short-term intentionality. Its reality is one of limited future and past, referred to the organic present.

Next, let us step down to the organizational levels below life.

In the world of massive matter that is, in the astronomical universe of galaxies dealt with both in Newtonian and in Einsteinian physics, causation is deterministic. It cannot and does not refer to a present, it is without preferred direction, it is one of pure succession.

In the universe of particle-waves—in the domain of Quantum Theory—causation is statistical, probabilistic.

Finally, the most primitive form of energy, that of the primeval universe, is a quantum foam in which no kind of causation can exist. It is best thought of as a world of absolute chaos or pure becoming.

Having descended all the way to the chaotic origins as well as foundations of the universe, we may begin climbing back up. We note that the cosmologically later forms of causations did not replace but were added to the earlier forms.[18] Therefore, there can be no examples of probabilistic causation without pure becoming. There can be no examples of deterministic causation without probabilistic causation and pure becoming. There can be no examples of short-term intentionality without deterministic, probabilistic and chaotic components. And, there can be no examples of human, long-term intentionality without the short-term intentionality of the body, deterministic causation of solid matter, probabilistic causation of particle-waves and of the ever present pure becoming.

In his *Space, Time, and Spacetime* Lawrence Sklar judged doubtful the whole point of searching for a physical theory of time different in detail from a causal theory of time.[19] The hierarchical theory agrees with this opinion and extends it. It differs, therefore, from earlier causal theories in that it recognizes an open-ended, nested hierarchy of level-specific forms of causations and relates each to a qualitatively different temporality.

Nootemporality or noetic time is the temporal universe of the mature human mind, manifest in long-term intentionality, working in the reality of open future and past, in the mental present.

Biotemporality or biological time is the temporal universe of living organisms, including man in his biological functions, manifest in short-term intentionality, working in the reality of limited horizons of future and past, in the organic present.

Eotemporality is the time of the "physicist's *t*." It is the temporal reality of the astronomical universe of massive matter, it is one of pure succession without preferred direction.

Prototemporality is the time of the world of particle-waves. It is undirected, non-continuous and non-passing. In a prototemporal world locations and instants are located only statistically.

Atemporality is the temporal universe of the primeval world, one of absolute chaos or pure becoming—lurking beneath all natural processes.

The canonical forms of causations and their temporalities constitute a nested hierarchy.[20] When considered as a family of phenomena, they suggest that when time arose from the primeval chaos, it did not hit the ground running as it were. Not as it was assumed to have done in Bishop Usher's calculations of 1686, when he concluded that the date of Creation was October 23, 4004 B.C. Rather, time evolved in the steps, sketched.

It follows from what was said that causal connectedness appropriate for the time of space-time intervals may only be probabilistic or deterministic as-well-as probabilistic. These forms of causations and their corresponding temporalities have no provisions for a present, none for distinguishing between future and past hence no provisions that for our idea of time's passage. An extension and examination of this assertion follows.

(iii) *The nature of time in relativity's space-time*

In 1908 Herbert Minkowski enlarged the mathematical toolkit available for the handling of Einstein's theory of 1905. Taking advantage, he wrote, of the "pre-established harmony between pure mathematics and physics," he registered his enthusiasm. "Henceforth space by itself, and time by itself, are doomed to fade away into mere shadows, and only a kind of union of the two will preserve an independent reality."[21] G. J. Whitrow, writing seventy years later, was more circumspect. He agreed with C. D. Broad's 1923 remark that the Restricted Theory of Relativity did not break down the distinction between space and time but only their isolation.[22]

The observers for whom the restricted theory assured invariance in the forms of the equations of mechanics and electrodynamics, were those associated with coordinate systems in uniform translation. The 1916 theory of relativity extended the audience to include observers in accelerated motion. This was accomplished by sculpting the geometry of space-time to provide—as seen with the wisdom of hindsight—trajectories for space-time intervals that could accommodate rates of change in velocities. For that end, the General Theory of Relativity replaced the Euclidean geometry of the space-time of the special theory by non-Euclidean geometry. For non-Euclidean geometry, about a century earlier, Gauss enthusiastically adopted the name, "astral geometry." The term intended to suggest that the distinction between Euclidean and non-Euclidean geometries will be manifest only at astronomical distances.[23]

I am going to sketch a representative token of the generalization of the restricted theory of relativity, then ask whether that generalization effects the conclusion reached earlier, namely, that the time of space-time can only be eo-, proto- or atemporal.

The representative token I mentioned is the modification of the equation for *ds*, the element of space-time interval. Its value was made to depend on its position in (structured) four-space. The new Pythagorean equation reads: $ds^2 = Adx^2 + Bdy^2 + Cdz^2 - Dc^2dt^2$. The coefficients A, B, C and D stand for complicated functions called metric tensors. They modify the projections of *ds* upon the four axes according to the inner structure of space-time that is assumed to be appropriate for the local region of interest or even to the universe at large.

Non-Euclidean spaces are classified according to the curves that provide the shortest distances between any two points in them. They are called geodesics. Since geodesics trace the shortest distances in the spaces involved, they are the straight lines of those spaces, verbal contradictions notwithstanding.[24] They are the routes, the natural tracks of moving objects. Since the geometry of a properly selected non-Euclidean space, together with the (modified) equation for space-time intervals *ds* specify the dynamics of planets, behavior which have been theretofore ascribed to gravitational forces, the geometry of space-time is said to have replaced those forces.[25]

The dynamics of general relativity, employing astral geometry, can predict phenomena that are unpredictable from Newtonian understanding. Therefore, by prevailing standards, it is judged superior to the earlier model. It also permits an extension of geometrical reasoning to the cosmos, as also already implied.. Einstein, for instance, speculated that the universe could neither be infinite nor have a finite boundary.[26] This is an exotic condition that resembles the idea of the coincidence of the contraries. Yet it becomes not only thinkable but correct, provided one is free to recognize that the geometry of the cosmos is non-Euclidean.

Although the space-time intervals of the general theory may represent accelerated motion as well as linear translation, they are still only traces that lie tenselessly in mathematical space as do their simpler versions in the restricted theory. It follows that the vast intellectual edifice of special and general relativity theory, this dance of a geometer god, cannot and does not say anything about the sources and nature of time's passage. In support of this claim, let me cite some well-qualified witnesses.

- The space-time of relativity theory, remarked Eddington, "is completely isotropic for all measurements; no direction can be picked out in it as fundamentally distinct from any other."[27] Just as directions in space have no preferences, the time of space-time has no preferred direction.
- There are some good reasons, wrote Herbert Dingle, why the theory can say nothing about "time in itself." Namely, it is concerned only with time at a distance.[28] We cannot find the roots of time's passage in relativity theory. We must bring our knowledge of passage, to it. Unless we know what to mean by the passing of time here and now, we cannot give any meaning to the passing of time at a distance.
- P. C. W. Davies noted that "The four dimensional space-time of physics makes no provision whatever for either a 'present moment' or a 'move-

ment' of time." Also, that it is a "remarkable fundamental fact of nature [that] *all known laws of physics are invariant under time reversal.*" [His italics][29]

- According to Einstein's friend and biographer, Paul A. Schilpp, "once Einstein said that the problem of the Now worried him seriously." That he thought that "there is something essential about the Now which is just outside the realm of science."[30] The belief, explicitly stated or tacitly assumed—as illustrated in the Einstein view cited—that if the roots of some phenomenon cannot be identified in physics then it must either be an illusion or else some not yet understood function of the mind, is broadly held among physicists. P. C. Davies, whose comments were already quoted, also stressed that "relativity physics has shifted the moving present out from the superstructure of the universe, into the minds of human beings, where it belongs."[31]
- "No formula of mathematical physics involving time implies that time passes," wrote David Park, "and indeed, I have not been able to think of a way in which the idea could be expressed, even if one wanted to, without introducing hypotheses incapable of experimental verification."[32]

Let me return to the space-time interval of $ds^2 = dx^2 + dy^2 + dz^2 - c^2dt^2$ and to its values being positive, zero or negative. A positive value is for motion that outruns that of light. Nothing in nature corresponds to such conditions. Zero value is for motion at the speed of light. This world is limited to photons. Clocks at that speed—if there could be any—would come to a halt. This leaves a negative value for ds^2 for all motion where a clock, carried by a moving body, will measure proper time. "Proper time is another name for local time, i.e. time which either is, or could be, recorded directly by a clock."[33] But, what may possibly be meant by readings on a clock in a world where neither present instants nor the passing of time may be given any meaning?

There is a curious blindness here. It is assumed that the notion of "time" must necessarily include nowness and passage, even though as far as we know, this is not the case in the physical world. For some enlightenment in his matter, I turn to an old Scottish folksong.

> I gave my love a cherry that had no stone…
> How can there be a cherry that has no stone?…
> A cherry when it's bloomin it has no stone.

Time, in its archaic, evolutionarily older, "bloomin" forms, retained in the lower integrative levels of nature, was without a present and had no direction. Before the evolutionary character of time was recognized in the hierarchical theory of time, there was no conceptual framework to accommodate temporalities which, compared to noetic time, appear incomplete. Let me illustrate the presence of the past of time by citing from a paper on time and quantum theory.

> [Time] which appears in the equations [of quantum theory] is not a quantum mechanical observable, which would be represented in the theory by an operator, but rather a parameter external to the microscopic system. This 'time' does not refer to something internal to the quantum system, but something measured by a laboratory device—a macroscopic clock.[34]

In the hierarchical theory time a quantum system is prototemporal, while the clock on the wall is a device for the quantification of nootemporality, of our experience of present, future, past and passage. By the principle of parsimony, this interpretation of known facts must be preferred.

If it is the case that the theories of relativity can make no provisions for our idea and experience of a "now" and of the flow of time, if they report about a world without nowness and directed time, then, instead of holding with Einstein that an understanding of the sources and nature of nowness and of time's passage are "just outside the realm of science," why don't we maintain only that they are outside the realm of physics and look elsewhere? I will suggest how such an understanding may indeed be found in the sciences of life, of the mind and of society, whence they are smuggled into physics and make us believe that the time of space-time has the same qualities as does organic and noetic time.

First, a backward glance upon inorganic matter.

We learn from the Special Theory of Relativity that, if we already know what is to be meant by "now" here, then we must conclude that "nowness" is not a cosmic but a local phenomenon. The theory, wrote Whitrow, "denies the universal simultaneity of spatially separated events." The simultaneity of events "becomes an indeterminate concept until a frame of reference (or observer) has been specified."[35] In the words of Einstein, "Every reference-body (coordinate system) has its own particular time; unless we are told the reference-body to which the statement of time refers, there is no meaning in a statement of the time of an event."[36] There is no cosmic "now" of which we—whoever and wherever "we" are—may become aware.

To help assimilate this anti-intuitive idea, let me restate its conclusions with the help of a narrative metaphor. The ordinary view of time may be represented by the movement of the underground cable of the San Francisco cable car. Driven by some distant and obscure machinery, the cable remains out of sight. We know that it moves because the cable cars attach themselves to it and are carried along for a ride. Likewise, time is ordinarily imagined as a cosmic motion of the "now," driven by natural or divine powers. Matter, life, man and society attach themselves to that "now" and are carried along for a ride.

Not so, says the restricted theory of relativity. There is no universal flow of time with a cosmic "now" to which anything could become attached. Nowness can only be defined locally. Once so defined, it may then serve as an anchor for the flow of time—for the future changing into the past—here. Beginning here, and with the help of relativity's instructions about synchronizing clocks, nowness may then be operationally defined at a distance. It may also be imagined as having served as anchor to the flow of time in the past, and as serving so in the future.

But: what defines the now, here?

For an answer I turn to chronobiology and employ the coordinated clockshop model of life.[37] We start by noting that the frequency spectrum of biological oscillators or cycles, sometimes called biological clocks, is spread across 24 orders of magnitude. In humans, from the skin's response to ultraviolet light at 10^{16} Hertz to circalunar and circannual periods of 10^6 and 10^7 seconds respectively, it covers 22 orders of magnitude.[38]

This stunning complement of bioclocks are much more than adaptive measures. They do not merely assist in the survival of the organism, as is often stressed, not any more the instrument of an orchestra assist the orchestra. Rather, those instruments make up the orchestra, their coordinated sounds constitute music. Likewise, the instant by instant coordination of biological oscillators constitutes the life process. A failure of that coordination is death.

I showed elsewhere how a coordinated assembly of molecular oscillators may become subject to natural selection, complexify, define self-directed purpose, distinguish between present and non-present conditions and in terms of its needs, distinguish between future and past. Based on these reasons, I also proposed that it is the life process that inserts a "now"—the organic present—into the nowless world of nonliving matter. Analogous reasoning leads to the conclusion that coordinating the trillions of neurons of the human brain, from instant to instant, defines the mental present and makes personal identity possible. The failure of this coordination is manifest in the impairment of personal identity that is, of selfhood. And again, the coordination of the behavior of men and women from instant to instant defines the collective present and makes the construction of collective identity of a tribe, society and civilization possible.[39] The failure of this coordination is manifest in the breakup of a society.

The mental present is not the evolutionarily oldest process that inserts nowness into the nowless world of inorganic matter. That task may be identified in the organic present. The mental present only enlarges the temporal horizons of the organic present. The collective or social present further enlarges and refines the horizons of the organic and mental presents and is responsible for the perception of history, in which religions, philosophies, ideologies and scientific cosmologies see the story of man and of the universe play out.

During the history of evolution futurity and pastness, referenced to the organic, mental and social presents, became established as parts of the organic, noetic and social realities of humans.[40]

That presentness, futurity, pastness and the flow of time are not found in physical formalism does *not* mean that physics fails to account for something it should account for. Nor does it mean that our experience of time's flow is an illusion or a figment of human imagination, as is a leprechaun. Rather, it means that the functions and structures of a world about which physics reports is one in which time has not yet evolved to the level of bio- and nootemporality.[41] Michael Heller, physicist and Catholic philosopher, has shown that the physical world is time orientable, that it allows for two directions of time, but it need not be so oriented, that it is complete and intelligible without preferred temporal direction.[42]

How are the ideas of nowness and of the flow of time smuggled into the time of space-time? The answer was given earlier. By ignoring that time was born in the primeval chaos and that it has evolved as a symptom or correlate of the structural and functional complexity of matter and hence assuming, wrongly, that the time of space-time has the nature of bio- and nootemporality.

Let me now propose a figure of speech to sum up what was learned. A light bulb has little to do with light until electric current is passed through it. Relativity's space-time has little to do with what we normally mean by time until nowness and the passage of time are smuggled into it from the processes of life, of the mind and of society. It is the task of an integrated study of time to disentangle and elucidate such details. The challenges of that task require the tools of critical interdisciplinarity.

2. Critical Interdisciplinarity

We learned that the time of space-time intervals may be prototemporal or else proto- and eotemporal but remains, in either case, sans present, sans future, sans past, sans passing. The space-time of relativity theory is not what Rupert Brooke had in mind when he volunteered that “Fish say... Somewhere behind Space and Time, is wetter water, slimier slime.” But, the fame of Einstein made “space-time” an intriguing and attractive notion and time in space-time remains identified—erroneously—with time “that has changed the auburn hair to white.”[43]

In *Time & Social Theory* Barbara Adam remarked that “Einstein's fusion of space and time has affected social theories and inspired a new perspective in geography, associated with Hägerstrand and the Lund School.”[44] I would like to offer a few illustrations in support of the citation.

- Time geography. “The essential unit of geography is not spatial, it lies in regions of time-space and in relation of such units to the larger spatio-temporal configurations.”[45] “To separate time from space is quite impossible.”[46] “In my own time-geographic studies, I have taken the liberty of talking about *time-space* when I refer to a locational coordinate system and view time and space as existential dimensions. The term *space-time* I use to denote a room resource in the settlement system, isomorphic to other resources, such as time and machine-time.”[47] “There is little sense to be had from making distinctions between space and time—there is only space-time. Attempts to privilege either time or space, suggesting that one or the other is the signature of an age, for example, make nonsense.”[48] It is “thus our belief that time and space cannot be separated from social studies.”[49]
- Experiential relativity. Research in phenomenological space-time suggests “spatial scale as a mediator in the experience of temporal duration....”[50]
- Neuroscience. “In this paper we have attempted to relate brain organization to space-time geometry. .... It appears that the parallel distributed structuro-functional features of neural networks do furnish the CNS with an innate a priori propensity to implement geometries.”[51]
- Anthropology. “This paper discusses the concept of spacetime in the context of some traditional notions of space and time in sociological and anthropological literature.” ... “The perspective proposed is that sociocultural processes are grounded in a broad set of physical phenomena and that space and time are constitutive of social events and actions. Spacetime contextualizes, orchestrates, and directs human endeavors in a necessary, if not sufficient degree. Thus, human centering does not merely operate in space and time; rather, human centering embodies its own spacetime.”[52]

A visit to the Worldwide Web suggests the broad appeal of the term, “space-time.“ See Appendix A. The simultaneous hits have not been sampled for their proximities or examined for the significance of their correlations. But their large numbers and the degree of similarity between the profiles of the English and German samplings suggests that, in many minds, there are affinities between the notion of "space-time" and the other terms.

The idea of combining space and time may serve as attractive nuisance. Wyndham Lewis, in his *Time and Western Man* asserted that it was "Bergson who put the hyphen between Space and Time." Lewis accused him of having done so to support of his idea of *durée* as the essence of creative change.[53] Be that as it may, the notion of "space-time" cannot claim universal usefulness based only on its successful role in Einstein's *Invariantentheorie*. Nor can any other unexamined idea, term or way of thought that surface in the ongoing, desirable and interesting mutual forays involving the sciences and the humanities. They should be examined with the help of critical interdisciplinarity. By critical interdisciplinarity I mean the need for professional familiarity with the material that enters an integrated study of a time-related theme, regardless of the position of that idea and of the theme in the corpus of human knowledge.

Such a need, I believe, is now in especially stark focus because of the intellectual ferment of the global community.[54] The information tsunami in which we live, combined with the proliferation of incompatible political and religious beliefs defended unto death are, as I see them, not the causes but the symptoms of a radical change in human values. Whatever the roots of the resulting creative and destructive chaos, the focusing of attention upon a single theme of universal interest—such as upon an integrated understanding of time—appears to be a useful tool in trying to assess the human condition of our days.[55]

The difficulties of an integrated understanding of time have been examined. They fall under three main headings.[56]

(i) Each of the many fields upon which such an understanding must draw, has its *jargon*. This paper deals with an example: that of the term, "space-time."

(ii) The modes of *testing for truth*—that is, the standards for what constitutes necessary and sufficient *proof*, differ greatly among fields of knowing. What for one field constitutes necessary and sufficient *proof*, may be judged from the perspective of another field as neither a necessary nor a sufficient demonstration of fact or truth. What by one set of standards is a salutary and correct argument may sometimes be held, by another set of standards, a useless waste of words.

(iii) Distinct modes of inquiries about man or the universe display different *personalities of knowledge*. Mathematics, embryology, the history of art

> and jurisprudence display different collective attitudes toward the past, the future—the world. Fields of knowing may are categories of self-expression that are rooted in different ways of life and different priorities in values. It is of course people who have personalities not bodies of knowledge. But men and women gather themselves according to their values and preferred ways of life and, thereby, create and maintain what may be called the personalities of professions.[57]

The differences in jargon, in methods of testing for truth and in the personalities of knowledge correlate with differences among the classes of processes and structures that various fields of knowing examine. Curved space-time is very unlike the taste buds of frogs, which is very unlike the history of Thirty Years War. A uniform method of testing for truth or the use of a single jargon or a single personality of knowledge would prohibit the division of labor that made the growth of civilizations possible. These pluralities should, therefore, not be judged as inhibiting the development of an integrated study of time but rather as guiding it. An integrated theory of time should not be required to eliminate the differences among ways of knowing but facilitate communication among them.

Is it possible to take a bird's-eye view of these and related problems that is, of interdisciplinarity itself?

Yes. In a unique and brilliant paper, "On Interdisciplinarity," Alexander Argyros noted that although "interest in interdisciplinarity is the herald of major paradigm shift in Western philosophical thought... its conditions of possibility remain largely unexplored."[58] What is necessary, he wrote, is "a fundamental shift away from the two dominant world views of the contemporary world—relativist textualism and reductive science—toward a paradigm which models the universe as an evolving, hierarchical, dynamic, and innovative system."[59]

As a family of suitable principles that correspond to the desired paradigm, he espoused the hierarchical theory of time because, he wrote, it is a "theory of systemic interdisciplinarity" that is able "to account for phenomena at both the natural and cultural levels, and to be in step with the best available scientific knowledge."[60] He describes the theory as one in evolutionary ontology. The propositions of the hierarchical theory of time may be found in Appendix B.

Argyros also adds that the theory's "systemic interdisciplinarity offers human beings the opportunity to encompass the regularities that characterize our world and... to respect them, finesse them or supplement them." Acknowledging the evolution of causation and of time itself and recognizing the related increase in degrees of freedom supports the integrity of the different tests for truth, the need for differences in methods of working, the use of specialized languages and the distinct personalities of knowledge. Thus, they support the integrities of the different fields of knowledge, a condition that must be maintained because, wrote Argyros, "interdisciplinarity is not... a substitute for disciplinary rigor but... an addition to it."[61]

For the extensive and intricate reasoning of evolutionary ontology and its appropriate epistemology the reader has to be referred to the literature.[62] Instead of trying to summarize its myriad details, I would like to represent them through a poem of the thirteenth century Persian poet, Jalal-al-Din al Rumi. He recognized evolutionary ontology in terms appropriate for his age and described it in poetic language.

> I died from mineral and plant became
> Died from the plant and took a sentient frame
> Died from the beast and donned a human dress
> When by my dying did I e'er grow less?[63]

Rumi's question may be answered in the spirit it was asked: "I never did." Critical interdisciplinarity ought to read this quatrain as a reminder of evolutionary ontology and embrace a corresponding evolutionary epistemology.

Appendix A

A SUGGESTIVE SURVEY OF THE POPULARITY OF THE TERM, "SPACE-TIME"

In English	*In German*
A sampling used for calibration. Total entries under the words:	
Shakespeare 14,800,000	
Einstein 11,200,000	
Space-time 952,000	Raumzeit 59,200
Simultaneous hits	
Space-time + Einstein 172,000	Raumzeit + Einstein 15,100
Space-time + geometry 153,000	Raumzeit + Geometrie 5,820
Space-time + magic 92,000	Raumzeit + Magie 832
Space-time + literature 74,000	Raumzeit + Literatur 9,300
Space-time + art 53,200	Raumzeit + Kunst 5,680
Space-time + sexuality 28,000	Raumzeit + Geschlechtlichkeit 713

Appendix B

PROPOSITIONS OF THE HIERARCHICAL THEORY OF TIME[64]

1. Nature comprises a number of distinct integrative levels.
2. These integrative levels form a hierarchically nested and evolutionarily open system.
3. This system is located along a scale of increasing complexity.
4. Each of the integrative levels is characterized by distinct, level-specific processes.
5. The level-specific processes function by different modes of causations.
6. They must be descried in different languages.
7. Each integrative level determines a qualitatively different temporality.
8. Each new integrative level adds new, unresolvable creative conflicts to those of the level or levels beneath it.

9. COORDINATED CLOCK SHOPS

Some 3,500 million years before Homo Faber began to fabricate clocks and started to worry about time at a distance, nature on earth began to fabricate, what came to be called, biological clocks. A biological clock is any reliably oscillating cell or system of cells that serves as a control of the timed functions of a living system. The spectrum of biological clocks, in the twenty-first century, is spread from an upper limit of 10^{16} Hz (the response of human skin to ultraviolet rays) to a lower limit of 10^{-7} Hz (the circannual period found in millions of species). It is an immense clock shop with clocks that span a frequency range of 10^{23} orders of magnitude.

According to a broadly distributed report written around the 6th or 5th c. B.C. "God said, 'Let the earth put forth vegetation, plants yielding seed, and fruit trees bearing fruit in which is their seed, each according to its kind, upon the earth.'"[1] From a backward glance, it seems that plants and fruit trees were then joined by animals and humans. We learned in Essay No. 1 that the number of organisms that ever lived, up to our epoch, may be estimated as between 10^{40} and 10^{50} distinct individuals. This many lives, considering also that the width of their oscillations is spread across 10^{23} orders of magnitude, amount to awe-inspiring success. What do traditions teach about its origin?

We may take a hint from an African-American spiritual called "Dem Dry Bones." Its inspiration is a prophetical book of the Old Testament, bearing the name of Ezekiel. It relates the prophet's visit to the valley of dry bones where he cried out, "Dem bones, dem bones, gonna walk around. . . ." To be able to walk around they had to organize themselves into stable, dynamic systems. The spiritual gives the details of that self-organization in terms appropriate for Ezekiel's place, epoch and audience. "The toe bone connected to the foot bone / The foot bone connected to the ankle bone / Ankle bone connected to the leg bone. . . ."

With the help of contemporary natural science and the hierarchical theory of time, the essay that follows visits the issue of the coming about the first internally coordinated shops of oscillators. Abiding by the mode of reasoning in natural science, it formulates its scenario in a manner uncommitted to any specific example. It sees biological clocks as not merely assisting life in its adaptive endeavors but comprising, making up living organisms. Employing this model, the essay traces the origin of life to natural selection in the cyclic domain, working on molecular aggregates. Applying thermodynamic considerations to the model of biogenesis sketched, leads to the understanding that life is identically equivalent to a dynamic equilibrium between growth and decay, maintained through an instant by instant coordination among complements of biological clocks.

That equilibrium is remarkably resilient to external and internal perturbations. It is often said that life is fleeting, ephemeral, easily cut off by sword or snuffed out by ill wind. This is true for individual organisms. But life as a process is neither ephemeral nor is it fleeting. It is robust and lasting. The organic skill of combining growth and decay from instant to instant and, through that combination, maintaining the continuity of the organic present, is 3.5 billion years old. This is three-fifth of the age of the universe.

TIME AND THE ORIGIN OF LIFE[2]

Abstract

This paper interprets the coming about of life entirely in temporal terms. It is an attempt to formulate a principle—that of biogenesis—in a manner that is uncommitted to any of its specific manifestations. Seeking such a general model of a process is a practice traditional in natural science.

The argument begins with a survey of the spectrum of biological cycles in species alive today. The findings suggest that oscillatory processes in living organisms are much more than adaptive measures. Instead, biological oscillations, observed across a frequency range of 24 orders of magnitude and synchronized from instant to instant, constitute the life process. In the phenomenal world, the inner synchronization is manifest as the organic present. It is thus that life creates a "now" in the presentless world of nonliving matter. The organic present so born and maintained allows for a distinction to be made between present and nonpresent conditions. In their turn, the nonpresent categories of time may be separated into futurity and pastness in terms of the present needs and available means of an organism.

When this model of the life process is reduced to its simplest schematic form, a set of minimal requirements obtain to which even the earliest molecular aggregates had to conform if they were to be called alive. According to this understanding, the perpetuation of life is seen as the passing along of the biological skills that are necessary for maintaining the organic present against external and internal perturbations.

The burden of the paper is the testing of the proposed model of biogenesis in terms of three criteria. Does a system so envisaged have the potentiality for complexifying through evolution by natural selection? Can it define self-directed purpose in terms of its needs, and with it, distinguish between future and past? Is the functioning of the model consistent with the principles of thermodynamics? The paper concludes by giving affirmative answers to these questions.

This paper outlines an understanding of life and its origin with the assistance of the hierarchical theory of time.[3] In so doing, it follows a type of reasoning which is traditional to natural science. It formulates an interpretation of a process in a manner that is independent of, and is uncommitted to, any of its specific manifestations.

The Notion of Biogenesis

Opinions about the origin of life have varied from epoch to epoch, and differ from culture to culture.[4] Creation legends suggest intellectual readiness to distinguish between living and nonliving objects and to identify the position of life with respect to the world at large. Early narratives about the origin of living things matured into reasoned arguments in the thoughts of the Presocratics such as when Anaximander in the 6th century B.C. reasoned, rather than declared, that animals came into being from slime (Kirk and Raven, 1975, p. 141). Aristotle in *The History of Animals* took the idea of spontaneous generation of primitive organisms for granted, as did almost everyone else until Pasteur succeeded in demonstrating that examples of presumed, spontaneous generation were instants of reproducing microorganisms.[5] Darwin imagined "some warm, little pond with all sorts of ammonia, phosphoric salts, light, heat [and] electricity in it" as a likely setting of the origin of life (from an editorial note in Darwin (1887) v. III, p. 18, from an otherwise unidentified letter by Charles Darwin written in 1871). In the 1930s Haldane and Oparin suggested that the appearance of life was preceded by a long period of chemical synthesis, during which polymerization proceeded at a slow pace, creating, eventually, the first prototypic cells (Oparin, 1966). Contemporary biologists, interested in the origin of life, study cosmochemistry and the geological and physical conditions of the biogenetic earth (for current work see the journal *Origin of Life* and the publications of the International Society for the Study of the Origin of Life).

The ancestry of their approach to biogenesis may be found in the thought of Democritus of Abdera, a Greek atomist who lived around the turn of the 5th century B.C. and maintained that atoms were very small and simple, alike in qualities but different in their relationships. He also believed that some groups of atoms are subtler than others. For instance, those of the soul were like those of fire: very small, smooth, and spherical, so as to secure for them the mobility that is necessary for penetration.[6] The molecules of life, as understood today, are small by ordinary measures, though they are gigantic in atomic terms, having molecular weights a billion times that of hydrogen. But they are neither round nor smooth. They resemble, instead, very thin and long solids.

In approaching the task of this paper, I take my cue from the dynamics of these molecules. I note that a DNA molecule is a system of hundreds of millions of atoms which continuously wiggle, vibrate, and oscillate at vibration rates which span the electromagnetic spectrum from radio waves to the infrared. Groups of oscillating patterns, quantized vibrations called phonons, wander around the molecule as if looking for a place to settle, but as long as the DNA remains an integral, functioning unit, the phonons never stop moving.[7] If, for whatever reason, the DNA falls apart, its collective oscillations vanish as water waves do when the water vanishes. These molecules can keep on carrying their coded messages only as long as their intricate dances are kept coherent according to the laws of chemistry and physics.

As a first step in trying to trace the origin of life in the temporal behavior of certain molecules, it will be necessary to understand the relationship between biological oscillations and the life process.

Biological Clocks and the Life Process

The terms *biological oscillations, biological rhythms*, and *biological clocks* are not used uniformly across the literature.[8] I am going to use them interchangeably and mean by them all cyclic phenomena that are involved in maintaining life. The spectrum of biological clocks in species alive today is spread across 24 orders of magnitude.[9] The human body possesses clocks across 22 orders of magnitude in frequency. The morphologies of the different clocks vary substantially, but they all share a common dynamics: they all oscillate at their particular frequencies.

With few exceptions (Bonner, 1974; Goodwin, 1976; Winfree, 1980) all definitions of biological oscillations imply that living organisms, having come about through some yet unidentified steps, have acquired through natural selection a store of cyclic variations as parts of their adaptive strategies. In other words, it is almost universally maintained that life is historically prior to the cyclic processes of life. I believe that this assumption is erroneous. It is analogous to claiming that the sounds of the instruments of an orchestra assist the orchestra in making music. But those sounds do not *assist* an orchestra in its performance, they constitute the music, provided that they are kept correlated from instant to instant according to selected principles of harmony.

Likewise, biological clocks do not merely assist life in its adaptive endeavor but, more fundamentally, they comprise—they make up—living organisms. In this view, life is seen as a process that consists of the instant to instant coordination of chemical and physical oscillations, according to principles that will help maintain them in mutually supporting rather than destructive relationships. I call this schematic representation *the coordinated clockshop model of life* and find it useful as a conceptual tool for understanding biogenesis. In pursuit of that understanding, we leave behind the 6 million named species alive today, and direct our attention backward along the history of life. During the journey we watch the contemporary forms of life devolve toward their common origin.

First the hominids vanish, then the primates, the mammals, the vertebrates, the invertebrates, and the protists (bacteria and algae) disappear. Then, far below the almost naked DNA known as prokaryotes, we arrive at the roots of the phylogenetic tree of life. At that point of the journey, as an exercise in purposeful curiosity, we take the imagined ancestor of all life forms, and place it under the microscope of our analytical capacities.

The Primeval Clockshop of Life

The clockshop beneath the microscope is a self-organizing system, able to maintain its dynamic balance against internal and external perturbations. It shares this capacity for homeostasis with all stable inorganic compounds, but with an important difference. The living clockshop has crossed a threshold of complexity beyond which the short-range atomic and molecular forces ceased to be adequate to insure ordering and integrity. For that reason, in addition to short-range forces, it had to have appropriate methods for long-range dynamic coordination, perhaps of the type I mentioned when discussing the DNA.

The need for coordination among molecules that are far apart in terms of atomic distances is a demand of life for the creation and maintenance of *simultaneities of necessity.* Here "necessity" refers to the securing of sufficient coherence to help hold the microscopic world of the giant molecule together. The referent of necessity is the integrity of the living molecular group, that is, of the life process itself. In the much simpler world of inorganic matter one can find only *simultaneities of chance.* By this term I mean any chance combination of elements and molecules which, by the laws of chemistry and physics, can maintain their stability until some other chance event disrupts it, but which do not demand long-range coherence. In the phenomenal world the condition I called simultaneities of necessity corresponds to the *organic* or *living present* of life, that is, to the instant by instant, nondestructive synchrony of the life process.

The organic present is the simplest form of the now. Just as life may be identified with the creation and maintenance of the organic present, so the mind may be identified with the creation and maintenance of the *mental present.* And again, a group of people may be said to constitute a society only if, and only as long as, they are able to create and maintain a *social present.* These three presents are hierarchically nested in that each higher one subsumes the one or ones beneath it and, for that reason, they are necessarily simultaneous. The reasoning given earlier for the introduction of organic present into the presentless world of physical time may therefore be extended to the evolution of a nested hierarchy of presents. This claim needs a bit of elaboration.

There is nothing in the nonliving world, as revealed through the equations of physics, that could be used to define a now.[10] But, future, past, and the metaphor of time's flow—which is the changing of future into past—make sense only with respect to a now. Consistently, there is nothing among physical processes that could be held responsible for the experience of the passage of time. As I shall show later, the secular increase of entropy in thermodynamically closed systems, often called the thermodynamic arrow of time, is also inadequate for the definition of a direction of passage. As one would expect from the logic of the situation, the absence of a definable now and that of directed time are complementary aspects of time in the physical world. This does not mean that the physical world is timeless, but only that its temporalities are undirected and that the metaphor of the flow of time cannot be applied to physical processes when they are considered in themselves.

It was the coming about of the organic present that broke the symmetry of physical time by making it possible for the information content of a complex system to acquire meaning. Specifically, the definition of the now allowed a distinction to be made between present and nonpresent conditions. It is with respect to nonpresent conditions that a particular type of meaning, called purpose, could arise. With the ideational category of purpose came the possibility of distinguishing among different kinds of nonpresent conditions. Those that related to the unfulfilled needs of an organism could be separated from those that pertained to fulfilled ones. Events and conditions which were not present but which could satisfy unmet needs were assigned to a category of time called the future and became targets of intentions. Nonpresent conditions responsible for the means that were useless for reaching future satisfaction were assigned to the past. The ceaseless rearrangement of intentions with respect to the future, narratives about the past, and sense impressions about the present is what we call the flow of time.

What is to be meant by the organic present being inserted into a presentless world? Let me offer a visual metaphor as an aid to its understanding. The San Francisco cable cars move by attaching themselves to an underground cable. The cable, which is hidden from sight, is moved by some distant and invisible machinery. The present of life does not attach itself to a cosmic flow of time whose motion is maintained by unknown powers. Instead, life creates those conditions and operational properties of matter which give rise to our experiences and ideas of the present and of the nonpresents.

Keeping these comments in mind, let us assume that miniature synchronized clockshops, as here envisaged, did come about. Then, inquire into the manner in which they were likely to have evolved under the selective forces of the temporal organization of their environments. If such a path permits organic evolution to have taken place the way we know it did, then the coordinated clockshop model of life has some merit, and in search for the origin of life, we may begin asking questions about those primeval clockshops.

The Evolution of the Primeval Clockshops

I submit that an assembly of molecular oscillators, complex enough to define an organic present, will necessarily become subject to natural selection under the pressure of the cyclic complement of environmental changes.

To begin in the beginning: there are good reasons to believe that our nearest nonliving ancestors resembled crystals.[11] Perhaps they occupied protective microscopic niches in geological formations, as blue-green algae occupy air spaces today in the rocks of the Dry Valleys of Antarctica. That environment was unfriendly to later forms of life: the days were 18 hours or shorter, the atmosphere was mostly hydrogen, the surface of the earth was bathed in ultraviolet radiation. Natural radioactivity was strong, there were continuous electric discharges, and hosts of obscure phenomena, such as sonoluminescence and possibly its inverse, sound produced by light. The ground continuously shook, boulders rolled, and the earth vibrated at many frequencies.

In that environment, improved adaptation meant the acquisition of such collective oscillatory behavior as would enhance the chances of survival of the system as a unit. I can imagine, for instance, a crystalline structure lodged in a crevice whose width oscillated at a stable rate. Crystals that could match that rate of oscillatory change would have a better chance for survival than other units which did not.

Since the regularities of these primeval clocks are unlikely to have exactly fitted the temporal niches of the oscillatory spectrum of the environment, there existed an error signal through which the environment could exert selection pressure on the clocks. The selective forces would favor those configurations which were able to provide better fits to the cyclic variations of the surroundings. By favoring certain frequencies over others, natural selection supplanted accident. During this selection process life was born when the molecules crossed a threshold of complexity such that, above it their integrity came to depend on long-range ordering, as discussed before. With that ordering came the capacity for defining self-interest with reference to an organic present.

These clockshops were bound to complexify for reasons similar to those which make the history of timekeepers show increasing complexity in the construction of clocks. Namely, each step taken to improve the precision with which an external cycle could be modeled—whether by a human clockmaker or by organic evolution—served as a means for recognizing more refined cyclic structuring in the environment. Each improvement opened up the system to the possibility of, and need for, finer tuning.

Imagine, for instance, that the proto-organisms evolved sufficiently to take advantage of the tides which carried the nutrients they needed. After the tidal rhythm was internalized, then expressed through external behavior, the diurnal and semidiurnal inequalities and the lunisolar variations of the tides began to serve as new agents of selection. The difference between what an organism was capable of doing and what the environment demanded it ought to be doing, was an early example of the perennial existential tension of the life process.

As the spectral analysis of environmental periodicities continued, the complement of internalized rhythms had to expand. Also, as the clocks of the clockshop widened their frequency spectrum, the inner coordination of the assembly had to become increasingly more precise. The increased precision is recognized in the evolutionary narrowing of the width of the organic present.[12] Also, the internal cycles created beat frequencies to which nothing in the external world needed to correspond. Some of these had to be control functions that assured. the continued viability (autonomy) of the organism. Henceforth, natural selection while acting upon the externally manifest (phenotypic) rhythms came to effect the internal (genotypic) rhythms.

The early Precambrian earth; upon which life was born, was a busy but stable place; its climate and geological conditions took millions of years to change. During these eons, some regions of the earth became populated with blue-green algae and prokaryotic bacteria. Their rate of evolution, though slow, was certainly faster than that of the nonliving earth because inanimate matter could change only through simultaneities of chance rather than simultaneities of need, defining a purpose. Somewhere along this history, the environment to which species had to adapt became biotic; the target of adaptation itself began to evolve. This novel condition made life subject to a Malthusian principle in evolutionary rates: in the intensifying evolutionary race, rapidly adapting species began to gain advantages over those that changed slowly. The same type of conflict between different evolutionary rates is being replayed in our epoch upon a higher integrative level. Namely, the living species on earth today, including our own bodies, are unable to change biologically at the rate required by the cultural changes that are being created by the collective power of human minds.

Let me turn to an early example of a conflict between different evolutionary rates and illustrate how such a conflict gave rise to something unpredictably new, namely, sexual reproduction.

To do so, I have to appeal to a principle in the theory of automata, formulated by von Neumann (1969, pp. 79–80). According to that principle, a system must reach a certain threshold of complexity before it can produce others of its own kind that are of equal or increased complexity and of higher potentialities. If this principle is taken to be valid for living systems, then only after a species has reached a certain level of complexity did it become advantageous for it to reproduce through heirs that developed instead of replicating through identical heirs, as do crystal-like structures which split or cells that duplicate by mitosis.[13]

That structural and functional complexity seems to have been reached during the early Paleozoic, perhaps 500 million years ago, resulting in the appearance of sexual reproduction. There is no general agreement on the immediate causes of the emergence of sexual reproduction. But a representative view, consistent with von Neumann's speculation, maintains that sexually reproducing species could adapt to a changing environment much more rapidly than could asexually reproducing ones.[14]

In the sexual division of biological labor, the task of securing the continuity of the organic present became a joint enterprise of two types of specialized cells. They. are the male and female germ cells: known collectively as gametes. The husbanding of the gametes became the task of large assemblies of somatic cells. Gametes and somatic cells bear different relations to time. Germ cells are potentially immortal, as are single-celled organisms that reproduce asexually by mitosis, and multicellular organisms that reproduce by splitting (after reproduction there are only daughter cells: there is no body left to be buried). In contrast, organisms that reproduce sexually by interchanging genetic material do age and do die by aging.[15]

Thus, while all living organisms may be killed, members of many species do not and cannot die by aging. Aging and death are, therefore, not necessary correlates of time and life, notwithstanding all the philosophical and religious traditions which contemplate time exclusively in relation to the dreams of everlasting. life. But all species live by temporal ordering referred to the organic present. It follows that time relates to life through the maintenance of the organic present, and not through the inevitability of aging and death.

In the course of organic evolution, with the appearance of the lower invertebrates, came the development of the nervous system, including the brain, the seat of its central command. It had, as one of its tasks, the coordination of the cyclic processes of organisms. Also, as the species became more complex and their behaviour less predictable, being able to anticipate the conduct of friend, foe, and food became advantageous and hence, one must assume, it was favored by natural selection. In other words, it is reasonable to assume that the ability ofthe mind to prepare for future contingencies has evolved in response to the need for anticipation. The mind made it possible for humans to create a world of imagined conditions and work out strategies through the manipulation of symbolic representations of reality, based on past experience.

The emergence of the mind reminds us of von Neumann's idea of complexification thresholds. Only above a certain complexity of the central nervous system was it possible for individuals of a species to assist other individuals in becoming as or more advanced in their skills than were their teachers. Crossing that threshold made it possible to create cumulative knowledge in the form of symbols and through them, pass on to later generations the fruits of acquired characteristics. This feat cannot be accomplished by biological means alone.

Let me sum up this section on the evolution of a primeval molecular assembly, schematically represented as acoordinated clockshop. The section traced the presumptive fate of such a clockshop, along a trajectory of necessary complexification. It found that the history of life, interpreted in terms of natural selection working upon that clockshop, coincided with what we know about the history of organic evolution, from biogenesis all the way to the mental functions of the human brain. In search for new perspectives upon the origin of life, we are justified, therefore, in seeking an understanding of the way in which the earliest miniature clockshops are likely to have come about.

The Origins of the Primeval Clockshops: A Perspective

Whenever a natural process is unavailable to direct human sense experience—as is the case with the creation of the universe or the structure of the atom or the origin of life—then the scientific method suggests the formulation of suitable abstract concepts. To make defensible guesses about the origins of coordinated, living clockshops, it will be necessary to employ one of these concepts. Conclusions reached through its use may later be translated back into statements about structures and functions of the phenomenal world.

In this case, that conceptual tool is the notion of entropy. For a definition of entropy we can do no better than to turn to one given by Professor

Whitrow: "The concept of entropy is a mathematical measure of the disorganization of a system. The idea first arose as a part of the theory of heat, but a similar notion can be associated with probability distributions of any kind" (Edwards, 1972; s.v. "Entropy").

I would like to amplify this definition as well as sketch the relation of entropy change to time and to the life process. Then, I would like to use whatever shall have been learned to arrive at a particular view of biogenesis in terms of the nature of time.

The measure of anything is a comparison expressed in numbers. The entropy of a system is a numerical comparison, usually between two different states of the same system. A measure of entropy is not intrinsic to a system, just as the number of length units is not a property of the length of a road: the numbers depend on the units used. Also, since there are many different processes to which the idea of organization may be applied, there are many different ways of defining entropy.[16] Furthermore, the reference state of a system to which zero entropy is assigned, remains arbitrary.

Consider, for instance, a container with two compartments separated by a wall, where each compartment contains a different gas. Remove the wall and watch the two noninteracting gases mix. If the initial condition is said to have had zero entropy then, at any later time, it should be possible to assign a value of nonzero entropy to the gas, measured in suitable units.

Consider next a freshly slaughtered sheep. Zero entropy may be assigned to its carcass as to its most highly organized state, because we know from experience that, beginning with that state, the body will decay. Again, it should be possible to measure the increase in the disorganization of the carcass, at least in principle.

Instead of a carcass, let us think of a living sheep. To give any meaning to its present degree of disorganization, one must have as a reference, a final and complete sheep. But there is no final sheep because evolution is open-ended. The task must, therefore, be reversed. One has to measure the degree of organization of the sheep with respect to one of its own, less organized states, such as when it was an embryo. Or, we may go further back and compare its present organization with a sampling of the primeval chaos. The new variable, necessary for the measurement, is called negentropy or information. Obviously, the gases with their increasing entropy, and the development of the sheep in terms of its increasing information content are oppositely directed changes.[17] Now, let us relate the concept of entropy time and organic evolution.

The long-term entropic behavior of systems that are thermodynamically isolated from the rest of the world are governed by a principle known as the

Second Law of Thermodynamics. There are many ways of stating that principle, depending on how entropy is being measured. But common to all of them is the rule that the total entropy of an isolated, or closed system can, in the long run, only increase.[18] There are yet other systems, said to be open, such as living sheep. These are not defined by a box within which they are enclosed, but by their geometrical boundaries. Being open means that energy and information from the rest of the world may freely cross those boundaries.

Thus, both closed and open systems assume an outside world from which they are—or are not—closed off.

The Second Law of Thermodynamics caught the attention of Eddington, who perceived in its statement about the universality of decay, the physical basis of the human sense of time. He associated the decay process with the human experience and idea of passage, and named it the arrow of time (Eddington, 1958, pp. 69, 79, 101). But, for the Second Law to be applicable, it is necessary to begin with systems that may become disorganized. A principle that governs decay must necessarily assume some other principles that govern growth. This simple reason is a powerful argument against received teachings on time and thermodynamics. Namely, Eddington's arrow may be attached, with equal justification, to the entropy increasing or entropy decreasing processes of the world, to the decaying carcass, no less than to the developing embryo, to the mixing of gases no less than to the emergence of macroscopic order from cosmic chaos. Since the association is arbitrary, neither growth nor decay, neither entropy decrease nor increase can serve, in itself, as the ultimate agency responsible for the direction of time and with it, the ultimate root of the passing of life.

We should have been able to reach the same conclusion, without knowing anything about entropy or thermodynamics, because no arrow in itself can define a direction, be it one in space or in time. Up needs down, right needs left, decay needs growth. The directedness of change we describe as the passage of time cannot come, therefore, either from decay or from growth considered in itself, because they both point in the same direction—and that direction may be arbitrarily selected. The cells of my body manufacture enzymes and create decay products in the same sense of time's passage.

The sources of time's flow must be sought, instead, in the purposeful behavior of living systems. It is the goal-directedness of life, referred to the organic present, that creates the distinction between what is judged as present and nonpresent, and divides the nonpresents into future and past, as discussed earlier. But then, how does one explain all of life having a common beat as it were, in respect to a shared organic present? The answer, which

I proposed elsewhere, is that families of living organisms share a common present only, and only to the extent that they communally establish that present through intra- and interspecies communication (see Fraser [1987, pp. 192–201; 1978, p. 90 and passim]).

The view that time acquires its directedness with the life process is an unpopular one for an age that is more impressed by the mechanical and physical than with the organic. Many clever proposals have been made to rescue the belief that the ultimate roots of time's passage must necessarily be found among the governing principles of physics.[19] Some proposals maintain that, at the time of its creation, the universe was given a great degree of ordering, and since then, history has but lighted fools to dusty death. Others argue for a continuous creation of macroscopic information from the chaos of the atemporal substratum of the world serving as the source of our sense of time, driven, ultimately, by the expansion of the universe. There are statements in the literature of time in physics that maintain that time is entirely a mental phenomenon, or else a mystery, and will forever remain beyond the reach of science.

But none of these ploys, erected on the Platonic foundations of physical thought, can account for the phenomenon of the present. For that reason, none may be used to define future, past, and the passage of time. As I asserted earlier, directed time was born with life itself as a phenomenon of inner coordination.

To integrate this view of life and directed time with the undirected time of the physical world, whence life arose, it will be necessary to consider the creation of the world as it is interpreted in the theory of time as conflict (Fraser, 1982, p. 104).

In all sciences, religions, and philosophies which concern themselves with the origin of the world, that origin is usually identified with the emergence of order out of chaos. At variance with this view, I submit that the aspect of Creation upon which we ought to concentrate is not that of order, but that of conflict between ordering and disordering. Ever since Creation, conflict has remained the fundamental dynamics of the universe (this is the basic tenet of *Time as Conflict* [Fraser, 1978]). The ordering and disordering processes of the inorganic world are uncorrelated. There are only few and only transient physical processes whose integrity depends on the coordinated decrease *and* increase of entropy. The undirected character of the three physical temporalities demonstrates a statistical averaging, or random mixing of growth and decay, conditions which are among the hallmarks of the inorganic world.

It is out of this matrix of directionless time that life first arose. The crucial prebiotic step of inorganic evolution was the appearance of nucleotides or, in any case, of large molecules whose structures and chemistry allowed them to be strung together into longer and longer polymers. In this process of expansion the significant variable was, as already stressed, that of complexity. Somewhere along the history of complexification, the ordering and disordering processes within the giant molecules have become so intricately interwoven that they could not be thought of anymore as separate—either theoretically or experimentally—without destroying the system. The mutual definition of the two opposing thermodynamic arrows is, therefore, a corollary of the creation of the organic present.[20] Life was born when an organic present came to be defined (as sketched) and thus introduced into the presentless and undirected temporalities of the physical world.

In the unity of opposing trends we may recognize a late and sophisticated example of orderability and disorderability, identified above with cosmic creation. This immanent dialectical contradiction is manifest in the existential tension of all life forms. In an easily recognizable manner, they are the conflicts between the needs of organisms in terms of their self-directed purpose, pitted against the possibilities determined by the environment.

It is often said that life is fleeting, ephemeral, easily cut off by sword or snuffed out by ill wind. But life is neither ephemeral nor is it fleeting. It is robust and lasting. The organic skill of combining growth and decay in the living present is now 3.5 billion years old, which is three-fifths the age of the universe.

Summary

What did this exercise in the natural philosophy of time reveal?

It demonstrated that it is possible to interpret life and its evolution in terms of time. To do so, it was necessary to abandon the idea of a qualitative homogeneity of time, and replace it by the recognition that time is comprised of a nested hierarchy of qualitatively different temporalities. It was also necessary to adopt the coordinated clockshop model, as the most general representation of life. In the course of reasoning it was learned that the coming about of life was—or seems to have been—a necessity, once molecular evolution crossed a certain threshold of complexity. Self-directed behavior that emerged and remained under the selective forces of the cyclic spectrum of the environment henceforth guaranteed a process of complexification and through it, the stability of life itself.

The arguments offered remain independent of any particular chemistry or spatial structuring of biological matter, and hence are sufficiently general not to limit the phenomenon of life to the particular path that it has taken on earth. Life was positioned, within the broadest conceivable horizons, which are those of the universe.

Anyone aware of the immense variety of organic forms may wonder how such a schematic understanding could assert anything about the community of God's vast zoo of life. The answer is that it is the task of science to identify the universal in the uncountably many of its particulars, and that the relationship between time and life constitutes one of those universals. Specifically, the reasoning suggests that from among the chemically and physically possible structures and functions, life has selected those that could serve in the creation and maintenance of the organic present.

I would like to close with a criticism of the ideas presented. It comes from Christopher Fry's "*The Lady Is Not for Burning*" (1950, p. 53).

> We have given you a world as contradictory
> As a female, as cabalistic as the male,
> A conscienceless hermaphrodite who plays
> Heaven off against hell, hell off against heaven,
> Revolving in the ballroom of the skies,
> Glittering with conflict as with diamonds;
> When all you ask us for, is cause and effect.

References

Aristotle *De Anima.*

—— *Historia Animalium,* 539b, Book V, Ch. 1.

Bernal, J. D. (1951), *The Physical Basis of Life.* London: Routledge & Kegan Paul.

—— (1967), *The Origins of Life.* Cleveland, IL: World.

Bonner, J. T. (1974), *On Development.* Cambridge, MA: Harvard University Press.

Cairns-Smith, A. G. (1981), Beginnings of organic evolution. In: *The Study of Time,* Vol. 4, ed. J. T. Fraser, N. Lawrence, & D. Park. New York: Springer.

Darwin, F., ed. (1887), *The Life and Letters of Charles Darwin.* London: John Murray.

Eddington, A. S. (1958), *The Nature of the Physical World.* Ann Arbor: University of Michigan Press.

Edwards, P., ed. (1972), *The Encyclopedia of Philosophy.* New York: Macmillan.

Encyclopedia Britannica (1983a), Biological rhythm. In: *Encyclopedia Britannica, Macropaedia,* 15th ed.

—— (1983b), Periodicity, biological. In: *Encyclopedia Britannica, Macropaedia,* 15th ed.

Fraser, J. T. (1978), *Time as Conflict.* Boston: Birkhäuser Verlag.

—— Lawrence, N., & Park, D., eds. (1981), *The Study of Time,* Vol. 4. New York: Springer.

—— (1982), *The Genesis and Evolution of Time.* Amherst: University of Massachusetts Press.

—— (1987), *Time, the Familiar Stranger.* Amherst: University of Massachusetts Press.

—— (1990), *Of Time, Passion, and Knowledge.* Princeton, NJ: Princeton University Press.

Fry, C. (1950), *The Lady Is Not for Burning.* New York: Oxford University Press.

Goodwin, B. C. (1976), *Analytical Physiology of Cells and Developing Organisms.* New York: Academic Press.
Kirk, G. C. & Raven, J. E. (1975), *The Presocratic Philosophers.* Cambridge, U.K.: Cambridge University Press.
Maddox, J. (1985), Editorial. The DNA as a kind of solid. *Nature*, 317:197.
—— (1992), *Nature*, 358:103.
Needham, J. (1944), Integrative levels: A reevaluation of the idea of progress. In: *Time the Refreshing River.* London: Allen & Unwin.
—— (1968), *Order and Life.* Cambridge, MA: MIT Press.
Oparin, I. A. (1966), *Life: Its Nature, Origins, and Development*, tr. A. Synge. New York: Academic Press.
Pasteur, L. (1864), Des générations spontanées. In: *Oeuvres de Pasteur*, Vol. 2. Fermentations et Générations dites Spontanées, ed. V.-R. Pasteur. Paris: Masson, 1922.
Schilpp, P. A., ed. (1963), *The Philosophy of Rudolf Carnap.* La Salle, Quebec: Open Court.
Smith, J. M. (1978), *The Evolution of Sex.* New York: Cambridge University Press.
Toothill, E., ed. (1981), *Dictionary of Biology.* New York: Facts on File.
von Neumann, J. (1969), *Theory of Self-Reproducing Automata*, ed. & completed A. W. Burns. Urbana: University of Illinois Press.
Wallace, R. A., King, J. L., & Sanders, G. P. (1986), *Biology: The Science of Life.* Glenview, IL: Scott, Foresman.
Winfree, A. T. (1980), *The Geometry of Biological Time.* New York: Springer.

10. FROM PUPPY LOVE TO FAITHFUL LOVE

The essay of this chapter is a constructive criticism of sociobiology, a science that seeks to identify the roots of human conduct in the social behavior found in the older species of the kingdom Animalia. The tool for the criticism is the hierarchical theory of time. Observing sociobiology with the help of an integrated study of time permits the placing of the significance of intraspecific selection for planning and memory into the broad perspective of "time that takes survey of all the world."

The ideas may be entered through the issue of nature versus nurture.

"She looks like my mother, she'll be a great and good woman," said the father, watching his year old daughter. "She'll be a fine woman," said the young mother, "not because she looks like your mother but because I drank lots of goat milk while I carried her and sang for her every day." Was it grandma's genes or mother's inner environment that made the little girl appear so promising? Are inherited qualities the primary determinants of the characters of men and women, of their abilities, of their health and perhaps even the directions of a their fate? Or are their environments the primary determinants?

Questions of this kind have been conveniently classed under such headings as heredity versus environment and—as already cited—nature versus nurture. This latter distinction may already be found in Shakespeare: "A devil, a born devil, on whose nature/Nurture can never stick..." says Prospero, the Duke of Milan, about Caliban, a savage.[1] The generalization of the comments by the Duke of Milan is due to the biologist Sir Francis Galton (1822–1911) and was born from his interest in eugenics. The science of eugenics studies the ways human health and mental capacities may be improved through artificial selection. The title of Galton's 1874 book says it all: *English Men of Science: their Nature and Nurture.*

Nature versus nurture became a nineteenth and twentieth century shorthand for the debate between two belief systems. One praised and blamed the individual for his or her fate, the other blamed and praised the environment, primarily, the community, the social setting. During the late twentieth century the nature-nurture dialogue precipitated around the understanding and misunderstanding of the science of sociobiology. The clear demonstration by sociobiology that many forms of social conduct have their roots in human biology, whence they arose through natural selection, remained incomplete, because sociobiology neglected the partial autonomy of social processes.

The essay that follows maintains that an understanding of social processes without considering their biological roots will remain fatally incomplete, as will interpretations that do not acknowledge the level-specific freedoms of the social process. The evolutionary significance and role of symbolic causes—the language peculiar to the social process—cannot be interpreted and appreciated without acknowledging the nested hierarchy of integrative levels that comprise humans and their societies. Faithful love of adults subsumes and integrates its evolutionarily and developmentally earlier forms.

TEMPORAL LEVELS: A FUNDAMENTAL SYNTHESIS

Abstract

The principles of a hierarchical theory of time are sketched and applied to problems in the study of biological and social structures. It has often been noted that the major integrative levels of nature, distinct in their degrees of complexity, are also distinct in the types of causation and lawfulness in terms of which they may be most efficiently discussed. It is proposed that time itself is so structured. Rather than a single one-way thrust in which the furnishings of the world equally partake, it is better understood as a hierarchy of temporalities appropriate to the different integrative levels. Methodologically, if we wish to identify the evolutionary policies that control the emergence of new stable structures and functions, we should regard the major steps in probiotic and organic evolution as developmental stages of time itself, rather than steps within the framework of a pre-existing absolute time which 'flows equably without relation to anything else'. This is an anti-intuitive stance whose value can be judged only in terms of its internal consistency and explanatory power. The theory is, therefore, tested.

Biogenesis is interpreted as the coming about of autonomous clock aggregates which, in the course of evolution, complexify into the cyclic order of life, by filling the available niches in the spectrum of environmental periodicities. The aging order of life is seen as a boundary condition to the spectral spread of the cyclic order. Man emerges from this matrix through his ability for symbolic transformation of experience, which makes possible intraspecific selection for planning, memory and language. As a social enterprise language is carried along, as it were, by its inertia and dictates the forms of temporal structures within which social issues are expected to be formulated and social functions understood. The concluding remarks pertain to the epistemic status and critical testability of the theory.

1. Introduction

This paper sketches a hierarchical theory of time and applies it to a number of problems in the study of biological and social structures.

The *theory of time as conflict* comprises a class of principles in the scientific and humanistic study of time (Fraser, 1975, 435–446; 1978a). Its concerns include the abstract time of physics, time manifest in life and organic evolution, and time relevant to man's individual and collective behavior. The basic idea is that time is not a single, one-way thrust in which the furnishings of the world partake equally, but a hierarchy of temporalities associated with a hierarchy of structurally stable integrative levels.

One of the surprising and counter-intuitive principles of the theory is that these temporalities themselves evolve under the pressures of certain level-specific conflicts and that this evolution is open-ended.

2. Species-specific Universes

Early this century the German theoretical biologist Jakob von Uexküll drew attention to the epistemological significance of the fact that an animal's receptors and effectors determine its world of possible stimuli and actions (Uexküll, 1909). He called these circumscribed portions of the environment that were effective for a given species the *Umwelt* of that species. Events and objects not in the specific universe of a species must be understood as not existing in its world. Thus, the ultraviolet patterns of certain butterflies exist for one another but not for vertebrates. Note here that the 'environment' of which the *Umwelt* of a species is a circumscribed portion is, in fact, the *Umwelt* of man. We know of the ultraviolet patterns only because we succeeded in enlarging our *Umwelt* through the use of exosomatic devices, extensions of our endosomatic capacities.

Taking advantage of the latent power of this principle, I extended it beyond its original boundaries so that in the study of time we may identify epistemology (whatever we are capable of knowing) with ontology (whatever we may assume to be the case). I retained the use of the word 'umwelt' but naturalized it in appearance.

The concept of umwelts is easily formulated for different sensory systems. In man, for instance, the auditory, tactile and visual realms continuously interact and yet they define distinct umwelts. From these worlds, determined by our senses, it is but a step to the umwelts of scientific instruments. The umwelt of a radio telescope comprises regions of the electromagnetic spectrum, distributed within ill-defined spatial boundaries. It is a crude and inarticulate world. Finally, certain portions of reality are open to us mostly or only through mathematical formalism. Sometimes our equations tell us about intrinsically probabilistic worlds wherein the exact instant of events has no meaning; or about other worlds wherein time may be said to 'flow' but, strangely enough, it has no preferred 'direction'. If we believe our equations as we believe our senses, then we must judge such revelations as acceptable truths of nature. They are subject to change as our knowledge of nature develops, but they are not in need of justification through experiential familiarity.

This greatly enlarged scheme of the idea of species-specific universes is the *extended umwelt principle.*

3. Level-specific Temporalities

When the extended umwelt principle is applied to time-related issues in physics, biology, psychology and sociology (Fraser, 1975) it points to the existence of several distinct levels of structural and functional complexities, that is, to a hierarchical view of nature (cf. Koestler & Smythies, 1969; Simon, 1969; Whyte, 1969; Pattee, 1973; among many others). For the study of time it appears sufficient to distinguish six major structurally stable integrative levels separated by five metastable interfaces. The higher levels subsume their own, rich hierarchical orders but each of the major levels, as a class of structures and functions, is distinct from all others.

We begin with physics and go on to biology. The substratum of the world is the (idealized) universe of radiation or, *pars pro toto*, the world of light. The next level is that of indistinguishable and hence completely interchangeable particles. Above it, as it were, we find the astronomical universe of heavy masses. Upon one of these masses we recognize life. One of the living species is man. Individuals of this species know how to transform their experiences into signs and signals that tie them into societies. I submit the following.

(1) Each of these structurally stable levels determines its characteristic causation, language and temporality.
(2) These temporalities are paradigmatic in that though they first appear on lower integrative levels, they are also identifiable among the temporal functions of higher order structures.
(3) The temporalities developed from an originally atemporal matrix. This claim appears self-contradictory; we have no vocabulary in which the evolution of time itself may be discussed in self-evident terms (Fraser, 1978a). We shall have to depend on the consistency and usefulness of the theory and let language take its course.

Following the advice of Bastin (1969) we commence by thinking about hierarchical structures abstractly. Then we shall give names to the six temporalities and dress the skeleton.

There are umwelts in which no meaning can be given to the idea of 'event', to conditions of 'before-after' or 'past-present-future'. We say that these umwelts are *atemporal*. Such worlds remind us of empty sets, not of mystical nothingness. Physics teaches that the umwelt of a photon is atemporal. Experiments have shown that fighting fish (*Betta splendens*) cannot resolve temporal structures of periods shorter than about 300 ms. The corresponding limit for human subjects is about 2 ms. These periods, for the respective subjects, are atemporal.

There are umwelts in which events may be identified but their temporal positions never precisely determined. We will call such worlds *prototemporal*. It was in Quantum Theory that it first became clear that temporal uncertainty can be a fundamental and irreducible feature of some levels of nature. But all sets of indistinguishable (hence interchangeable) elements may determine features of prototemporality: populations of mice or men, genes in a gene pool, workers in a factory. Three hundred accidents for Labor Day weekend may be a reliable forecast but the instants (and places) of the mishaps cannot be predicted. Connections among prototemporal events may only be specified probabilistically.

In certain umwelts time may be said to 'flow' yet it does not have a 'direction'. The paradigm of such *eotemporal* umwelts (for Eos, goddess of dawn) is the world of classical mechanics and electrodynamics where the existence of directionless time was first noted. But all purely oscillatory processes are eotemporal: vibrations of macro-molecules, rhythms studied in chronobiology, social cycles. In the symbolic domain, historical time based on myths of eternal return are eotemporal. Connections among eotemporal events may be specified as necessities and hence the characteristic eotemporal causation is deterministic.

The three lower temporalities, strange as this must sound to those not familiar with the study of time in physics, have no features whatsoever to which the idea of 'nowness' could be attached. The crucial break in the development of temporalities came with biogenesis; specifically, with the need of organisms to provide for continuous, internal synchronization of biochemical processes. As we shall see, in terms of the instant by instant control of compatible and incompatible conditions nowness emerged from the now-less world of inanimate matter. Futurity and pastness then became meaningful because future and past make sense only in reference to a present. Thus there evolved the *biotemporal* umwelt peculiar to life. But time's arrow was born with a small head. Early forms of life are almost purely cyclic (eotemporal). Later, with the evolution of aging, the past-future polarization in ontogeny became sharper, beginnings and endings became meaningful in terms of the autonomy of the organism, though not always without ambiguity. The single-path, deterministic causation of the eotemporal world gave rise to multiple causation characteristic of life, as well as to final causation (goal-seeking behavior).

Human communication, especially language, determines the *nootemporal* umwelt of man. Human beings 'have broken through the old vertebrate restraints... by acquiring the intelligence to consult the past and plan for the future' (Wilson, 1975). In the nootemporal umwelt the arrow of time has an unarguable direction that derives from man's knowledge of his inevitable personal aging and death. The biotemporal 'now' opens up to the mental present of man with its continuously varying boundaries, while the plurality óf biological causations expands to include human freedom.

We may talk about *sociotemporal* umwelts for human groups if they are sufficiently coordinated to act in self-interest. The epistemic difficulties in identifying the hallmarks of social time are great. Partly, we are both the observers and the observed (as already for the nootemporal umwelt); partly, as we shall see later, there are reasons to believe that in dealing with societal time we shall need a new language whose vocabulary and syntax derive entirely from social and not from individual experience.

4. Level-specific Languages

Arguing for and against differences between physical and biological lawfulness is as old as the history of biology itself. Let us narrow our concerns to the past decade only. Mayr (1968) and Elsasser (1969) stressed the radical change in the number and character of causal pathways as we enter the study of living matter, and the corollary changes in type of lawfulness. Koestler & Smythies (1969) made the non-reducibility of biological law the theme of a conference. Fox (1972) reflected on the evolution of levels of evolution and the changes in the focus and method of selection. Grobstein (1973) spoke of set—superset transitions between levels of lawfulness, Goodwin (1978) about the control of pattern formation from above the molecular-genetic level.

The structuralist theory of time offers a framework for a hierarchical epistemology that can accommodate, as we shall see, both the continuities and discontinuities of lawfulness that bind and separate the stable levels of nature. Let us review and comment upon the hierarchy of causations.

- Causation has no meaning for atemporal umwelts.
- Prototemporal events are joined by probabilistic causation.
- Eotemporal events are joined by deterministic causation.
- Biotemporal events are joined by multiple and by final causation. Rigid programming gives way to dynamic programming.
- Characteristically human actions are connected through symbolic causes, known as ideas. The possibility of choice among ideas and corresponding actions has been called free will. I prefer the concept of human freedom. Ideas are also the means of sociobiological evolution for they can produce

responses to imaginary challenges: an imagined bear can increase defensive readiness against real bears.

- Socially important events are joined through historic causation. Since in a hierarchical epistemology each level is under multiple control, historic causation subsumes all lower level connectivities. However, its hallmark is the capacity of man to change his social institutions in response to symbolic causes. A unicorn in the garden can change history by increasing defensive readiness against unicorns.

Each of these levels may be said to have its peculiar language, in which we include all signs, signals and symbols employed for communication. For human society, taken in this broad context, 'language' includes art, artifact, body cues, speech and written language. It is not surprising, therefore, that in the umwelt of man, the languages necessary to deal with physics, biology, psychology and sociology incorporate distinct jargons.

Writing about biological hierarchy, Pattee (1970) remarked that the postulate of stable life forms is equivalent to the postulate of interfaces and that 'each side of the interface requires a special language. The lower level language is necessary to give what we might call the legal details, but the upper level is needed to classify what is significant.' When all the structurally stable levels are considered together, not only the biological hierarchy, a case can be made for a generalized asymmetry of level-specific languages. Looking down from above, the laws of a lower integrative level, expressed in the appropriate language, appear obvious; looking sideways, they are intelligible; looking up, from a lower umwelt to a higher one, the language appears unintelligible and hence mysterious. Thus, human action in terms of human biology and psychology is intelligible though its understanding is badly incomplete. But morphogenesis in terms of physics alone is unintelligible. Likewise, many sociologists maintain (e.g. Lukes, 1973; Mendelbaum, 1973) that certain societal concepts are not reducible to concepts of individual behavior. The interfaces, we might say, are semipermeable to the languages which they separate.

5. The Emergence of Stable Structures

There are, or at one time had to be, structures intermediate between those that comprise the known stable integrative levels. But such intermediate structures and functions either do not exist today or else are very rare.

Again, we begin with physics. We know of no form of energy intermediate between the atemporal (light) and the prototemporal (particle). Also, matter is either particulate (prototemporal) or massive (eotemporal) by preference as it were, for almost all matter of the universe is in radiation, particles or galaxies. Going from the eotemporal to the biotemporal we note that biogenesis is not now taking place spontaneously. Likewise, going from the bio- to the nootemporal umwelts, we have no living species intermediate between ape and man.

I have examined what we know and what we may reasonably assume about the interfaces, because it is from them that new, stable structures have arisen in the course of evolution (Fraser, 1978*a*). As a class they seem to follow certain common policies.

They tend to be *metastable*: either relapse into the lower level or rapidly evolve into the higher, stable level. But they sometimes survive in modified, slowly changing forms. Thus, if we believe Cairns-Smith (1971) the structural stability of our crystalline ancestors (on the top of the eotemporal world) survived in the stability, though not in the chemistry of the DNA molecule, on the bottom of the biotemporal world. Likewise, the behavioral pattern of the species that was between ape and man survived in a class of behavioral functions on the bottom of the nootemporal world. In depth psychology these functions make up the unconscious of man.

Across the interfaces we observe *very strict selection processes*. Going from the eotemporal to the biotemporal world, we find that the percentage composition by elements of living matter does not correspond either to that of the universe or to that of the earth. Going from the biotemporal to the nootemporal world, out of the millions of species only a very few, possibly only one, evolved the brain whose umwelt is nootemporal. Going from the biology of the individual to that of society, it appears likely that if our tribal plurality is to change into a planned global society, only a very few of the multitudes of social structures will guide the global colony of man.

Countering the discontinuities, we can identify continuities in *new uses for old structures and forms*. Crossing from matter to life, as I shall argue later, the self-organizing capacity of matter that produces ordered structures continues in the self-organizing capacity of living matter that produces ordered functions. Crossing from animal to man, the examples are numerous. For instance, the auditory loop which, in animals, helps blend individual sensitivity into collective sensitivities evolved and made possible human language, the foremost tool of personal identity in man. Crossing from the individual to society, technology is a collectively expanded form of endosomatic functions which became, in its own right, an instrument of collective survival.

An aside on methodology and quantitative reasoning is appropriate at this point.

It is tempting to look to catastrophe theory as a possible quantitative means for handling transitions between stable integrative levels, across the metastable interfaces. Careful examination reveals, however, that the theory in its present form cannot accommodate these examples of evolutionary novelty, even in principle. Namely, the coming about of a new, structurally stable integrative level corresponds to an increase in the codimensions and coranks of the manifolds upon which the developments of level-specific systems are to be written, as well as to the coming about of new attractors. The new codimensions, coranks and attractors represent specifications to which nothing in the earlier spaces can correspond. What is needed, therefore, is a super-catastrophe theory which is inductive in character, as distinct from a generalized catastrophe theory which is deductive. A super-catastrophe theory would have to map transitions from lower order sample spaces to higher order ones and from sample surfaces of certain character to more complicated sample surfaces. During the super-catastrophes the higher order spaces and sample surfaces would have to be created rather than presumed. Current catastrophe theory cannot permit this as far as time goes, for the theory is written in conventional physical space-time and therefore it is limited by the physical assumptions hidden in that concept (Fraser, 1978*a*, pp. 176–177).

Let us return now to the common policies of metastable interfaces.

A corollary to the enrichment of languages across the interfaces is the existence on each stable level of certain functions and structures which, since they cannot be specified in the language of that level, remain *undetermined.* The laws of physics constrain but do not determine the paths of development along Waddington's epigenetic landscape; biology constrains but does not determine the specific forms of human creativity. Out of the undetermined regions emerge the laws and languages of the stable structures of new integrative levels and with them, new temporalities.

The problems of the following section deal with this and related issues. I will maintain that the evolutionary emergence of new stable structures of increased complexity is best understood as developmental steps along the evolution of time itself, rather than evolutionary steps in the framework of a pre-existing 'absolute, true and mathematical time [which] of itself and from its nature flows equably without regard to anything else' (Newton, 1687).

6. Applications of the Theory

6.1. *Biogenesis*

Goodwin (1963, 1970, 1976*a*) and Bonner (1974) are two among many biologists who regard oscillatory behavior as the fundamental dynamic mode of living systems. Let us assume that oscillatory behavior did indeed characterize life from the very beginning and seek a model of biogenesis as a step in the development of temporalities, specifically, as the coming about of biotemporality from the eotemporal umwelt of the pre-biotic world.

Darwin (1871), while speculating about the origins of life, imagined 'some warm little pond, with all sorts of ammonia, phosphoric salts, light, heat, electricity' in it. Let some such mixture be our primordial soup. Needham (1936, 1944) wrote about mesoforms which occur between successive levels of organization and noted that between living and non-living matter the realm of the crystalline represents the highest degree of organization. Bernal (1967) called for a generalized crystallography as the key to the biology involved in the origins of life. More recently Cairns-Smith (1971, 1975) suggested that the ancestors of life were, in fact, crystals with structures that stood in for the later DNA–RNA–protein system of biochemistry. Let us assume that the biogenetic landscape did include crystals which occupied protective, microscopic niches in whatever solid geological forms existed, as blue-green algae colonize air spaces (today) in rocks in the Dry Valleys of the Antarctic. Their energetic environment included optical radiation, high frequency solar radiation, heat, electric discharge and supersonic waves (Fox & Dose, 1972).

Let us now detour to thermodynamics. Beginning with mechanical systems, continuing with chemical systems close to equilibrium and then going on to physical chemistry, a case can be made for the claim that while the history of closed thermodynamic systems is, indeed, controlled by the Second Law of Thermodynamics, for many conditions that resemble those of living processes, the final state of equilibrium is reached along paths of minimal rates of entropy increase (Fraser, 1978*a*, p. 64). This is a generalized form of Prigogine's celebrated principle of minimal entropy production (Prigogine, 1955, 1973). Let this principle be schematically represented as the vector sum of two opposing arrows: an entropy growth and an entropy minimizing one. Their difference is the net entropy increase.

Goodwin (1970), continuing the work of others, argued that autonomous cyclic operations can occur only for non-linear systems with steady states displaced from equilibrium. Let us next assume that Prigogine's extension of his principle of minimal entropy production is valid for such conditions. But then we may identify the thermodynamics of biogenesis with the lengthening of the minimizing arrow into the entropy-absorbing (self-organizing, growth) arrow of living matter. This transition represents both a continuity and a discontinuity. Non-living matter can only minimize rates of entropy increase; living matter can do better, by altogether reversing the entropy increase trend within the walls of energy-permeable bags known as open systems.

Perhaps some of the crystals of the biogenetic landscape were able to absorb oscillating energy at some frequencies and use it to maintain oscillations at other frequencies that fitted the frequency niches of their immediate environment. Those crystals that could do so had a better chance for maintaining their structural integrity than those that could not. The accidental modeling of environmental periodicities had to have a statistical spread. Such a spread made possible the working of environmental selection pressures and we may thus imagine that natural selection supplanted chance as order in structure was replaced by order in function.

A system of coupled oscillators can endure only if its individual clocks remain, instant after instant, mutually supporting rather than mutually destructive. [On this, see the extensive literature of chronobiology which conceives of pathogenesis as temporal mismanagement; also Richter (1965) and Cohen (1967).] The logistics of microscopic living clock shops had to include specifications as to what may or must and what should not happen simultaneously if collective viability was to be maintained. It is in this demand for inner coordination that we may identify the coming about of nowness in the background of the pure succession of physics. Simultaneity or its absence is also important for certain steps in physical and chemical evolution. But the prebiotic world has no structures whose autonomy may or may not be lost through temporal mismanagement.

There is a much-debated principle in the theory of automata, formulated by von Neumann (1951) according to which a system must reach a certain threshold of complexity before it can produce others of its kind of equal or increased complexity. Below that it can produce only systems of lesser complexity. If this principle is taken to be valid for biology, it then follows that reproductive capacities need not be coeval with the life process. Only after the early oscillating units reached a certain level of complexity was it advantageous to reproduce through heirs that develop (rather than to replicate identical heirs) because only then could the offspring be potentially better adapted than the parent (cf. the 'cost of reproduction' argument of Bonner, 1974). The evolutionary priority of living functions to the reproductive ones suggests a period in early organic evolution when the umwelt of life was almost eotemporal with a very broad 'now' whose boundaries then became increasingly narrower and better defined.

6.2. *The cyclic and aging orders of life*

In the model of biogenesis just proposed, improved fitness consists of improved copying of the spectrum of environmental cycles. By ever finer tuning, each new clock is an adaptive measure which increases the viability of the organism, as evolution fills the temporal niches of geological, astronomical and physical rhythms. For this model the characteristic variable of evolutionary progress is complexification (cf. Saunders & Ho, 1976).

Increasing complexification surely led to the internal generation of new cycles to which nothing in the external world needed to correspond. Some of these had to be control functions which assured the continued viability (autonomy) of the primitive organism. Natural selection, henceforth, while acting upon the externally manifest ('phenotypal') rhythms did, in fact, effect the internal ('genotypal') rhythms. Thus, under the continued environmental pressure of cyclic change, evolved the cyclic order of life whose dynamic stability came to be best describable in terms of the dynamics of limit cycles rather than through the statics of limiting point-states (Goodwin, 1970; Winfree, 1975).

The cyclic spectrum of biological clocks known today is very broad. Human skin responds to ultraviolet rays at 10^{16} Hz; retinal cells to 10^{15} Hz; most organisms pick up heat at 10^4 Hz. The periods of neural signals are between 3–10 s. Probably all living things, down to the genes, show circadian rhythms. Lunar periods are widely spread as are circannual rhythms and some bamboos flower every 7–8 years. The morphology of oscillators across this band of 24 orders of magnitude must vary greatly and thus must force upon living matter a division of labor by specialization, a multitude of life forms and, by the argument of Simon (1969), a hierarchical organization.

Not only in the cyclic functions of the individual, but also in the cyclic behavior of populations, we can observe an increasingly finer tuning of rhythms and the corresponding evolution of appropriate implementation. For instance, populations respond to external perturbations by a set of adjustment mechanisms that resemble relaxation oscillators which differ in their activation times (Slobodkin & Rapoport, 1974). And, collective rhythms in general, tend to fill the available niches in the frequency spectrum of the environment (Cloudsley-Thompson, 1961).

The upper frequency boundaries of biological clocks have probably been reached in the ultraviolet clock. The lower boundary would be a clock whose changes are almost linear. This limit, I suggest, is the aging process. Its organic implementation involves an almost purely cyclic, 'immortal' germ cell and a 'throw-away' soma. A fine gradation of time's arrowheads may thus be identified among the living, from the polynucleotides which, though not immortal, can hardly be said to die, to primates which are surely mortal.

In reproduction, aging and death—in the life cycle—organic evolution retained a record of time's rites of passage. The sociotemporal world of societies and the nootemporal world of individuals are assured of continuity through periodic regression to the temporality of a single cell. But the integrity of life process remains carefully guarded. Life cycles are not connected through 'naked genes', which would amount to a return almost to the eotemporal world of physics, but through genes in a cellular environment that contains the records of past evolution (Bonner, 1974).

During the evolution of the cyclic order of life, some of the autogenic cycles (to which no prior external cycles needed to correspond) were likely to have become externally manifest as behavioral cycles. Thus the very success of organic evolution enriched the spectrum of rhythms to which the species had to adapt. It is safe to assume that, because of the multiple and final causation available to living matter, the behavior of advanced organisms became progressively less predictable. For this reason behavioral controls also had to become progressively less satisfactory, at least for some species in some regions on earth.

Some time after the appearance of the human brain and before the appearance of the first human groups, one or more forms of genus *Homo* learned to work out future behavioral strategies based on past, individual experience, through the use of a symbolic system that includes human language. This evolutionary step reminds us again of von Neumann's complexification threshold. Above a certain level of the complexity of his brain, an individual of the species *H. sapiens* was able to assist, through the dynamics of human teaching and learning, conspecific individuals to become better adapted than himself. Through language it became possible to create a store of collectively generated cumulative knowledge, comprised of acquired characteristics. Thus, from the biotemporal umwelt emerged the nootemporal world of man with its peculiar dynamics. Though the signs, signals and symbols of this world, just as human actions themselves, are limited by physical and biological restraints, the direction in which the noetic world evolves is not determined by the laws and regularities of lower integrative levels.

6.3. *Phenomenology of the immensely complex*

The minimal set of essential responses to token stimuli found in the most primitive instinct-reflex mechanisms represents the action of a central nervous cord of hundreds of thousands of neurons. An organism with a fully elaborated central nervous system and brain, containing 10^5–10^8 neurons, can learn to handle particularities of its environment, such as remembering places. In organisms with 10^9–10^{10} neurons behavior is not rigidly programmed and socialization is prolonged. The key social feature of this grade is 'a perception of history' (Wilson, 1975, p. 151). I would prefer to describe their temporal umwelt as including non-cyclic planning and individual memory, and reserve the perception of history to man. But man is still in the same league with the chimpanzee and the baboon when it comes to the number of neurons. For this reason, the difference in mental capacity between man and the higher apes cannot be accounted for by the numbers of neurons alone, nor simply by difference in brain size (Lenneberg, 1964).

Let us construct a measure of complexity by identifying it with the number of distinct states that an autonomous system may assume. A cubic centimeter of water contains perhaps 10^{22} molecules. Borrowing a term from Elsasser's 'statistic and the concept of immensity' (1966, p. 69) we may say that the number of microstates of such an assemblage, that is, the number of ways in which it may be partitioned, is 'immense'. But this immensity has little structural or functional significance for none of the microstates will be very different from any other microstate of that sample, or even of a sample of 10^{32} molecules of water. A cubic centimeter of water is not an autonomous system.

Beer (1967) calculated that as a digital device, with each axon on or off, the number of various states of the human brain is perhaps 10^{109}. (The number of different sequences in a chess game, which is a set of well-formed rules, is about 10^{120}.) If we assume that each of the brain states may correspond to a different (global) engram, then we can claim that the fully developed human brain is the most complex system known, even if the figures are off by a few thousand billion. But a number of a similar order of magnitude must also obtain for the brains of baboons. Therefore, the complexity (as defined) of the un-engrammed brain, whatever engrams might eventually turn out to be, is only a genetic potentiality. It could evolve into what it is now only if its capacity to form, retain and generate engrams was strongly selected for. Let us try to identify one set of selection pressures appropriate for that task.

Like a sprinter in training, we shall have to return to the starting line a number of times before we can go on and reach our goal.

When talking about the expanding clock shop model of organic evolution, I maintained that sooner or later autogenic rhythms had to appear, with no corresponding external rhythms. One would also expect that with an increasing engram complement, the brain would begin to generate new engrams to which nothing in the external world need to correspond. On the internal landscape of the mind we would thus expect to find images (actors, symbols, signals) corresponding to real as well as to imaginary objects and events.

As the behavior of the living environment became less predictable, sooner or later there had to arise the necessity for a new and peculiar symbol on the inner landscape. It stood for an object, for a constant companion, whose presence made food, friend and foe display predictable behavioral changes. One may speculate that in the course of evolutionary development the symbol for this companion and the symbols for the sensate body merged into a single symbol standing for selfhood.[2]

Consider now certain limiting processes and structures with which we are familiar. Examples are the immensely large (objects of the size of the universe); the very small (elementary particles); the very rapid (speeds close to that of light); the very cold (temperatures near absolute zero); possibly the very hot (the cosmic fireball). All these limiting processes or conditions have their peculiar laws which, though not reached discontinuously from human-sized conditions, are nevertheless unpredictable from them.

I propose that the complexity of the brain of historical man represents a limiting condition in biology comparable to those in physics, just listed. In this view, the human brain is identically equivalent to the 'immensely complex' and, for that reason, we should neither be surprised nor disturbed if its laws are not easily extrapolated from the laws of other biological structures.

I further submit that selfhood, conscious experience, human freedom, human creativity comprise the phenomenology of the immensely complex. If a comprehensive science of man is to include concerns with selfhood, conscious experience and the capacity to create human time, then it is with the phenomenology of the immensely complex that the work must begin. Whether such a science could still be called sociobiology depends on whether human sociobiology will recognize and be able to accommodate the high degree of autonomy which must be assigned to the noetic and societal functions of man.

Now we may attempt to answer the question about selection that would strongly favor the evolution of immense complexity of the brain.

Consider first that the self is partly a sensate and partly an imaginary object. It is 'out there' with all other objects but it is also an imaginary construction put together from impressions gained from the behavior of others. (If it would not be a symbolic continuity, no meaning could be attached to the utterance 'I don't want to die' if it is coming from a dying person.) Consider next that I cannot explore myself visually or by smell as completely as I can explore others, but I can hear myself probably as completely as I can hear others. It is perhaps for this reason that the spoken word became the most immediate way of representing the self—such as by naming—and also representing everything else that constellates about the self: the passage of human time, responsibility of choice and the certainty of death.

I would envisage the emergence of man as an autocatalytic and hence, necessarily, rapid process triggered by the discovery of selfhood and language. In turn, language and selfhood made possible the working of powerful intraspecific selection for long term planning, long term memory and the creation and manipulation of symbols. In brief, it was a self-reinforcing selection process for the advantage of knowing human time. Structurally, the transition belongs in the class of metastable interfaces discussed earlier.

6.4. *Linguistic inertia*

Let us grant, with Waddington (1961), that human evolution was made possible by, and depends on, human language. The reason why the sociobiological multiplier effect and network expansion have been so tremendously effective in the history of human societies is to be sought in the flexibility and freedom of language. The shot fired at Concord in 1775 was not really heard around the world, but words about it were.

Thom (1970) compared the words of natural language with Waddington's creodes. I find this analogy inappropriate because creodes—preferred trajectories in phase space of epigenesis—can lead only to such organic forms as are permitted by the laws of chemistry and physics. Human language is not so limited. By giving process descriptions of the brain (Fraser, 1978*a*, p. 117), it can specify structures, functions and events to which nothing in reality need correspond, either actually or potentially. As Steiner maintains in his theory of translation (1975) 'language is the main instrument in man's refusal to accept the world as it is'. It can express not only potential novelties but also impossible futures and counterfactual pasts. Mythical animals that live in man's umwelt did not evolve through natural selection.

Let us now turn to the moving forces of social evolution. Wilson identifies two of them: ecological pressure and phylogenetic inertia (1975). To get the full benefit of the latter idea, let us define inertia not simply as resistance to change but as a resistance to changing the direction of prevailing motion.

Consider now that all living things are subject to Newtonian inertia. But this, though important, is not very interesting because living matter has certain freedoms that non living matter does not, and hence, possesses its peculiar conservativism, including phylogenetic inertia. Likewise, language is subject to phylogenetic inertia. But this, though significant, is not that interesting because human language has certain freedoms that animal languages do not and hence it possesses its peculiar conservativism. For this reason, in human sociobiology, *linguistic inertia* must be added to the other two as the third, distinctly human organizing and ordering force, and mover of social evolution.

By linguistic inertia I mean the tendency of language and, by extension, the tendency of the total human communication network including art and artifact, to resist changes in the direction of the cultural process. The tremendous inertia of accumulated knowledge and value judgment embodied in tradition, deposited in human language and dictating the preferred views of time, is manifest throughout political and intellectual history (Fraser, 1975). It is complementary to the inborn 'authority bearing' trait of man and together with that trait is responsible for 'beliefs of the particular tone we call ethical' (Waddington, 1961).

6.5. *The evolutionary office of symbolic causes*

How may an individual select a path of action from the practically infinite store of actions which he can imagine? Traditional responses to the individual's need for guidance have been classed under such headings as ethical behavior, praiseworthy conduct or correct morality. As far as one can tell, these instructions have always been under collective control. I will not attempt to deal with specific instructions but ask, instead: what is the evolutionary role of ethical legislation?

Sociobiologists seek the evolutionary sources of human ethics in altruistic behavior, observable throughout the animal kingdom. Altruism is defined as behavior that decreases the fitness of an individual so as to favor the fitness of others. On this scale, colonial invertebrates form the most perfect society because the genetic identity among individuals makes possible the evolution of unlimited altruism (Wilson, 1975). As electrons in quantum mechanics, genetically identical individuals are (ideally) indistinguishable and therefore completely interchangeable. For the very same reason, however, we cannot speak of self-sacrificing behavior. There are no selves, as distinct from other selves, that may be sacrificed. It is true that certain behavioral responses among the higher social species do suggest something of a selfhood, but we find no implementation that could possibly raise the importance of identity to the level which it assumes in human social evolution. The gap is so vast that it cannot be meaningfully measured and so we will assume man's selfhood to be one of the distinct hallmarks of the nootemporal and sociotemporal worlds.

In higher animals, where the absence of genetic identity precluded the evolution of unlimited altruism, we find such practices as reciprocal altruism, which is an exchange of favors. Man's species-specific knowledge of time extends this practice to deferred reciprocal altruism: invest now, pay later; thus there is a degree of continuity across the biotemporal-nootemporal interface. But we also find a significant discontinuity. The highly developed skill of our species to create and use symbols allows for the practice of self-sacrifice on behalf of symbolic causes.

Queen bees attack and even kill their fertile daughters, yet in no sense can they be said to commit a moral or immoral deed; neither do salmons commit self-slaughter after spawning. They have no choice and they have no selves. Mother rabbits often protect their kits even at the price of their lives. They have great genetic investment in them and they are programmed to protect that investment. But Abraham, who was surely programmed to defend and protect Isaac, was, nevertheless, ready to kill him.

To get away from the quagmire of metaphysics, I will assume that he did have a choice. I assume, in fact, that human individuals often have such choices and that acting on behalf of symbolic causes, they frequently make fateful decisions. Their acts may be as dramatic as the Socratic self-sacrifice or as undramatic as a decision to favor a particular style of art. Insofar as such decisions pertain to symbolic causes I will class them as 'moral choices', quite divorced from any particular school of ethical teachings. Moral choice may thus be seen as the peculiarly human form of altruism.

We may now rephrase the earlier question: what is the evolutionary office of moral choice? Why did social evolution allow for man to be willing to be disadvantaged for favoring a style of art, to be willing to die for abstract ideas annunciated by people far away or in ages past, conveyed through art, artifact and language, all of which roll along under their collective inertia?

Consider now that scientific truth suggests a body of eternal and hence discoverable verities, as does the ordinary use of the concept 'ethical law'. But the epistemic status of moral choice is quite different from that of scientific law: each decision is an act of creation. The discovery of the laws of planetary orbits did not change those orbits. But an engraving of an image by Cro-Magnon man on a bison horn or the self-sacrifice of Socrates did change the path of social evolution, some in small, some in large measure. The engraving on the horn and the Socratic code of behavior were not eternal verities 'there' to be discovered; they were created determinants of undetermined futures.

If this epistemic distinction is admitted as valid, then the answer to our question comes to mind. The evolutionary office of moral choice, including the self-sacrifice of man for sustaining symbolic causes, is to provide the means whereby the direction of social evolution, carried by linguistic inertia, may be altered. Whereas altruism in animal and man maintains the momentum of phylogenetic inertia, moral choice in man makes possible the changing of the functions and structures of social institutions.

6.6. *The temporal environment of a global colony of man*

As do biological clocks, so societal rhythms model geological and astronomical cycles. They also generate rhythms peculiar to the culture and economy of the group (Zerubavel, 1976, 1977, 1978). A dramatic extension of the domain of social cycles may be observed in our days, described by Thrift (1977) as 'flextime' and by Melbin (1977) as 'the colonization of time'. Melbin has shown, for instance, that night-time social life in large cities resembles the social life of former land frontiers (Melbin, 1978). To this we may add many other changes of the same type, such as the disappearance of the privileged position of Sundays in the social rhythm of commerce. The 'colonization of time' is contiguous with the expansion of the cyclic order of life discussed earlier. With the availability of artificial daylight, the circadian rhythms of individuals are likely to get out of phase with the sun, while the circadian rhythms themselves remain entrained.[3] The same scattering may be observed regarding days of rest.

The colonization of time has aspects of sociobiological preadaptive measures. It gets individuals ready to serve society in an ever finer tuning of the division of labor. The process is contiguous with the division of labor that accompanied the evolution of the cyclic order of life.

In the dynamics of social cycles the autogenic rhythms of society act as forcing functions for individual rhythms. The better organized a society, the less significance there is to individual preferences. When you are told to rise, you rise; when told to die, you die. Under the pressure of increasingly coherent world-wide rhythms demanded by industrial production and distribution, and with the assistance of complexifying networks of communication, the world community seems to be changing in the direction of a global colony of man. Because of the increased mobility of people there is a dramatic rise in the rate of gene flow around the world. Invoking the idea of group selection, Wilson (1975, p. 575) noted that this flow presages a decrease in group-selected genes and, therefore, it is likely to lead to a decrease or even loss of human conflict and with it, of human creativity.

Because of the vast autonomy of the noetic and societal integrative levels, it is hasty to draw far-reaching conclusions on the presumptive fate of human creativity, based on genetic arguments alone. Yet the warning is valuable because mankind is unique both biologically and culturally.

We learned earlier that the individual self is a symbolic continuity developed through interaction with other members of the human species. Group identity is also such a symbolic continuity. It is a 'we–they syndrome', 'a living thing that grows, changes and strives or withers according to the rise and decline of its own vitality and the conditions in which it exists' (Isaacs, 1975). Since there is no other humanity with which it may be compared, the identity of a global society would enter the class of such other concepts as God or the universe. Paraphrasing Steiner (1975): 'no encounter, no name'. Not only the biological mismanagement but also the cultural and ideological experiments of a world-wide family of man would go unchallenged by external forces. The social, economic and ideological swings of a global community may easily become oscillatory (cf. Watt, 1978) and, by crossing a certain critical threshold, send mankind back to tribal fragmentation.

I would like to make a few observations regarding linguistic changes that parallel in the symbolic domain of the phenotype what gene flow around the world does for the genotype.

Under the pressure of common socio-economic needs around the earth, we observe a rapprochment among tongues, a trend for standardization. This practice has been eroding the distinct emotional logics of languages, their characteristic colors as it were, deriving from their different histories. Part of the standardization is an increasing stress on present political change and a de-emphasis of historical continuity. This, in turn, tends to inhibit intellectual inquisitiveness unless it leads to immediate results. The division of labor attendant to a global community of man demands exchangeability of people within professions, reminding us of Wilson's insect societies (1975, p. 379) which live in 'impersonal intimacy' by recognizing castes but not individual nestmates. The umwelt of exchangeable individuals resembles the prototemporal and, hence, one would expect that in a highly organized human society of specialists, questions regarding individual destiny and human time would be increasingly difficult to formulate. This trend is already detectable in the ways that issues of time and the modern self are handled in literature (Sebba, 1972; Ungvari, 1972). As does the colonization of time, so the narrowing of temporal horizons of daily language has aspects of sociobiological measures, preadaptive to the emergence of a global colony of man.

But against the pressures that favor a single, planned society, we find a proliferation of tribal interests, an explosive increase in pluralism, a powerful centrifugal force. It is difficult to imagine a social structure that could include both: the increasing demands for distinctness and the increasing call for uniformity. Our epoch is a 'time of confused and chaotic passage' (Isaacs, 1975). In the structural theory of time described in this paper, this passage has the hallmarks of metastable states comparable to the conditions that accompanied the emergence of life and the emergence of man (Fraser, 1978*b*).

7. Concluding Remarks

In a biology of social structures, what comprises the temporal environment to which social institutions need to adapt, if they are to be understood as evolving by natural selection?

I tried to show that evolving life, while adapting to pre-existing temporal patterns, creates its own temporal environment which is markedly different from the time of physical process. Then I argued that in social evolution, while human societies adapt to pre-existing temporal patterns, they also create their own temporal umwelts which are markedly different from the temporal umwelts of human individuals and of non-human societies. Temporalities may thus be seen as evolving no less than biological and social structures.

The hierarchical approach to the evolving temporalities seems to have enough flexibility to accommodate the numerous methodologies that must enter the study of time in human sociobiology; it also has the coherence to allow for an understanding of phenomenal reality as a single process. A fundamental synthesis of temporal passage may thus be effected, provided we are prepared to give up the belief that the experiential time of our species possesses an a priori, necessary, and universal validity for all levels of complexity that inorganic and organic evolution created. Whether or not this is a reasonable step to take may be decided only by critical testing of the usefulness and consistency of the principles sketched in this paper.

References

Bastin, T. (1969). A general property of hierarchies. In *Towards a Theoretical Biology* (C. H. Waddington, Ed.) Vol. 2, p. 252. Chicago: Aldine.
Beer, S. (1967). *Decision and Control.* New York: Wiley.
Bernal, J. D. (1967). *The Origins of Life*. Cleveland: World.
Bonner, J. T. (1974). *On Development*. Cambridge, Mass.: Harvard University Press.

Cairns-Smith, A. G. (1971). *The Life Puzzle: on Crystals and Organisms and on the Possibility of a Crystal as an Ancestor*. Edinburgh: Oliver and Boyd.
—— (1975). *Proc. R. Soc.* B 189, 249.
Cloudsley-Thompson, J. L. (1961). *Rhythmic Activity in Animal Physiology and Behaviour*. New York: Academic Press.
Cohen, J. (1967). *Physiological Time in Health and Disease*. Springfield, Ill.: Thomas.
Darwin, C. (1871). *The Life and Letters of Charles Darwin* (F. Darwin, Ed.). London: Murray, 1887. Editorial note to letter to Hooker, 29 March 1863.
Elsasser, W. M. (1966). *Atom and Organism: a New Approach to Theoretical Biology*. Princeton: Princeton University Press.
—— (1969). *Am. Sei.* 57, 502.
Fox, T. O. (1972). Evolution of levels of evolution. In *Molecular Evolution: Prebiological and Biological* (D. L. Rohlfing & A. I. Oparin, Eds). New York: Plenum Press.
Fox, S. W. & Dose, K. (1972). *Molecular Evolution and the Origin of Life*. San Francisco: W. H. Freeman.
Fraser, J. T. (1975). *Of Time, Passion, and Knowledge: Reflections on the Strategy of Existence*. New York: Braziller.
—— (1978a). *Time as Conflict: a Scientific and Humanistic Study*. Basel: Birkhäuser & Brookfield, Vermont: Renouf USA, p. 419.
—— (1978b). The individual and society. In *The Study of Time*, Vol. III, 419 (J. T. Fraser, N. Lawrence & D. Park, Eds). New York and Heidelberg: Springer-Verlag.
Goodwin, B. C. (1963). *Temporal Organization in Cells*. New York: Academic Press.
—— (1970). Biological stability. In *Towards a Theoretical Biology* (C. H. Waddington, Ed.) Vol. 3, p. 1. Chicago: Aldine.
—— (1976a). *Analytical Physiology of Cells and Developing Organisms*. New York: Academic Press.
—— (1976b). *Theoria to Theory* 10, 33.
—— (1978). *Theoria to Theory* 11, 247. Also *J. Social Biol. Struct.* 1, 117.
Grobstein, C. (1973). Hierarchical order and neogenesis. In *Hierarchy Theory* (H. H. Pattee, Ed.). New York: Braziller.
Isaacs, H. R. (1975). *Idols of the Tribe: Group Identity and Political Change*. New York: Harper & Row.
Koestler, A. & Smythies, J. R. (Eds) (1969). *Beyond Reductionism*. New York: Macmillan.
Lenneberg, E. H. (1964). A biological perspective of language. In *New Directions in the Study of Language* (E. H. Lenneberg, Ed.). Cambridge, Mass.: The M.I.T. Press.
Lukes, S. (1973). Methodological individualism reconsidered. In *The Philosophy of Social Explanation* (S. Ryan, Ed.). Oxford: Oxford University Press.
Mayr, E. (1968). Cause and effect in biology. In *Towards a Theoretical Biology* (C. H. Waddington, Ed.), Vol. 1, p. 42. Chicago: Aldine.
Meibin, M. (1977). The colonization of time. In *Timing Space and Spacing Time in Social Organization* (T. Carlstein, D. Parkes & N. Thrift, Eds). London: Arnold.
—— (1978). *Am. Sociol. Rev.* 41, 3.
Mendelbaum, M. (1973). Societal facts. In *The Philosophy of Social Explanation* (S. Ryan, Ed.). Oxford: Oxford University Press.
Needham, J. (1936). *Order and Life*. Cambridge, Mass.: The M.I.T. Press, reprinted 1968.
—— (1944). *Time: the Refreshing River*. London: Allen & Unwin.
Newton, I. (1687). *Sir Isaac Newton's Mathematical Principles of Natural Philosophy* (A. Motte, Tr., F. Cajori, Ed.). Berkeley: University of California Press.
Pattee, H. (1970). The problem of biological hierarchy. In *Towards a Theoretical Biology* (C. H. Waddington, Ed.), Vol. 3, p. 117. Chicago: Aldine.
—— (1973). *Hierarchy Theory*. New York: Braziller.
Prigogine, I. (1955). *Introduction to Thermodynamics of Irreversible Processes*. New York: Wiley.
—— (1973). Time, irreversibility, and structure. In *The Physicist's Conception of Nature* (J. Mehra, Ed.). Boston: Reidel.

Richter, C. P. (1965). *Biological Clocks in Medicine and Psychiatry*. Springfield, Ill.: Thomas.
Saunders, P. T. & Ho, M. W. (1976). J. *theor. Biol.* 63, 375.
Sebba, G. (1972). Time and the modern self: Descartes, Rousseau, Beckett. In *The Study of Time*, Vol. I, p. 452 (J. T. Fraser, F. C. Haber & G. H. Müller, Eds). New York and Heidelberg: Springer-Verlag.
Simon, H. A. (1969). The architecture of complexity. In *The Sciences of the Artificial* (H. A. Simon, Ed.). Cambridge, Mass.: The M.I.T. Press.
Slobodkin, L. B. & Rapoport, A. (1974). *Quart. Rev. Biol.* 49, 181.
Steiner, G. (1975). *After Babel: Aspects of Language and Translation*. Oxford: Oxford University Press.
Thom, R. (1970). Topological models in biology. In *Towards a Theoretical Biology* (C. H. Waddington, Ed.) Vol. 3, p. 89. Chicago: Aldine.
Thrift, N. (1977). *Progr. Hum. Geog.* 1, 413.
Uexküll, J. J. (1909). *Umwelt und Innewelt der Tiere*. Berlin: Springer-Verlag. See also his *Theoretical Biology* (1926). New York: Harcourt, Brace & Co. Also, with G. Kriszat (1934), A stroll through the worlds of animals and man. In *Instinctive Behavior* (C. H. Schiller, Tr. and Ed.). New York: International Universities Press, 1957.
Ungvari, T. (1972). Time and the modern self: a change in dramatic form. In *The Study of Time* Vol. I, p. 470 (J. T. Fraser, F. C. Haber & G. H. Müller, Eds). New York and Heidelberg: Springer-Verlag.
Von Neumann, J. (1951). The general and logical theory of automata. In *Cerebral Mechanisms and Behavior* (L. Jeffres, Ed.). New York: Haffner.
Waddington, C. H. (1961). *The Ethical Animal*. New York: Atheneum.
Watt, K. E. F. (1978). *J. Social Biol. Struct.* 1, 53.
Whyte, L. L., Wilson, A. G. & Wilson, D. (Eds) (1969). *Hierarchical Structures*. New York: American Elsevier.
Wilson, E. O. (1975). *Sociobiology: the New Synthesis*. Cambridge, Mass.: The Belknap Press.
Winfree, A. T. (1975). *Nature* 259, 343.
Zerubavel, E. (1976). *Sociol. Inquiry* 46 (2), 87.
—— (1977). *Am. Sociol. Rev.* 42, 868.
—— (1978). The Benedictine ethic and the spirit of scheduling. Presented at the Meeting of the International Society for the Study of Civilizations. Unpublished paper.

11. LOGOS AT THE EDGE OF THE COSMOS

Much has been said in the preceding chapters about the unique position of thinking humans in the cosmos at large. In those reflections we encountered and reencountered the idea of logos, a word that signified the coherence or ordering principle of the cosmos, the reason innate in all things.

What do we know about the coherence, the ordering principles of our own, innate logos? Its hallmarks include the capacity to create symbolic transformations of experience, the ability to speak a language and to construct and maintain personal identities.

To explore the nature of this inner, human logos in relation to time, we reenlist Galatea and the Wife of Lot. The journey of the Wife of Lot was "a collapse from the nootemporal umwelt through the biotemporal and eotemporal worlds to the prototemporal as she returned to incoherent dust." The journey of Galatea was that of emergence along the evolutionary umwelts from the prototemporal to the noetic. "We may imagine the lava cooling, the stone assuming form, acquiring life, eventually achieving selfhood and the capacity to fear, hope, remember, anticipate, hate and love." What I described in that chapter as the "aesthetic adventure" was the human search for knowledge, our voyage along the nested hierarchy of temporalities. It involved a ceaseless descent into and ascent out of the depths of the mind.

In the same chapter where we met the two ladies, we also watched Plato's slaves walk—then run—to their freedom. Their path, directed toward the sunlit world, led along an upward slant of the cave. "The wall of the cave, which is also the mind of modern man, retained the history of nature.... From [that] wall—from their own minds—the marching brotherhood read the history of time."

The essay of this chapter supplies some of the details of that reading about the history of time in the mind of man. But now we are not listening to a brotherhood talking mathematics but to the words, to the reasoning of a commuter as it were, into and out of the deepest levels of the mind. Everyone is included regardless of race, religion or county of origin. The aesthetic adventure is well-defined: it is a continuous, self-scanning survey of the brain, a process known as consciousness. The scanning, that inner survey, is carried out by the newer parts of our brain. It consists of monitoring, of testing the level-specific realities simultaneously present in the mind. What makes the survey possible is the fact that the nootemporal umwelt of the modern mind subsumes all the archaic reality assessments by the older parts of the brain, including different judgments about future, past and the present.

TEMPORAL LEVELS AND REALITY TESTING

The External and Internal Reality of Time

In psychoanalysis the concept of reality testing refers to the capacity to appraise the external world in its generally accepted meanings. It signifies the ability to distinguish between perceptions and thought or fantasy. This division, normally well defined, gets modified or breaks down in the case of mental illness. Whereas philosophers have been seeking normative criteria for a categorical definition of reality, psychoanalysts are concerned with the mental process whereby the patient establishes a working concept of his self in relation to others and the world at large. For this reason there is no need in psychoanalytic theory to assume a final, absolute reality independent of man. It is more appropriate to talk about visions of reality (Schafer, 1970) inherent in, or at least traditional to psychoanalytic thought.

In its broadest sense, reality testing is a method of adaptation to the environment (Hartmann, 1956). But life, even in its most primitive forms, is characterized by drives for bringing about environmental changes favourable to the inner needs of the organism. Adaptation between the species and the environment is, therefore, always mutual: simple life forms manipulate their environment by mechanical and chemical means, man through the power of the symbolic transformation of experience. Because of this mutuality, a model of the environment as independent of the adaptive process itself, is only a heuristic approximation. Likewise, the reality to which man may be said to adapt comprises both internal and external aspects which, as I shall argue, mutually define each other.

The importance of inner reality has been implicit in psychoanalytic theory and practice and its significance has been receiving increased attention (e.g. Obendorf, 1941; Chessick, 1957; Hurvich, 1970; Schafer, 1970). Yet, when psychoanalysts deal with time in the psychoanalytic situation, they assume the existence of an objective temporal matrix in which psychic processes take place and to which the subjective time of the patient is gradually attuned. A desirable result of therapy is an identification of the subjective time of the patient with the assumed, objective physical time. Psychoanalytic theory may, therefore, be seen as endorsing that view of time which was proposed by Isaac Newton in 1687. He wrote that 'absolute, true and mathematical time, of itself, and from its own nature, flows equably without relation to anything external, [that is, anything other than itself] and by another name is called duration...' But this understanding of time must be rejected, though not at all for reasons that come from post-Newtonian physics[1] but rather because of reasons that stem from evolutionary biology (Fraser, 1978*c*) and from psychoanalytic insight.

It is generally assumed in psychoanalytic thought that the mind is capable of distinguishing between, on the one hand, what comprises material objects, external conditions and other persons and, on the other hand, elements of dreams, wishes and fantasies. Reality testing may then be conceived of as a method of experimentation through which a person learns to separate ideas from perceptions. According to this interpretation, future and past must, therefore, be classed as ideational experiences which are for that reason unreal, while perception remains limited to the present, which is real. But then, by what means do we learn to judge whether images of future or past stand for certainties, likelihoods or impossibilities? Assuming a strict dichotomy between ideas and perceptions appears to be an insufficient theoretical basis for the understanding of how reality testing works with respect to time. Perhaps we are faced with a spectrum of realities which differ among themselves in qualitative ways.

Consider, however, that the mind perceives much more than is actually sensed. Perceptions are best understood as expectations derived from memory, guided and screened by conscious and unconscious fantasies and modified by sense impressions. Reality testing, although at first approach it seems to pertain only to the present, demands the mobilization of a broad spectrum of feelings and ideas which relate to the human sense of time. In acute mental illness, for instance, confusion between inner and outer reality correlate significantly with temporal disorganization (Melges & Freeman, 1977) suggesting that the distinguishing features of human time may be correlative with the definition of the self. Accordingly, this paper examines the status of time as a fundamental dimension of reality testing and stresses the mutuality between the external and internal aspects of time.

The dynamics of internal-external mutuality, as it pertains to living organisms, found scientific expression early this century in the work of Jakob von Uexküll (1909). He drew attention to the epistemic significance of the fact that an animal's receptors and effectors determine its world of possible stimuli and actions. He called such a species-specific world the *Umwelt* of the species. Events and things that may be identified by members of a species constitute the reality of that species; what is external to that umwelt must be regarded as non-existent for the species. In modern psychology *Umwelt* is defined as 'the circumscribed portion of the environment that is meaningful and effective for a given species and that changes its significance in accordance with the mood operative at a given moment' (English & English, 1964). Note that the 'environment' of which the umwelt of a species is 'the circumscribed portion' is the umwelt of man.

Taking advantage of the latent power of von Uexküll's notions, I have formulated an *extended umwelt principle* in which the originally biological concept is enlarged to include, in the case of man, the conscious and unconscious manifestations of the mind as well as the totality of our exosomatic paraphernalia. I called this broadened world the noetic umwelt of man.

I shall argue that it is the evolutionary heritage of the mind to comprise a hierarchy of distinct umwelts, of which the noetic is the highest. Also, that each higher level subsumes those beneath it in a consecutive, nested, series while adding certain new functions and freedoms of its own. It follows that through our noetic faculties we should be able to identify, without much ambiguity, the physical and biological functions of nature, including those of our bodies. But an individual does not command yet a higher umwelt of which the noetic manifestations of his mind would be a circumscribed portion. It further follows, therefore, that certain mental processes will be found intrinsically self-referential and hence, whatever constitutes noetic reality must also remain, in some respects, undetermined. Noetic time will also remain ambiguous, even if some of its properties may be embedded in Hartmann's (1956) 'average expectable environment' as parts of a 'consensus about reality, a consensual validation about the nature of the fixed world "out there",' to borrow Wallerstein's (1973) apt phrase.

We will study the developmental and maturational aspects of time experience and emphasize its facultative and functional elements. This will lead us to a structural-dynamic theory of time that can accommodate the potentiality for regressive reactivation of primitive forms of mental umwelts, along a hierarchy of distinct temporal levels. Such journeys into the depth of the mind are important in psychoanalysis in the context of the all-pervasive influence of instinctual conflicts and anxieties. Using the framework of this new theory of time, we will find that reality testing, when it comes to time, ought not be thought of as the exploration of an external, objective temporal organization but must signify, instead, the capacity to create and maintain a peculiar mental set. We will try to explore just what functions of the mind belong in that set.

Temporal Levels vs. Timelessness

This section offers a critique of the concept 'timelessness' in terms of a new theory of time (Fraser, 1975, 1978*a*). It employs the idea of temporal levels, a concept first found useful in studies of the hierarchy of lawfulness that characterizes the stable integrative levels of nature: radiation, elementary particles, massive bodies and living matter. The temporalities of nature are paradigmatic. Although they may be recognized as emerging in the history of complexifying matter, they may also be identified in the structural organization of the mind.

Temporal levels

The concept of temporal levels will be introduced through visual metaphors.

Let us imagine the picture of an arrow on a sheet of paper and let the arrow stand for our experience of passage from cradle to grave. The head and the tail are clearly drawn so as to represent a clear distinction between future (with its anxieties and certainty of death) and past (with its regrets and shades of guilt feelings). The two categories combine with sense perception in the *mental present* with its continuously shifting boundaries, depending on attention, fantasy, and the relative investment of psychic energy among future, past, and present objects. The shaft of the arrow may represent the mental present. An umwelt characterized by these features we call *nootemporal.* This is the world of symbolic causes, the tensions between what is desired and what is possible, the stage upon which the human drama is played.

But the head of the arrow may also be badly drawn, amounting to no more than ambiguous limits to the shaft. The picture, then, is a visual metaphor for *biotemporal* umwelts. Here the mental present of the noetic world reduces to a 'creature present', a category of time experience more or less identical with a perceptual set but without appreciable mental content. Along its edges, as it were, there are elements of the immediate future and past, without long-term expectation or memory. As far as one can conjecture, this is the umwelt of advanced forms of pre-human life, of early man, of the human infant prior to the consolidation of the self and of the higher portions of the archaic levels of the fully developed mind. Thus, for instance, behaviour controlled by the pleasure principle demands immediate satisfaction because its reality is that of the biotemporal world near the soma. Since in this umwelt distant future does not exist, delay in satisfaction amounts to complete refusal and hence total frustration.

If both the head and the tail of the arrow are missing we are left with the shaft, a line, an image of *eotemporal* umwelts (named for Eos, goddess of dawn). Here, although not everything happens at once, the directional quality of time is absent. Since instants are not delimited by future and past, eotemporal experience is that of the philosopher's *nunc stance* or abiding present. Because of the absence of defineable 'now', temporal units cannot be specified and hence time has, so to say, a rubber scale.

All cyclic processes determine eotemporal umwelts for they provide variations but not a preferred direction. This is the world of rhythmic tensions and relaxations associated with the breast period, the universe of heartbeats, rocking chairs and repetition. Endless time, eternity, and the abiding present are different ways of talking about eotemporal conditions. They may represent memories of childhood when time was future- and past-less. They may describe experiences of regression to conditions where boundaries of the self are ill-defined. They may represent the influence of unconscious wishes, themselves eotemporal, operative on the noetic level of time apprehension (Arlow, 1973). Eotemporality is also the predominant umwelt of the manifest content of dreams, of the virtual immortality of impressions that have been sunk into the unconscious by repression and of wishful impulses which have never passed beyond the unconscious.

The shaft of the arrow may further disintegrate into slivers of wood, representing fragments of time. Such a picture symbolizes *prototemporal* umwelts (for 'proto-', the first of a series). These are worlds of fragmented time. Events hang together, if at all, only very loosely, such as in psychotic states when the continuity of the self is disrupted but not totally absent. This may also be the time of remembered dream content before secondary elaboration, and of the worlds of certain fantasies.

If even the fragments of the arrow are lost, we are left with a blank sheet of paper symbolizing *atemporal* umwelts. As the most primitive form of time, it blends into space. Atemporality is used in dream work for the representation of events which, in the nootemporal world, would be classed as at great distances in time. Experientially, atemporality may also represent complete schizophrenic chaos or the way we imagine the inner world of fugue in senility.

We shall repeatedly employ these categories of time in the context of various psychoanalytic issues. It is appropriate, therefore, to remark upon certain systematic relationships among them, especially since in actual experience what one observes is usually a dynamic interaction of one level with the others.

We already learned that each of the temporal umwelts may be associated with a stable integrative level of nature and that each temporality subsumes the levels of time beneath it, while adding some specific features of its own. But not only temporalities are hierarchically nested. So are connectivities. Thus, no connectivities (causations) exist in atemporal chaos. Events in prototemporal worlds are connected by statistical laws which, from the psychoanalytic viewpoint, are forms of magic causation. Connexions among eotemporal events are deterministic: effects may follow or precede their causes. This is also a form of magic causation, one that informs myth, fantasy and dream alike. Only in the biotemporal can final causation (intentionality) have any meaning. It appears in terms of the self-interest of the organism which prompts it to display instinctive behaviour directed toward a purpose. Finally, only for the nootemporal umwelt of the mature mind does human freedom have any meaning. Stemming from the over-determination of the elements of psychic life, it constitutes the highest type of connectivity among events.

It follows from the hierarchical and nested nature of temporalities and levels of causation that, as the patient attunes himself to an (assumed) external time, he must learn to master the contributions to his nootemporal umwelt of all the latent temporalities and also learn to live with all the archaic types of causations which coexist with human freedom.

Timelessness

From the perspective of the nootemporal, all lower umwelts should appear as, in some ways, deficient with regard to temporal attributes. This relative poverty has given rise to the custom of describing temporal experiences appropriate to the lower umwelts as those of timelessness.

In 'The unconscious' (1915) Freud wrote that 'the processes of the system *Ucs.* are timeless; i.e. they are not ordered temporally, are not altered by the passage of time; they have no reference to time at all. Reference to time is bound up... with the system *Cs.*'

For 'timeless' we may now read 'deficient in certain temporal attributes when compared with the time of the system *Cs.*' The absence of temporal ordering suggests the fragmented conditions of the prototemporal. 'Not altered by the passage of time' signifies very slow rates of change. The umwelt of the id, being close to the umwelt of the soma, behaves in many respects like the soma. The stability of the system *Ucs.* reminds us of the stability (very slow change) of certain biological forms rather than of the rapid adaptive changes available to the conscious mind, which deals with symbolic transformations only. That unconscious processes have no reference to time is an issue that was examined by Bonaparte (1940). She maintained that the reason for this absence is that there is nothing in the id that would correspond to any idea, including that of time.

In 'Beyond the pleasure principle' (1920) Freud remarked that 'as a result of certain psychoanalytic discoveries, we are today in a position to embark on a discussion of the Kantian theorem that time and space are "necessary forms of thought"'. He held that, in so far as the unconscious is timeless, the Kantian dictum cannot be maintained. We may now refine Freud's comments in terms of the heavily evolutionary stance of the hierarchical theory of time. The nootemporal umwelt is, indeed, the characteristic mode of perception for man and hence, in that sense only, it is a necessary form of thought. However, unlike Kant's pre-Darwinian *a priori*, the noetic umwelt of man (i) subsumes all lower temporal umwelts and (ii) is in no way a necessary terminal state of perception. Organic evolution is open-ended.

In 1930, referring to his correspondence with the French novelist Romain Rolland, Freud wrote that there is 'a feeling which he [Rolland] would like to call a sensation of "eternity", a feeling of something limitless, unbounded—as it were "oceanic"'. Freud goes on reasoning that this feeling constitutes a regression of the mind into a state where the boundaries of the self become ill-defined. We shall learn later that noetic time is experienced only when there is a tension between (i) the continuity which is the oneness of the self and (ii) the unpredictable, many-sided matrix of the external and internal (somatic) world. Under certain conditions this tension may be lessened, the eotemporal with its abiding present becomes the dominant temporality of the mind and the patient reports an experience of oceanic feeling.

We may conclude that what has traditionally been described as timelessness possesses, in fact, a distinct structure of temporalities. Experiences of timelessness have often been called transtemporal states. Yet, they do not constitute very advanced but, rather, primitive states of the mind.

Time and the Uncanny

In 'Civilization and its discontents' (1930) Freud formulated a strong recapitulation principle for the mind. He asserted that whereas in the body earlier evolutionary phases are usually absorbed in and supplanted by later phases, the mind preserves unchanged the stage of its evolutionary development. 'The fact remains that only in the mind is such a preservation of the earlier stages alongside the final form possible' and 'we are not in a position to represent this phenomenon in pictorial terms'.

Brief reflection will reveal that the two quotes are corollary specifications. Namely, consecutive stages of a changing structure, occupying a certain space, cannot be retained in that space. But any number of processes may take place simultaneously, occupying the same span of time. (The music of several acts of an opera could be played simultaneously, but the play could not be so acted within the confines of the stage.) The most efficient, and sometimes the only possible way of retaining historical stages are process and not state descriptions. It follows that the mind, which is a process description of the brain (Fraser, 1978*a*) and hence comprises temporal functions, cannot be represented in pictorial, that is, spatial terms.

If this all be so, then, subsumed in the nootemporal umwelt of the modern mind, we should be able to identify the temporal umwelts that constitute archaic realities. To demonstrate that the early forms of time are indeed continuously present in the mind, we turn to the experience of the uncanny.

In a paper first published in 1919 ('The uncanny') Freud asked: what kind of things appear uncanny? Through comparative word analysis he first established that the 'uncanny' is associated with such other ideas as dismal, foreign, sinister, demonic, concealed, secret—yet it also often displays origins in concepts that stand for the familiar, the well-known, the comfortable. Taking his cue from the linguistic ambivalences, he then examined literary pieces which are known to bring forth the feeling of the uncanny. He found that common to them was an uncertainty regarding the position of a thing or event on the scale of reality. An organism that may be animal or human, a man who may be alive or dead, a house that looks familiar yet we know we have never seen it before, all create feeling of the eerie. Freud's final conjecture was that 'the "uncanny" is that class of the terrifying which leads back to something long known to us, once familiar'. In 1934 Bergler reasserted that the feeling of the uncanny is usually brought about when old beliefs, ready to cease upon confirmation, seem to be confirmed, when unconscious fantasies are brought to the conscious by conditions which seem to correspond to them. The feeling of the eerie, then, suggests the return of the repressed.

Continuing earlier work on this subject (Fraser, 1966*b*, 1975, 1978*a*) I will now try to demonstrate that the feeling of the uncanny, as it accompanies certain time-related experiences, constitutes *prima facie* evidence for the presence in the unconscious of pre-noetic temporalities.

The Picture of Dorian Gray by Oscar Wilde is a paradigm of literary compositions which manipulate time so as to produce a feeling of the uncanny. The plot turns upon the exchange of temporal features between a young man and his portrait. The portrait ages, while the man remains untouched by the passage of time. Eventually the inversion of umwelts becomes unbearable and the still young-looking hero thrusts a knife through the canvas, which shows an old man. The hero is found dead with the knife in his heart, an old, withered man; the portrait is once again that of a man of 'exquisite youth and beauty'.

Unageing people and live paintings were once among the furnishings of the child's reality. But perceiving the world in such terms becomes a taboo as the child's dominant umwelt comes to approximate that of the mature adult. The early temporal umwelts are delegated to the class of unacceptable truths and hence, repressed. Just why children renounce and/or are coerced into renouncing their primitive temporal umwelts (as they are forced to renounce their oedipal wishes) is not clear. Developmental child psychology generally neglects the psychoanalytic dimension of growth and hence gives little guidance. One may speculate that the collectives of man cannot tolerate the sense of individual invincibility which is a correlate of the child's experience of time. Society prefers to command that authority and reintroduce such feelings for the benefit of the group. Be that as it may, Wilde makes the reader believe in the validity of a world in which identities and the direction of time are only loosely defined, if at all. The repressed temporal modes of perception then return to consciousness.

Consider next the feeling of the eerie—the 'getting-the-shivers'—that sometimes accompanies the observation of crowd behaviour. Whether celebrating Eros around a maypole or Thanatos around a guillotine, crowd scenes often suggest the presence of something hidden: divine, demonic, foreign. People appear to be moved by unknowable supernatural powers. A suitable description of conditions which produce that feeling may be found in the words of the Marsellaise:

> Allons, enfants de la patrie!
> Le jour de gloire est arrivé!

The participant becomes an *enfant de la patrie*, a child of the almighty fatherland. His self definition lessens as he turns into an indistinguishable member of the mob whose actions may only be described statistically. Drives appropriate to the bio-, eo- and prototemporal umwelts cathect the mental representation of objects and produce the corresponding temporal feelings: that of the abiding present and/or fragmented time, the eternal *jour de gloire*. The significance of future and past lessens or vanishes. The temporal world of the child emerges into consciousness as a rediscovered reality.

Let us turn to the eeriness of the double. 'The quality of [its] uncanniness can only come from the fact of [it] being a creation dating back to a very early mental stage, long since surmounted—a stage, incidentally, at which it wore a more friendly aspect. The "double" has become a thing of terror, just as after the collapse of their religion, the gods [in Heine's *Die Götter im Exile*] turned into demons' (Freud, 1919, p. 236). Freud saw in the double a protection of the self against death through a permanent, unageing form of that self. The temporal umwelt of the experience of the double is again that of the child. Time is that of the abiding present, or else is passing in both directions and hence people may be young or old but they age no more than does the picture of Dorian Gray.

The uncanny as a source of unpleasure may turn into its twin, the uncanny as satisfying and a source of pleasure. An example is the oceanic feeling that may accompany the discovery of a *Doppelgänger* [2] in a member of the opposite sex. For such occasions the return to childhood is a happy one and, surely, of some adaptive value. One way to analyse the associated feelings involves the identification of the double with the imago of self representation. Plato recognized the presence of this identification (and we may recognize its significance in defying individual death) in the ritual of love. He wrote in the *Symposium* that

> we are like pieces of the coins that children break in half for keepsakes—making two out of one like the flatfish—and each of us is forever seeking the half that will tally with himself... And so all this to-do is a relic from that original state of ours when we were whole, and now, when we are longing for and following after that primeval wholeness, we say we are in love.

In love, the return to earlier temporal umwelts is a pleasurable and useful journey.

We may note here that the double experience in the life of a non-psychotic child belongs to the period of time after the child has grasped the significance of his vulnerability to death. The fantasy of the double represents a playing with time and death in the service of establishing mastery over the passing of the self. If we permit ourselves the use of a recapitulation theory, we may speculate that the normal fantasy of the double corresponds to that stage in the evolution of our species when selfhood, noetic time and the awareness of death became mutually reinforcing elements in the making of the mind.

From a demonstration that the archaic umwelts of time are always present in the mind, let us turn to the archaeology of futurity, pastness and presentness, the three categories of noetic time. Will the uncanny give us some guidance in this exploration?

Consider the taboos erected to prevent the return of the repressed. In the animistic world of the infant there was no distinction between living and non-living objects but subsequent animism, as an aspect of reality, was repressed. A person who may be a thing (or vice versa) is not only uncanny but dealing with him (or it) is controlled by taboos: dolls are not to be loved as we love people, persons are not to be handled as dolls. The region between life and no-life is filled with mysterious beings, as is the region between this world and the other world. Creatures that belong in these ill-defined areas are subject to inordinate praise or abuse.

There also exist strong taboos against tampering with temporal categories. People for whom future and past seem to appear as present are usually invested with highly loaded emotions. It is as if the mind would wish to protect its great discovery and asset, perhaps its reason for being: the discovery of the adaptive advantages of future, past and present.

The most intense feeling of the uncanny and the strongest taboos relate to presumed non-inferential knowledge of the future. Unexplainable foresight was always noted: prophets were praised or condemned, but seldom ignored. Arlow (1951) sees the power of the prophet in his ability to give structure to the nascent unconscious fantasies held in common by the members of a group. He argues convincingly that the determining dimensions of the group experience lie in the past and it is the past which the prophet reads accurately, formulating from them visions of the future. But the stuff that prophets sell after having meditated upon the past, is the claim of knowing about the future what to others is unknowable. Prophecy keeps its audience through the *Angstlust* of time travel, not unlike science fiction.

Cohen (1964) in his survey of divination lists over a hundred methods of fortune telling. They stand witness to the keen interest in the future by people of epochs when serious questioning of the past was all but unknown.

Non-inferential knowledge of the past may also appear uncanny but seldom with the same degree of intensity as that of the future. Seers of the past are less fascinating than prophets and are controlled by taboos only if they disturb otherwise cathected feelings of sex, death, or ideology dealing with either. Finally, knowledge of current, distant happenings, though maybe interesting, can hardly be imagined as bringing forth revolutions as may the utterances of the prophet. We conclude that there appears to be a gradation in the intensity of the uncanny and the strictness of taboos, according to the sequence: future, past and present.

Kafka & Bolger (1949) observed that the speed with which a reasonably realistic description of the future can be accomplished in writing a future autobiography is a function of many factors, among which the lack of unresolved conflicts ranks first (see also: Melges, 1972). We may evaluate this finding in terms of Freud's conjecture (1950) that there are 'good grounds for suspecting that the arousing of the repressed... follows the law of development. Further, that repression proceeds backwards from what is recent, and affects the latest events first'. Whatever was repressed the earliest, lies the deepest. One cannot go to the future without clearing a path to it through the removal of obstacles in the past, operating in the present. In the removal of past obstacles we recognize the programme of psychoanalytic therapy.

The suggestion comes to mind that the relative depths in the unconscious of certain feelings ordinarily attached to futurity, pastness and presentness reflect the evolutionary sequence in which these perceptual categories of noetic time evolved in phylogeny and, as I shall argue, evolve in ontogeny.

In a psychoanalytic study of artistic vision and hearing, Ehrenzweig (1953) maintained that every act of visual perception recapitulates the ontogenic development of the visual motor-pattern of the child. Hence, each act of visual perception runs through infantile stages of dreamlike structures before the final images are formed that emerge into consciousness. We must assume he concludes, that consciousness itself continuously oscillates among different levels of form differentiation.

I would like to postulate that some such mechanism is also at work in all acts that involve the noetic sense of time. Before temporal judgments are presented to consciousness, the mind already scanned the archaic temporal umwelts in the unconscious as well as in the preconscious, with its concern about future and past. What emerges is a complex of feelings and thoughts about time that has already been screened, modified and, to a large extent directed by mental exploration. This hierarchy of temporal umwelts is, then, the peculiar mental set which reality testing explores.

Time and the Dream Work

Time in its relation to dream is frequently mentioned in the psychoanalytic literature and is often singled out for special attention. With few exceptions, however (e.g. Simonson, 1928; Calhoun, 1976) the model of time assumed is, once again, that of an objective, man-independent aspect of nature which the dream is said to distort, or of which it displays a lesser amount. We will now examine the role of time in dream work from the perspective of the evolutionary levels of time.

In the 'New introductory lectures' (1933) Freud observed that in the manifest content of the dream 'both smallness and remoteness in space have the same significance: what is meant is remoteness in time…' This remark is preceded by an aside (with reference to a dream wherein the multiple appearance of the same figure represented events that occurred repeatedly) that 'the dream is not taking very much on itself in expressing frequency by multiplicity'. Freud was pointing to the German language which retained the developmental primacy of space perception to time perception. *Häufigkeit*, or 'multiplicity in time' derives from *haufen*, 'to hoard', that is, provide multiplicity in space. In our terms, since the temporal umwelt of the manifest dream content does not include the potentiality of long term memory or expectation, the dream symbol may only be placed into an atemporal umwelt, which is space.

In 'The interpretation of dreams' (1900) Freud wrote that 'a dream is able to compress into a very short space of time an amount of perceptual matter far greater than the amount of ideational matter that can be dealt with by our waking mind'. This fact exemplifies the rubber-scale quality of temporalities below the biological, a characteristic which, as we have seen, follows from the absence of a defineable 'now' in the eotemporal—of a defineable instant between future and past. Freud adds that the dream 'reproduces logical connections by simultaneity'. Here we may quote again his earlier comment that 'the dream is not taking very much on itself'. For, logical relationships constitute abstractions, assumed to be unchanging with time. Mathematics, geometry and logic are the sciences of space *par excellence*, serving for their disciples as intellectual havens from the conflicts of instinctual demands which help generate the higher temporalities.

Let us try to identify the contributions of the different temporalities to the manifest content of dreams. By Freud's strong recapitulation principle, they ought to be simultaneously present. Atemporality is represented by space. Prototemporality is represented by chaotic segmentation of dream images. Eotemporality is recognized as deterministic causation and its corollary, the two-wayness of time. This is the feeling of that curious fore- and aft-knowledge which resembles the listening to a composition already well known. Arlow (1959) observed that all *déjà* experiences may be analysed as if they were dreams. In our terminology, their umwelts are eotemporal.

The biotemporal contribution to dreams appears through final causation that is, intentionality. Underneath the dream action, the instinctual wishes force the appearance of futurity, in which the forbidden satisfaction may yet be obtained. The dreamer is threatened with coming face to face with hitherto repressed, unconscious wishes, inconsistent with superego demands. The dreamer then wakens or, in the words of Heideggerian philosophers, he is 'thrown' into historical (noetic) time of future, past and present.

In 1913 Freud noted that the time of day in a particular dream stood for the dreamer's age at an emotionally important junction of her life. Gross (1949) examined dreams in which precise numbers referred to time. He proposed that they are usually associated with current, highly charged conflicts. This view was upheld and further elaborated by Hartocollis (1978). One may be tempted to regard the numbers functioning as screens, but this would assign too much sophistication to the lower mental umwelts. It is probably more correct to recognize that there is nothing in the lower umwelts that can correspond to superego demands and object relationships, which are the usual sources of highly charged conflicts. The best the dream work can do is to represent such conflicts by number. Conflicts appropriate to the noetic language (imagery) are thus reduced to the vocabulary of dream umwelts. We may find some support for this hypothesis in examples given by Freud (1900) where time was represented by money counted. Whatever else the dream metaphor of money may signify, the counting in numerals changes highly charged emotions into neutral, atemporal relationships.

It cannot be by chance that when people turn strongly inward by way of meditation, or by concentrating on abstract tasks, they appear to be sleeping. The dominant umwelt of their minds are pre-noetic and their bodily behaviour reflects this fact.

If we assume with Arlow (1966) and others that dreaming is an adaptively useful way of periodic depersonalization, we must then also insist that the efficiency of dream work resides in the periodic regression it provides for the mind into temporal umwelts without future and past. In keeping with what has been said before, I am suggesting that the maintenance of noetic time represents a burden to the mind, for it is a corollary of maintaining a permanent conflict between the self and the non-self. Half a century ago Simonson (1928) remarked that the absence of 'waking time' (our nootemporality) not only facilitates the dynamics of the dream work but in fact, makes that work possible in the first place. Dreams must, and do, make their statements in forms appropriate for archaic temporal umwelts.

It follows that when, in the analytic process, the past is recreated as a living present, the patient must overcome two sets of difficulties. One is the resistance of psychological forces which would prefer to leave certain memories repressed. The other resides in the fact that bringing the contents of the unconscious into consciousness demands the dressing of the skeleton with the flesh of noetic time, a task whose difficulty is proportional to the depth whence the skeleton is being resurrected. The archaic languages—temporalities—of memories and instincts represent conservative forces which ally themselves with defence mechanisms in general.

The Ontogeny and Phylogeny of Human Time

In 1913 Ferenczi outlined the stages in the development of the sense of reality in terms of ego instincts, from the primary process stage to the stage of secondary process typical of waking thought. Although later work greatly enriched our knowledge of the course of this development (Hurvich,

1970) Ferenczi's ontogenic specifications are still admitted as valid. We shall examine its stages in terms of the idea of nested, level-specific temporal umwelts.

Ferenczi speculated that the embryo 'must get from his existence the impression that he is in fact omnipotent' for he 'has nothing left to wish for'. It is difficult to admit a necessary and complete happiness in the womb because the embryo must suffer the vicissitudes of biological development as do all other parts of the mother's body. Granting, however, the great privileges of the foetus, we may still agree with Ferenczi and put our claim thus: in the umwelt of the embryo, causation may be given little or no meaning because the universe is identically equivalent to the organism. Omnipotence is the affective aspect of the absence of causal connexions, otherwise characteristic of atemporal worlds.

Through and after the trauma of birth, the feeling of unconditional omnipotence changes to one of magical-hallucinatory omnipotence, which is a gradual enlargement of the store of means whereby the relative peacefulness of the womb may be partly recreated. First magic thoughts, then magic gestures bring forth the desired results. The infant's experiential world grows from one of prototemporal causation (magic, statistical) to eotemporal causation (deterministic) mediated by his kinaesthetic powers. Space gets filled with objects which maintain their identities for increasingly longer periods of time. Eventually, though much later, the self of the child becomes the paradigm of such enduring identities.

In Ferenczi's stages, the various feelings of omnipotence are followed by an animistic period, an infusion of life into all matter and space. The umwelt of the infant becomes biotemporal as intentionality gets associated with all objects. Here we witness the origins of the intimate connexion between the body of the self and all other objects, giving rise to the capacity for symbolic transformation of experience. As the child learns to speak, he endows the spoken word with the power which was earlier attributed to wishing and imagination. The nootemporal umwelt then begins to emerge, defined and maintained by human language (Fraser, 1978*c*).

In 1920 Freud remarked that the history of the earth and its relation to the sun must have left its mark upon the 'living vesicle' or, in contemporary terms, upon the procaryotic cell. Evidence gathered during the last decade demonstrates that the most fundamental of all cycles, the circadian rhythm, is indeed endogenous in all life forms down to and including the gene. The broad spectrum of astronomical, geological and physical cycles which may be identified in living matter may be used to support an idea of biogenesis which sees the coming about of primordial life as an extension of the cyclic order of inanimate nature (Fraser, 1978*a*). According to that theory, life arises in the form of biological clock aggregates which, through their demand for internal coherence create presentness in the eotemporal umwelt of directionless change. Even the human infant, albeit of immense complexity, is mostly a co-ordinated biological clockshop driven and regulated by instinctual mechanisms. He is at the end of a journey which began in the prototemporal umwelt of giant DNA molecules and at the beginning of another journey in which he will have to create and learn to live with both the old and the new kinds of realities.

It has been suggested by Gifford (1960, 1978), Hartocollis (1974) and others that the origins of man's sense of time go back to the periodicities of hunger and satiety of the infant, to the kinaesthetic sense of periodic elimination and to the adaptation to the day-night cycle. Yates (1935) has given evidence that certain difficulties in the appreciation of musical rhythm may, indeed, originate at the breast period. But in so far as noetic time is *qualitatively distinct* from the temporalities of the infant, it might be more accurate to say that the rhythmicity of early experience gets embedded in the later sense of time, but is not equivalent to an early form thereof.

The development of the sense of time, beginning with rhythmic experience, is suggested by the opening words of a co-operative survey (Fraser, 1966*a*) of the scientific and humanistic views of time.

> Slowly but quite perceptibly, the young child learns the rhythms of life. With each sunrise and sunset he adapts himself more intimately to the enduring and to the changing aspects of his environment. As the seasons pass he begins his lifelong search for personal identity through a complex mental process involving expectation and memory... Through the life of this man, as through the life of all men, personal identity becomes intelligible and communicable to others because of the existence of a subtle private and communal understanding of an ordering principle.... This principle is couched in terms of an idea called *time*...

There exists an impressive body of evidence regarding the sequence in which linguistic designations for the categories of noetic time appear in the child (e.g. Stern, 1924; Ames, 1946; Orme, 1969; Friedman, 1978). The first category of time-words which the child can verbally manipulate refers to the present. But this is not the mental present of the adult but a biotemporal present ruled by the demands of the pleasure principle for immediacy. At around 24 months the child commands future-time words, at around 30 months past-time words and then a growing vocabulary of all these, but with a greater variety of future-time words than past-time words.

Futurity in the noetic umwelt is not identical, however, with the periodic expectations of a cyclic world. The sense of the future may develop from the cyclic matrix only after effective regulatory blocking of wishes for immediate satisfaction have been learned, as primary narcissism is enlarged to secondary narcissism. In a 1962 paper Loewald suggested that the superego may be thought of as protecting and forming the category of future. Since the future is the earliest emergent among the categories of noetic time, it is surely formed and guided by parental and societal authority, more so than the other categories. Even in mature life we claim to direct our future actions according to societal demands of praiseworthy conduct. But similar controls also hold sway over the past, though to a lesser extent. The continuous rewriting of history according to the dominant shared fantasies of each epoch demonstrate such influence.

Having begun his transition from heteronomy to autonomy and having learned to live with periods of separation from gratifying objects, the child discovers the past while imagining future threats and satisfactions. Later, the boundaries of futurity, pastness and presentness begin to expand under the pressure of certain mutually reinforcing experiences. Among these we may list the conscious recognition of affects, the separation of the self from the non-self, and ideas about the inexorability of individual death. Lastly, conscious and unconscious fantasies place need satisfaction into an expanding future, relying on actual and fantasied past events. The fully developed mental present does not appear until after the self acquired a history informed of regrets and feelings of guilt, and a future loaded with anxieties and informed of death. In terms of the universal need and fear that attends the knowledge of time, Shakespeare's Sonnet LXIV may serve as a delineation of human time experience:

> Ruin hath taught me thus to ruminate—
> That Time will come and take my love away.
> This thought is as a death which cannot choose
> But weep to have that which it fears to lose.

Let us assume with Ferenczi (1913) that the sense of reality in modern man represents a succession of repressions to which mankind was compelled. Let us also recall Freud's insistence that the mature mind retains in it the developmental steps of its phylogeny. Finally, let us remember that when we ranked the intensity of uncanny experiences and the strengths of related social taboos, we have found reasons to believe that futurity is the earliest emergent category of noetic time, followed by pastness and the mental present. We may then ask: what dangers reside in the discoveries of future and past that would mobilize defence mechanisms and lead to a sequence of repressions, manifest in the phylogenic and ontogenic sequence of future, past, mental present?

Among the anxieties that help establish the future as a category of noetic time, the certainty of individual death is surely the most universal and significant. Whereas awareness of passing does have immense adaptive advantages in preparing for the future in terms of past experience, a permanent mortal terror surely does not. We are not supposed to think continuously about the suffering and circumstances of our death. Accordingly, there exists a corpus of individual and collective measures against keeping death in mind, so much so, that people must often be reminded that they will die. I wish to speculate that no sooner than futurity came to include the certainty of death, a mental process of selective repression evolved relative to future objects.

What is there in the past that threatens the ego and must, therefore, be excluded from consciousness by repression? There is no single event comparable to death, but there exists a constellation of threatening feelings, foremost among them the oedipal situation. Although the vicissitudes of oedipal conflicts are necessary for the healthy development of the ego, a continuous reminder of them would hamper, rather than assist adaptation. Accordingly, there exists a corpus of individual and collective measures which help keep them out of mind, so much so that people must often be reminded that they ought to love their parents. I wish to speculate that no sooner than the past came to include the mastering of oepidal conflicts, a mental process of selective repression evolved relative to past objects.

We thus witness the compartmentalization and selective repression of anxiety and guilt feelings. These affects on the one hand and futurity and pastness on the other hand, have become mutually generating modalities of the mind.

Let us now re-approach the problem of the development of human time from a different perspective.

Masler (1973) concluded from anthropological data that time is associated with objects perceived as changing, yet retaining identity. In this suggestion we recognize the dialectic of Western thought about time and reality, reaching back at least twenty-five centuries (Fraser, 1975). The Eleatic philosophers have already asked: does reality comprise permanence (being, predictability) with change (becoming, unpredictability) as a figment of imagination, or is reality ceaseless becoming with permanence but a construct of the mind?

In the concept of being we recognize a projection of the continuity and indivisibility of the self, prime reference of permanence in the human psyche. In its abstract form, it stands for the experience of time as totally predictable.

In the concept of becoming we recognize a projection of all experiences which function independently of our wishes and fantasies. In its abstract form it stands for time as totally unpredictable. The noetic experience of time includes both projections, even though they are logically incompatible.

In philosophy, the being-becoming debate has been generally conceived of as an either-or proposition. From the psychoanalytic viewpoint the debate itself is a projection of the ceaseless conflict and mutual definition of the self and the non-self. If self definition vanishes so does the conflict—and with it, noetic time. In terms of the hierarchy of temporal umwelts in the mind, 'being' reminds us of the deterministic causation of the eotemporal, 'becoming' of the degrees of unpredictabilities of the lower temporalities. In the well-integrated ego these two aspects of time are not mutually exclusive but function in a hierarchical balance.

The general sequence of phylogenic and ontogenic development of time, is, thus, not from chaos toward order as conventional beliefs generally hold. Rather, it is from chaos toward conflicts of increasingly better defined articulation, along a hierarchy of temporal umwelts.

Issues in the Psychodynamics of Temporal Levels

After introductory thoughts, this section discusses (1) changes in the apparent speed of time, (2) an example of temporal disorganization in mental illness, (3) the question of time in the psychoanalytic setting and (4) the moods characteristic of the temporal umwelts. It concludes (5) with a metapsychological aside on the theory of time as conflict.

Freud (1920) introduced the idea of psychological projection by comparing the system *Cs.* with a living cell encased in a semipermeable membrane. This kind of structuring enables the cell (or by analogy, the system *Cs.*) to distinguish between what is external and what is internal. It also makes possible certain strategies, such as the projection upon the external world of internal dangers which, if taken as internal, would cause extreme unpleasure. In an isomorphic argument Federn (1952) spoke of ego boundaries as constituting the flexible periphery of the ego.

Using the same imagery we may claim that the central nervous system, including the brain, is the sense organ of human time. We may support this claim as follows. The task of the brain is the working out of strategies for and directing the behaviour of the soma. But strategies and behaviour are functions and processes, not structures. For this and many other reasons, the mind of man, usually credited with the work of the brain, should be regarded as something purely temporal (Fraser, 1975, 1978*a*). Though 'mind' is a noun, it ought to be a verb for the function of the brain is the minding of the body.

Unlike other sense organs, however, whose office is to keep survey of the world external to the body, the purview of the mind includes the body itself as well as the universe external to it. The feeling of internal time tends to predominate, however, for the nervous system is close to—it is part of—the cyclic and ageing orders of the body, shielded from external processes through the functions of our senses (cf. Freud, 1920, Sec. IV).

We have already established that the mind retained its developmental stages. It follows that all temporal levels appropriate to nature must have some corresponding umwelts in the human mind, because of the evolutionary journey of living matter from biogenesis to man. Whether time is purely external or purely internal to man is a very frequent question in philosophy. But stated in the form of a simple dichotomy it is too simplistic and hence it cannot be answered. However, we may refine the issue by formulating it in terms of temporal levels.

Consider, therefore, that atemporal, proto- and eotemporal conditions may be identified throughout the physical world, including among the functions of the body itself. It follows that the hallmarks of lower temporalities may be projected upon the external world and upon the body image, with stunning appropriateness. The hard sciences bear witness to this fact. Biotemporality may be identified only for life processes. When the hallmarks of biotemporality are projected upon the external world they are found to be appropriate only for our own bodies and for other living organisms. The limited usefulness of animism is one demonstration of this level-specific aspect of nature.

Nootemporality is unique to man. When projected upon the non-human world it seems to make little sense. Religions, philosophies and even the sciences have been trying to force noetic qualities upon nature at large, or else advocate various ways in which we may live without such projections. It is in the noetic world that temporal issues become overdetermined and hence ambiguous. Furthermore, there are epistemic reasons why it may not be possible to determine what aspects of human experience may eventually be made intelligible, and which aspects will remain inexplicable in terms of languages now available for dealing with the relation of the individual to society (Fraser, 1978*b*).

The psychodynamics of temporal levels is, thus, a syndrome of immensely complex processes, innocuously hidden in the statement that we feel time pass. The sketches that follow cannot define, but only imply, this complexity.

The passage of time

It has long been assumed that feelings about the rate at which time passes are useful guides to the understanding of emotional states, that they correlate with personality[3] and may also be symptomatic of certain psychopathic conditions. The metaphors used to describe the passage of time, at least in Western languages, tend to be motional: time moves too rapidly, too slowly, or else it stops.

But 'speed of time' metaphors tend to be misleading and ambiguous terms (Melges & Fougerousse, 1966). Also, they correspond to different affective textures depending on whether they refer to conditions expected, remembered or experienced. They stand for feelings whose origins are increasingly more difficult to identify as we contemplate the future, the past or the present.

Studies on the assessment of future time are numerous (an entry may be had through Orme, 1969 and Doob, 1971) but psychoanalytically oriented works, such as Melges (1972) appear to be rare. When contemplating the future, we seem to be dealing with the evaluation of fantasies, present needs and unresolved conflicts in terms of the perceived amount of freedom available for attaining the goals of drives. When guilt delimits that freedom then we feel with Macbeth that 'To-morrow, and to-morrow, and to-morrow, Creeps in this petty pace from day to day, To the last syllable of recorded time'. But to men in joy 'What is six winters? They are quickly gone... But grief makes one hour ten' (*Richard II*). The relative simplicity of future perfect tenses, in languages where they exist, suggest the relative simplicity of future as a category of noetic time.

For dealing with the past, many languages provide distinct grammatical forms. Thus, in most Western tongues it is possible to distinguish between processes judged complete (perfect or passive) and incomplete (imperfect or active). Comparison of even a few languages reveals, however, the arbitrariness of this division. The ambiguity is rooted in the fact that past events *qua* past events exist if, and only if they live on in memory and are therefore, even by grammatical definition, incomplete and active.

Whether last summer's vacation or last winter's hospital confinement appears now as having passed slowly or rapidly is a function of present investment of psychic energy in mental representation of past feelings, directed by present needs and filtered by present fantasies. But the ambivalence of language is great even without mental filtering. A past experience of oceanic feeling may be equally well described as one in which time appeared to have stopped or else it flew with great rapidity.

In discussing the experiential speed of time's passage, the model of time employed by physician and patient alike is, again, that of an objective, man-independent time which flows 'by itself and by its own nature' and whose speed the patient misjudges.

It is true that no clock may be said to be fast, slow or correct in itself but only with respect to a reference clock. In scientific measurements of time the authority of the reference clock must, in its turn, be validated. This is usually done by abstract reasoning, such as by reference to a law of nature. The same three-cornered dynamics (two clocks and a theorem) may also be identified in reports about variations in the experienced speed of time.

It is the thesis of this subsection that the three elements which determine the rate at which time seems to pass are (i) the affective dimensions of temporal experience (time felt); (ii) the rationalized or intellectual dimension of time (time understood) and (iii) a set of drives, fantasies and superego demands which mediate between time felt and time understood. It is this third component that guides the decision as to which of the 'clocks' (time felt or time understood) is to be regarded as the dominant form of reality. We may now try to tie this scheme to some known functions of the mind.

Freud, writing on narcissism (1914) remarked that 'at the bottom, the self-criticism of conscience coincides with self-observation on which it is based. Thus, the activity of the mind which has taken over the function of conscience has also placed itself at the service of internal research...'. In clinical cases of depersonalization this internal research process is disturbed. Arlow (1966) has given examples which illustrate the disturbance. Beneath the feeling of losing their sense of reality, his patients report dream-like detachments from their own and other peoples' actions; there was a split in the self representation into a participating and observing self. The major sense of identity was usually with the observing self, connected distantly to the participating self, which was experienced with a lesser sense of reality. Arlow suggests that clinical cases, such as the ones he examined, 'may be understood as representing the dissociation of the function of immediate experiencing from the function of self-observation'.

Let us assume that the observing and participating selves are always present in the mind representing, respectively, phylogenetically older and younger modes of perceiving reality. In the absence of psychopathic conditions the participating self is associated with secondary process functions: it prefers to deal with ideational issues, its umwelt comprises well-defined categories of future, past and present. It understands time. The observing self is associated mostly with primary process functions: it prefers to deal with time in its instinctual and affective aspects. Its umwelt comprises all the pre-noetic temporalities. It feels time. The two selves, or actors, communicate through fantasy, both conscious and unconscious. These fantasies are the 'theorems' which determine what is to be regarded as real and where the major sense of reality is to be placed. The selection is generally made so as to protect the executive integrity of the ego and thus secure a dynamic balance between the observing and participating selves. This is what we mean when we say that the task of the mind is to direct behaviour so as to promote the survival of the species in the evolutionary enterprise. However, in the presence of overwhelming danger, internal or external, fantasy ceases to be the mediating agent between the two portions of the self and becomes, instead, the vehicle that conveys the danger which then precipitates an overt split. The patient then cries out with Hamlet: 'The time is out of joint; O cursed spite/That ever I was born to set it right'.

In the case of Hamlet, the vicissitudes of murder and the duplicity of court life hyper-cathected the objects in the temporal umwelt of the observing self and thus challenged the executive powers of the participating self. To rescue the freedom of the participating self (acting in a world of future, past and present) the murderous impulses are projected onto another entity. This other entity exudes an atmosphere of inevitability that is, it functions in a world without future and past.

> If Hamlet from himself be ta'en away,
> And when he's not himself does wrong Laertes,
> Then Hamlet does it not; Hamlet denies it.
> Who does it then? His madness.

The tension between Hamlet's two selves left its healthy operating range, the intensity of the struggle became unmanageable and the conflict became chronic. He began and continued to regress toward the conflictless umwelt of the omnipotent child. His dying words were about the eotemporal, abiding present: 'The rest is silence'.

We have already learned that, from the perspective of the nootemporal, all lower temporalities appear incomplete, unshaped. Let us assume that the inverse also holds: looking up, as it were, each higher umwelt appears increasingly directed and structured. Then, time too slow, undirected, unshaped suggests the perspective of the noetic self looking down, as it were. Time too fast, too structured suggests the pre-noetic self looking up. The feeling that time is out of joint suggests an unmanageable distribution in the allotment of psychic charges among the different temporal umwelts. Not surprisingly, disturbances in the sense of time tend to be correlated with confusion regarding the structuring of reality (Melges & Freeman, 1977).

In the following I shall sketch some further applications of the facultative scheme discussed, as it relates to problems in the apparent speed of time. The ideational content of time experience (the contributions of the id, the ego and the superego) is to be understood as modulating and refining the shifts in psychic energy and alternations in levels of functioning.

Thus, for instance, it has often been noted that time in old age tends to pass with increasing rapidity. Perhaps this is a way of saying that the major sense of identity has shifted to the observing self with its biotemporal and lower umwelts. As long as reality is still identified with the noetic umwelt of secondary process functions, with a well-defined structure of future, past and present, the dominant self will behold the world as passing all too quickly.

It has also been noted that the time of early adolescence often passes too slowly, though we must add that this experience is heavily biased by societal conditioning. The apparently slow passage of time may be a way of saying that the major sense of identity is becoming vested in an emerging, noetic self. However, as long as reality is still identified with the lower umwelts of the child, time will be felt as passing too slowly.

Among non-clinical examples, changes in the experienced speed of time have been studied from the perspective of the prisoner by Farber (1944), Engel (1963), Des Pres (1976) and others. In benign, experimental isolation and under the self-imposed regimen of a strong ego, the relative balance between the observing and participating selves may be maintained. The underestimation of the total number of days spent in confinement is likely to have a physiological basis: the endogenous, free-running circadian organization of man tends toward periods longer than 24 hours. (The literature is extensive. See, e.g. Aschoff, 1965.) However, if fantasy is unable to secure the balance among the different levels of reality and time, the instinctual, somatic levels of time may become hypercathected and threaten the noetic, participating self. With reality vested in the noetic world but identity in the instinctual one, a conscious awareness of a frightfully slow passage of time may become manifest. Deterioration of perceptual faculties follows, including hallucination accompanied by free-floating anxiety. The subject may then reach for, what quite appropriately has been called, the panic button.

Under extreme conditions of deprivation, in contrast to benign isolation, the major portion of psychic energies must be invested in the lower umwelts of the mind, so as to serve bodily survival. If not hindered by unresolved conflicts, fantasy can accommodate a shift of the major sense of identity from the participating to the observing self. Secondary process activity is then maintained at a level no greater than necessary for survival. The experience of time becomes that of surviving from instant to instant in an abiding present. But if unresolved conflicts prevent the noetic self from yielding its claim for dominant identity and determination of reality, the prisoner may then identify with the jailer, the prey with the predator. Time will then be felt as passing with unbearable slowness.

Alternately, fantasy may manage to invert the evolutionary roles with respect to temporal umwelts of the two selves. While the participating (ideational) self directs behaviour in a future- and past-less world, the observing (instinctual, primary process) self projects itself into a world where eotemporal eternity is the form of dominant reality. The survival of the captive is again aided, this time by what is known in common language as hope.

In certain clinically significant cases the superego may play the role of the captor that enforces isolation. The noetic umwelt is identified as the only level of reality and hence, authority in matters relating to time. Noetic time may even be invoked as an aid in the control of instinctual demands (Dooley, 1941). Time felt, judged from the perspective of time intellectually understood, will appear as insufficiently structured and flowing too slowly. In a curious example of the return of the repressed, behaviour may become fragmented and hectic, appropriate to the 'chaos, a cauldron of excitement' which Freud saw in the id. The patient's actions will then reflect the temporality of the id, whose demands for recognition the whole strategy was supposed to have negated.

Burnham (1969) dealt at length with the role of time in the need-fear dilemma of schizophrenic patients.

The inordinate need for contact with others, whose presence the patient feels as reassuring, suggests a dominant reality with biotemporal horizons: what is not here-and-now, does not exist. The associated emotions are those of an unstructured void which the patient is driven to fill by activities, often of heroic proportions. Whether time is described as flowing too rapidly or too slowly, or both, is irrelevant. It is 'out of joint'.

The often equally inordinate fear of the reassuring objects may lead to a defensive denial of the passage of time. The associated emotions are those of dread, an unpleasure of patterned and filled time. Again, it is irrelevant whether time is described as too fast or too slow. There is an unbalance between time felt and time understood, which is one of the many ways schizophrenic reactions may sometimes be described. (For the evaluation of certain schizophrenic dysfunctions in terms of reality testing and time, see Seeman, 1976.) The associated experience of derealization and depersonalization are corollaries of the unbalance in the evolutionary hierarchy of the temporal umwelts of the mind.

Lastly, it should be mentioned that changes in the experienced speed of time have been examined in the context of artistic and religious experience, under the influence of drugs and hypnosis and in states of heightened sensitivity such as in love or in mortal danger. (Cooper & Erickson, 1950; Melges *et al.*, 1970; Doob, 1971; Yaker *et al.*, 1971; Melges *et al.*, 1972; Aaronson, 1972; Melges *et al.*, 1974.) It would be impractical and it is unnecessary to sketch the details of this work beyond mentioning that phenomenologically and symptomatically they resemble the paradigmatic cases already discussed.

In summary, an interpretation of changes in the experienced speed of time appears feasible in many cases in terms of a three-cornered dynamics. Its elements are time understood (a secondary process function involving noetic time), time felt (a primary process function involving bio-, eo-, proto- and atemporality) and conscious as well as unconscious fantasies which mediate between the two.

An example of temporal disorganization

Melges & Freeman (1977) have found a substantial positive correlation between temporal disorganization and symptoms of inner-outer confusion in 58 acutely ill psychiatric patients.

Temporal disorganization was measured by 20 questions deriving from five categories of statements. I will reproduce the five categories below, followed in parentheses by a statement of the issues, as seen from the point of view of temporal levels.

'Rate-duration changes.' (The question of apparent speed of time. In which temporal umwelt does the patient place his major sense of identity? In which his major sense of reality?)

'Tracking difficulties.' (The question of continuity. What seems to be the dominant temporal umwelt of the patient in terms of temporal continuity?)

'Temporal indistinction.' (Categories of future, past and present. What is the dominant umwelt of the patient in terms of the organization of his mental present?)

'Impaired goal-directedness.' (Questions of causation. What is the patient's dominant umwelt in terms of connectivities?)

'Desynchronization.' (Consistency of the dominant umwelt. To what degree does the patient 'travel' among the temporal umwelts of his mind?)

Symptoms of inner-outer confusion were measured by questions derived from five categories of statements. They were phenomenological descriptions of the ways in which the patients separated perceptions from ideas. In so far as perceptions refer to the presumed world of the non-self, whereas ideas are taken to operate within the self, the answers to the questions suggested the boundaries of the patient's self and the sharpness of its definition.

The authors stressed that, if memories and expectations cease to be attached to futurity and pastness then they might be experienced as current, perceptive phenomena. This reminds us of the question asked in the beginning of this paper: if reality testing is but the separation of perception from imagination and fantasy, then how do we evaluate future and past?

The correlation factor obtained between total scores of the two sets of variables was an impressive .73, significant at the .001 level. The authors warn us, however, that correlation does not necessarily signify cause and effect; that must be separately tested by demonstrating mutually influential interaction between the variables. They regard their findings as 'covarying changes over time', thereby tacitly assuming the exclusive reality of an external, man-independent time that flows by itself.

What we have learned about time counsels a change in perspective. For the purposes of testing, identifying the psychiatrist's nootemporal umwelt with reality is heuristically justifiable. But for interpreting the data a hierarchical, level-specific structure of time and reality appears to be more useful. The suggestion comes to mind that the correlation found by the tests, while certainly not coincidence, is also not cause and effect. It suggests, instead, that degrees of self-definition modes of connectivities and varying perceptions of reality are corollary manifestations of the structuring of human time.

The psychoanalytic journey

Namnum (1972) has stressed the importance of contrast between the 'timelessness' of the analytic hour and the strict control of time regarding the limits of that hour. We may say that they are counterpoints of the melody that maps the reality of the world. But time (more precisely, nootemporality) and timelessness (more precisely, atemporality) are but the highest and lowest members of a hierarchy. I would like to develop the idea of Namnum in terms of temporal levels.

Freud's advice to the physician treating a difficult case was 'to behave as timelessly as the Ucs. itself, if he wishes to learn anything or achieve anything' (1918). The recommendation is to imitate the temporal character of the umwelt of the unconscious. Being near the soma, its rates of change, as I have already remarked, resemble those of biological rather than ideational processes.[4] This strategy may assist and encourage the patient to begin and recreate his past as a living present. It implies empathy for the patient's task in a situation unique to the psychoanalytic setting.

Whereas archaeologists explore the past of a noetic present, psychoanalysts and their patients are confronted in their explorations with *qualitative* changes in the temporal framework itself. For, as one gets closer to infantile memories, the temporal umwelt itself in which the memories have been recorded, changes. One of the profound difficulties in piercing infantile amnesia surely resides in the fact that a nootemporal framework, in whose language the traumatic events must eventually be reported, did not exist at the time of the event. The task, therefore, involves a process of creation so that upon the termination of the analysis the developmental hierarchy of the temporal umwelts may be well ordered in the patient's mind—both in their conscious and unconscious dimensions.

Consider now that the task of conflict resolution in terms of ego psychology consists of penetrating pathological defences and revealing the elements of successive layers of conflicts. With thorough working through of the insights so gained, the analysand becomes accepting of his instinctual drive derivatives—his private hierarchy of realities—and finds appropriate channels for their expression. From the perspective of temporal levels I would like to represent this, the psychoanalytic journey, through the symbolism of two mythical figures.

The first figure does not have a name of her own: she is Lot's wife. Having transgressed the laws of her Creator she was changed to a pillar of salt. We may fancy the change from a woman of mind and body to one of mindless life, then to dead matter. Whatever the collective fantasy that underlies the myth, we may take it to represent the mental crippling under the pressure of various inner conflicts, a state of catalepsy, a loss of the joy of being human. It is also the painful journey of the analysand into the world of ossified wounds.

The other figure is that of Galatea, fashioned by Pygmalion from milk-white marble, laid on his bed, brought to life by Aphrodite, the goddess of love. She became wife to Pygmalion and mother of a son and daughter. Galatea journeyed along the umwelts from the lower temporalities up to the noetic. On the way she acquired selfhood and the capacity for fear and hope. Following the steps of psychosexual development, she learned to live with certain unresolvable conflicts which include anxiety and feelings of guilt.

The task of analysis is to descend with Lot's wife into the primordial, unfree temporalities that are always present in the mind and which constitute the framework of primary process thinking. From there the analysis is to aid the patient in the reconstruction of the hierarchy of inner realities and thus return with her as Galatea.

We find a fine example of this journey in Orgel's report (1965) on the uses and meaning of time for a young woman patient. When the patient first sought psychoanalytic assistance she resembled Lot's wife for 'she maintained a continuous blank slate—on which the writing of the moving finger of time was erased as soon as an impression was recorded'. Her prior descent into this state had to be systematically explored and then the dynamic structure of temporal umwelts rebuilt. On her way to becoming Galatea 'the beginnings of a belief in the flow of time, and especially in the reality of time farther and farther away from the present in both forward and backward direction, coincided in the analysis with a gradual giving up of... compulsive attribution of causality to unrelated events'. As the temporal umwelts fell in their places, so did their appropriate modes of connectivities.

The moods of the temporal umwelts

Writing about fantasy and reality testing, Arlow (1969) remarked that 'it is easy to see how moods may be evoked by perceptions of reality, in the sense that real experience stimulates the emergence of specific memories and systems of fantasy'. Since the perception of reality, as ordinarily understood, subsumes a hierarchy of archaic realities, one should be able to identify experiences which stimulate the emergence of umwelt-specific moods. We cannot expect them to appear in abstract purity, however, but only as dominant emotional states or colours of feeling.

Let us first sketch the hallmarks of temporalities: chaos for the atemporal; fragmentation for the prototemporal; directionless continuity for the eotemporal; directed continuity for the biotemporal; future-past-present, self *vs.* other, guilt, anxiety and human freedom for the nootemporal.

The following are representative examples from the arts which I have found as inducing, in my experience, the peculiar moods that correspond to hallmarks of temporalities.

Aleatory paintings, novels whose pages may be freely exchanged (Spencer, 1971), radio and television programmes put together from unrelated fragments exude the atmosphere of the atemporal and the prototemporal, depending on their degree of incoherence. So does the second movement of Alban Berg's *Lyric Suit,* appropriately named 'Desolato'. In these symbolic objects later *vs.* earlier, here *vs.* there can be given no meaning for they possess no intrinsic reference to which the direction of human time may be tied. Atemporal and prototemporal conditions may appeal to fantasies lodged in the lowest temporal umwelts, at the bottom of primary process functions.

Whether their moods are felt as pleasurable or unpleasurable should depend entirely on the psychic content.

The music of drum cultures (rhythmic continuity without a preferred direction) evokes an eotemporal mood. So do cubist paintings with their combination on canvas of various views of the same object, but views which are perceivable only by using extended periods of time. The cubist doctrine demands that the painter's temporal world resemble the abiding present of the eotemporal.

The following dialogue from Beckett's *Endgame* also exudes eotemporal mood, sounding rather the way one may fancy a monologue in the unconscious:

> Hamm: What time is it?
> Clov: The same as usual.
> Hamm: Have you looked?
> Clov: Yes.
> Hamm: Well?
> Clov: Zero.
> Hamm: It's need rain.
> Clov: It won't rain.

There is a degree of organization in this dialogue: not everything happens at once, the universe is not one of chaos. But there is no development: the answers are interchangeable. One may hypothesize that the eotemporal mood is the characteristic feeling of that level of the mind wherein object constancy has already been established but the systematic relationship among constant objects, has not.

For the small child, the eotemporal present represents ultimate reality. Schematically, it is the world of countable objects and events which are, however, not orderable along a temporal direction (Fraser, 1978*a*). The concept of repetition describes such umwelts. The spikes of a fence along which one walks create a world of countable objects with no intrinsic preference for the direction of counting. Passing the hand from spike to spike is like listening to the beat of drums. For adults, the mood thus evoked may represent a return to childhood, a regression along the temporal umwelts.

Bartemeier (1941) described a counting compulsion of a patient who felt oppressed by (noetic) time. Predictably, she preferred to contemplate geometrical objects. She might or might not have known that such objects constitute Plato's timeless Forms, a world into which one may escape from the tensions of noetic conflicts. Park & Youderian (1974) described an autistic child for whom number seems to have been the highest symbolic form through which she could express feeling and meaning. Those objects of human experience which demanded more advanced, qualitative kinds of meaning, perhaps because they appeared threatening to her were, it seems, banished from her mind.

Passage from the eotemporal to the biotemporal and finally the nootemporal mood is suggested by the following quote from *Remembrances of Things Past* by Marcel Proust, master of the twilight zones between the self and the non-self:

> my sleep was so heavy as completely to relax consciousness; for then I lost all sense of the place... and when I awoke... not knowing where I was, I could not be sure at first who I was, I had only the most rudimentary sense of existence, such as may lurk and flicker in the depths of an animal's consciousness; but then... memory... would come down like a rope let down from heaven to draw me out of the abyss of not-being... in a flash I would traverse and surmount centuries of civilization, and out of half visualized succession of oil lamps, followed by shirts with turned down collars, would put together by degrees the component parts of my ego.

The pervasive characteristic of the nootemporal mood is that of the unresolvable conflict between growth and decay, perhaps Eros and Thanatos. This element may be identified not only in the graphic arts and literature but also in music. In a psychoanalytic study of the dying composer Gustav Mahler, Feder (1978) appealed to the moods of Mahler's late compositions to demonstrate the fear of death's anticipated regression: from the freedom of the noetic to the unfreedom of matter. Of special interest to us is a symphony composed a year before Mahler's death. In the portion discussed by Feder, the melodic line grinds to a standstill as muffled drum strokes come to carry the eo- and prototemporal moods of musical experience. From the directed melody to the undirected drum, it is a journey of Lot's wife from life to death.

In a paper on time and the ego, Cohn (1957) described beautifully the power of music in inducing temporal moods.

> The deepest regression that appeals to the body most broadly, like music, provides the time quality of companionship with the vital organs, and though the ego's vigilance seems to cease here in sleep, it still retains the one last function—dream symbolism—that registers the basic recognition that we, the living, are time (p. 189).

I have argued elsewhere (Fraser, 1975, 1978*a*) that the power of art resides in the capacity of the artist to roam freely among the moods of temporalities and then report about his journey through such metaphors (cf. Arlow, 1977) as are able to stimulate the emergence of fantasies, appropriate for the stages of the journey. The similarity of experience, though not of method, between the artist's round trip and the guided journey of psychoanalytic therapy is part of the intellectual heritage of the profession. The present argument points to the roots of this similarity in the psychodynamics and moods of temporal levels.

A metapsychological aside upon a theory of conflict

This paper employs the conceptual tools of a new understanding of time. The theory of time as a hierarchy of unresolvable, creative conflicts was first sketched in Fraser, 1970; worked out in detail in 1975, then extended and summed up in 1978*a*. Though the theory comprises not one but a class of principles in the scientific and humanistic study of time, its theoretical basis may be correctly represented by one of Freud's remarks in 'Beyond the pleasure principle' (1920).

> No substitute of reactive formations and no sublimations will suffice to remove the repressed instinct's persisting tension; and it is the difference in amount between the pleasure of satisfaction which is *demanded* and that which is actually *achieved* that provides the driving factor which will permit of no halting at any position attained, but, in the poet's words, 'untamed, forever forward presses'[5] (p. 42).

The difference between the imagined (or demanded) and the encountered (or achieved) is recognized by the theory of time as conflict as man's existential tension. It is then postulated that existential, or constitutive tension is the fundamental driving force of nature, of which the conflict of the human mind is but one example.

The theory points to the stable integrative levels of nature and identifies the unresolvable conflicts of each. By 'conflict' in this broad context is meant the coexistence of two opposing trends, regularities or groups of laws in terms of which the processes and structures of each integrative level may be explicable. By 'unresolvable' is meant that, by means indigenous to an integrative level, the level-specific constitutive conflicts may only be maintained (and thereby the continued integrity of the level secured) or else eliminated (and thereby the level collapsed into the one beneath it) but not, in any sense, solved.

It is shown that the known methods of inorganic and organic evolution will, on the average, increase rather then decrease constitutive tension. In the history of nature the increasing constitutive tensions have sometimes created conditions that were unmanageable by means available on an integrative level. In such cases the unresolvable conflicts of that level have occasionally provided the motive force for the emergence of a new integrative level which, in some sense, did resolve the conflicts of the lower level. However, from its very inception, each of the new levels could once again be identified with certain unresolvable conflicts of its own.

Thus the theory traces and describes the mechanisms of emergence from an atemporal matrix of light, the subsequent integrative levels of elementary particles, massive matter, life and mind.

In this metapsychological scheme the human mind, identified with its peculiar conflicts between passion and knowledge (shorthand for instinctual demands and intellectual understanding) seems in our epoch to be entering a transitory period comparable, earlier in history, to the emergence of life from matter, and mind from life (Fraser, 1978*b, c*).

Summary

Psychoanalytic practice has tacitly assumed the existence of an absolute and objective temporal matrix in which physical, biological and psychological processes take place and to which the subjective time of the patient in therapy gets gradually attuned. But psychoanalytic thought does not demand an ontology of absolute frameworks independent of man but only an epistemology in which several visions of reality are traditional.

This paper reformulates the concept of time so as to accommodate three demands which appear to be mutually exclusive. The new concept of time (1) eliminates the need for assuming that it is a feature of the world independent of man, (2) retains the advantage of the heuristic assumption that, to first approximation, there does exist an objective passage of time and (3) does not contradict the teachings of other sciences about time.

Careful attention to the feeling of the uncanny as it relates to time, together with considerations of the development of the sense of time suggests that our conscious evaluation of noetic time is directed and weighted by unconscious, archaic modes of time perception. These phylogenetically old forms of the sense of reality determine the temporal character of the unconscious.

The dynamics of psychic energy invested in the various levels of time seems to determine the rates at which time is experienced as passing. The feeling that time passes at the right rate probably corresponds to such combinations of perceived causations and temporalities which do not threaten the integrity of the ego. If, because of internal or external danger or physiological dysfunction that balance changes, so does the sense of reality and with it, the affective texture of experienced time.

Judged from this perspective, the reliability of reality testing is a measure of the degree to which the patient is able to accommodate the archaic reality of primitive causations and temporalities, lodged in his fantasy, memory and dreams, and integrate them with the ambiguities of human freedom.

For their detailed comments I am deeply indebted to Jacob A. Arlow and H. Robert Blank.

Translations of Summary

La pratique psychanalytique a supposé tacitement l'existence d'une matrice temporelle absolue et objective dans laquelle ont lieu les processus physique, biologique et psychologique et avec laquelle se met graduellement d'accord le temps subjectif du patient en thérapie. Mais la pensée psychanalytique ne demande pas une ontologie de charpentes absolues indépendantes de l'homme mais seulement une epistemologie dans laquelle sont traditionnels plusieurs visions de la realité.

Cet article reformule le concept du temps pour adapter trois exigencies qui semblent s'exclure entre elles. Le nouveau concept du temps (1) élimine le besoin de supposer que c'est une caractéristique du monde indépendante de l'homme, (2) retient l'avantage de la supposition heuristique que, dans une première approximation, il exist un passage objectif du temps et (3) ne pas être en contradiction avec les enseignements des autres sciences à propos du temps.

L'attention soigneuse au sentiment du sinistre associé au temps, avec des considérations du developpement du sens du temps suggère que notre évaluation consciente du temps abstrait est dirigée et pesée pas des modalités inconscientes et archaiques de la perception du temps: ces formes phylogénétiquement anciennes du sens de la realité determinent le caractère temporel de l'inconscient.

La dynamique de l'énergie psychique investie dans les differents niveaux du temps semble determiner la vitesse avec laquelle on experimente le passage du temps. Le sentiment du temps qui passe à la vitesse correcte correspond probablement aux combinaisons des causations perçues et à la temporalité, ce qui ne menace pas l'integrité du moi. Si par un danger intérieur ou extérieur ou d'une dysfonction physiologique qui équilibre des changements, la même chose sera faite par le sens de la realité et avec lui, la texture affective du temps expérimenté.

Jugée dans cette perspective, la confiance sur l'épreuve de la réalité est une mesure du degré avec lequel le patient peut adapter la réalité archaique des causations primitives et temporalité, logée dans ses fantasmes, mémoires et rêves, et les intégrer aux ambiguités de la liberté humaine.

Die psychoanalytische Praxis hat stillschweigend die Existenz einer absoluten und objektiven zeitlichen Matrize angenommen, in der sich physische, biologische und psychologische Prozesse ereignen und der sich die subjektive Zeit eines Patienten in Therapie allmählich anpasst. Psychoanalytisches Denken verlangt jedoch keine Ontologie von absoluten Bezugsrahmen, die vom Menschen unabhängig ist, sondern nur eine Epistemologie, in der verschiedene Visionen der Realität landläufig sind.

Dieser Artikel gibt dem Begriff der Zeit eine neue Formulierung, um drei Forderungen gerecht zu werden, die sich gegenseitig auszuschliessen scheinen. Der neue Begriff der Zeit (1) eliminiert die Notwendigkeit der Annahme, dass sie ein von der Menschenwelt unabhängiges Phänomen ist, (2) behält den Vorteil der heuristischen Annahme bei, dass ein objektiver Zeitablauf—wenigstens im Sinne einer ersten Annäherung—existiert, und (3) widerspricht den Aussagen anderer Wissenschaften über die Zeit nicht.

Sorgfältige Aufmerksamkeit auf das Gefühl des Unheimlichen und dessen Beziehung zur Zeit im Zusammenhang mit der Entwicklung des Zeitsinnes erweckt den Eindruck, dass unsere bewusste Einschätzung von subjektiver Zeit durch unbewusste, archaische Wahrnehmungsweisen der Zeit gesteuert und beeinflusst wird. Diese phylogenetisch alten Formen des Realitätssinnes bestimmen den temporalen Charakter des Unbewussten.

Die Dynamik der auf den verschiedenen Zeitstufen investierten psychischen Energie scheint die Einschätzung zu bestimmen, auf Grund derer der Ablauf der Zeit erlebt wird. Das Gefühl, dass die Zeit mit der richtigen Geschwindigkeit abläuft, ist wahrschleinlich im Einklang mit solchen Kombinationen von wahrgenommenen Ursachen und Zeitbedingtheiten, die die Integrität des Ichs nicht bedrohen. Falls sich dieses Gleichgewicht auf Grund einer inneren oder äusseren Gefahr oder infolge eines physiologischen Fehlfunktionierens verändert, so ändert sich damit auch der Realitätssinn und mit ihm die affektive Qualität der erlebten Zeit.

Aus dieser Perspektive gesehen, ist die Verlässlichkeit der Realitätsprüfung ein Anhaltspunkt über den Grad, in dem ein Patient in der Lage ist, die archaische Realität primitiver Ursachen und Zeitbedingtheiten, die seinen Phantasien, Erinnerungen und Träumen anhaftet, in sich aufzunehmen und sie mit den Zwiespältigkeiten menschlicher Freiheit zu integrieren.

La práctica psicoanalitica ha asumido tácitamente la existencia de una matriz absoluta y objetiva en la que tienen lugar los procesos fisicos, biológicos y psicológicos, y a la que el tiempo subjetivo del paciente en la terapia se va adaptando poco a poco. Pero el pensamiento picoanalítico no requiere una ontología de armazones absolutos independiente del hombre, sino sólo una epistemología con varios esquemas tradicionales de la realidad.

Esta comunicación reformula el concepto del tiempo a fin de dar cabida a tres exigencias que parecen ser mutuamente exclusivas. El nuevo concepto del tiempo (1) elimina la necesidad de asumir que éste sea un rasgo del mundo independiente del hombre, (2) retiene la ventaja de la suposición heurística de que, a primera vista, existe un paso objetivo del tiempo, y (3) no contradice las enseñanzas de otras ciencias respecto al tiempo.

La consideración atenta del sentimiento de extrañeza en relación con el tiempo así como del desarrollo del sentido del tiempo sugiere que nuestra evaluación consciente del tiempo noético lleva la dirección y el peso de modos de percepción del tiempo arcaicos e inconscientes. Estas formas filogenéticamente viejas del sentido de la realidad determinan el carácter temporal del inconsciente.

Le dinámica de energia psiquica invertida en los varios niveles del tiempo parece ser lo que determina a qué ritmo se experimenta el pasar del tiempo. La sensación de que el tiempo pasa al ritmo apropiado quizá corresponda a combinaciones de causaciones percibidas y temporalidades que no amenazan la integridad del ego. Si, por causa de un peligro interno o externo, o de una disfunción fisiológica ese equilibrio varía, también cambia el sentido de la realidad, y con él la textura del tiempo experimentado.

Juzgada desde esta perspectiva, la fiabilidad de la prueba de la realidad es una medida del grado en que el paciente es capaz de dar cabida a la realidad arcaica de causaciones primitivas y temporalidades que están alojadas en su fantasia, en su memoria y en sus sueños, y de integrarlas en las ambigüedades de la libertad humana.

References

Aaronson, B. S. (1972). Time, time stance and existence. In J. T. Fraser *et al.* (eds.), *The Study of Time I.* New York and Berlin: Springer-Verlag.

Abraham, G. (1976). The sense and concept of time in psychoanalysis. *Int. Rev. Psycho-Anal.* 3, 461–472.

Ames, L. B. (1946). The development of the sense of time in the young child. *J. Genet. Psychol.* 68, 97–125.

Arlow, J. A. (1951). The consecration of the prophet. *Psychoanal. Q.* 20, 374–397.

—— (1959). The structure of the déjà-vu experience. *J. Am. Psychoanal. Ass.* 7, 611–631.

—— (1966). Depersonalization and derealization. In R. Loewenstein *et al.* (eds.), *Psychoanalysis—a General Psychology.* New York: Int. Univ. Press.

—— (1969). Fantasy, memory, and reality testing. *Psychoanal. Q.* 38, 28–51.

—— (1973). Disturbances of the Sense of Time. Freud Anniversary Lecture. New York Psychoanalytic Institute, (in press).

—— (1977). Metaphor and the Psychoanalytic Situation. Loewenstein Memorial Lecture. Unpublished.

Aschoff, J. (1965). Circadian rhythms in man. *Science* 148, 1427–1432.
Bartemeier, L. (1941). A counting compulsion: contribution to the unconscious meaning of time. *Int. J. Psycho-Anal.* 22, 301–309.
Bergler, E. (1934). The psycho-analysis of the uncanny. *Int. J. Psycho-Anal.* 15, 215–244.
Bonaparte, M. (1940). Time and the unconscious. *Int. J. Psycho-Anal* 21, 427–468.
Brandon, S. G. F. (1966). Time and the destiny of man. In J. T. Fraser (ed.), *The Voices of Time.* New York: Braziller.
Brown, N. O. (1959). *Life against Death—the Psychoanalytic Meaning of History.* New York: Randon House.
Burnham, D. L. (1969). Time in personality and social structure: an aspect of the hospital careers of schizophrenic patients. In D. L. Burnham *et al.* (eds.), *Schizophrenia and the Need-Fear Dilemma.* New York: Int. Univ. Press.
Calhoun, C. J. (1976). Continuity and change: the significance of time in the organization of experience. *Int. Rev. Psycho-Anal.* 3, 291–304.
Chessick, R. D. (1957). The sense of reality, time, and creative inspiration. *Am. Imago.* 14, 317–331.
Cohen, J. (1964). *Behaviour in Uncertainty.* London: George Allen & Unwin.
Cohn, F. C. (1957). Time and the ego. *Psychoanal. Q.* 26, 168–189.
Cooper, L. F. & M. H. Erickson (1950). Time distortion in hypnosis II. *Bull. Georgetown U. Med. Ctr.* 4, 50–61.
Des Pres, T. (1976). *The Survivor, an Anatomy of Life in the Death Camp.* New York: Oxford Univ. Press.
Doob, L. W. (1971). *Patterning of Time.* New Haven and London: Yale Univ. Press.
Dooley, L. (1941). The concept of time in defence of ego integrity. *Psychiatry* 4, 13–23.
Ehrenzweig, A. (1953). *The Psycho-Analysis of Artistic Vision and Hearing.* New York: Braziller, 1965.
Engel, S. W. (1963). Die Zeit des Gefangenen. In G. Schaltenbrand (ed.), *Zeit in nervenärztlicher Sicht.* Stuttgart: Emke.
English, H. & English, P. (1964). *A Comprehensive Dictionary of Psychological and Psychoanalytical Terms.* New York: David McKay.
Farber, M. L. (1944). Suffering and time perspective of the prisoner. In K. Lewin *et al.* (eds.), *Authority and Frustration.* Iowa City: Univ. of Iowa Press.
Feder, S. (1978). Gustav Mahler, dying. *Int. Rev. Psycho-Anal.* 5, 125–148.
Federn, P. (1952). *Ego Psychology and the Psychoses.* New York: Basic Books.
Ferenczi, S. (1913). Stages in the development of the sense of reality. In *Sex in Psychoanalysis.* New York: Dover, 1956.
Fraser, J. T. (ed.) (1966*a*). *The Voices of Time.* New York: Braziller.
—— (1966*b*). Comments on time and the uncanny. In J. T. Fraser (ed.), *The Voices of Time.* New York: Braziller.
—— (1970). Time as a hierarchy of creative conflicts. *Studium Generale* 23, 597–689.
—— (1975). *Of Time, Passion, and Knowledge: Reflections on the Strategy of Existence.* New York: Braziller.
—— (1978*a*). *Time as Conflict—a Scientific and Humanistic Study.* Basel & Boston: Birkhauser.
—— (1978*b*). The individual and society. In J. T. Fraser *et al.* (eds.), *The Study of Time III.* New York & Berlin: Springer-Verlag.
—— (1978*c*). Temporal levels: sociobiological aspects of a fundamental synthesis. *J. Social. Biol. Struct.* 1, 339–355.
Freud, S. (1900). The interpretation of dreams. *S.E.* 4.
—— (1913). Observations and examples from analytic practice. *S.E.* 13.
—— (1914). On narcissism, an introduction. *S.E.* 14.
—— (1915). The unconscious. *S.E.* 14.
—— (1918). From the history of an infantile neurosis. *S.E.* 17.
—— (1919). The 'uncanny'. *S.E.* 17.

—— (1920). Beyond the pleasure principle. *S.E.* 18.
—— (1930). Civilization and its discontents. *S.E.* 21.
—— (1933). New introductory lectures on psycho-analysis. *S.E.* 22.
—— (1950). Extracts from the Fliess papers. Draft M. *S.E.* 1.
Friedman, W. J. (1978). Development of time concepts in children. *Advances in Child Development and Behavior* 12, 267–297.
Gifford, S. (1960). Sleep, time, and the early ego. *J. Am. Psychoanal. Ass.* 8, 5–42.
—— (1978). Aspects of time perception in infancy and old age. (Unpublished.)
Gross, A. (1949). The sense of time in dreams. *Psychoanal. Q.* 18, 466–470.
Hartmann, H. (1956). Notes on the reality principle. In *Essays on Ego Psychology*. New York: Int. Univ. Press, 1964.
Hartocollis, P. (1974). Origins of time. *Psychoanal. Q.* 43, 243–261.
—— (1978). Time and the dream. Paper read at the Fall, 1978 meeting of the Am. Psychoan. Ass. (Unpublished.)
Hurvich, M. (1970). On the concept of reality testing. *Int. J. Psycho-Anal.* 51, 299–312.
Kafka, J. S. & Bolger, H. (1949). Notes on the clinical use of future autobiographies. *Rorschach Research Exchange* 13, 341–346.
Loewald, H. W. (1962). The superego and the ego-ideal: II. Superego and time. *Int. J. Psycho-Anal.* 43, 264–268.
Masler, E. G. (1973). The subjective perception of two aspects of time. *Int. J. Psycho-Anal.* 54, 425–429.
Melges, F. T. *et al.* (1970). Temporal disintegration and depersonalization during marihuna intoxication. *Archs. Gen. Psychiat.* 23, 204–210.
—— (1972). Future oriented psychotherapy. *Am. J. Psychother.* 26, 22–33.
—— *et al.* (1972). Marihuana and alcohol. *Archs. Gen. Psychiat.* 27, 812–815.
—— *et al.* (1974). Temporal disorganization and delusional-like ideation. *Archs. Gen. Psychiat.* 30, 855–861.
Melges, F. T. & Fougerousse, C. E. (1966). Time sense, emotions, and acute mental illness. *J. Psychiat. Res.* 4, 127–140.
Melges, F. T. & Freeman, A. M. (1977). Temporal disorganization and inner-outer confusion in acute mental illness. *Am. J. Psychiat.* 134, 874–877.
Namnum, A. (1972). Time in psychoanalytic technique. *J. Am. Psychoanal. Assn.* 20, 736–750.
Newton, I. (1687). *Mathematical Principles of Natural Philosophy.* F. Cajori (tr. & ed.). Berkeley: Univ. Calif. Press, 1973.
Oberndorf, C. P. (1941). Time—its relation to reality and purpose. *Psychoanal. Rev.* 28, 139–155.
Orgel, S. (1965). On time and timelessness. *J. Am. Psychoanal. Assn.* 13, 102–121.
Orme, J. E. (1969). *Time Experience and Behaviour.* London & New York: American Elsevier.
Park, D. & Youderian, P. (1974). Light and number: ordering principles in the world of an autistic child. *J. Autism and Childhood Schizophrenia* 4, 313–323.
Schafer, R. (1970). The psychoanalytic vision of reality. *Int. J. Psycho-Anal.* 51, 279–297.
Schneider, D. E. (1948). Time-space and the growth of the sense of reality. *Psychoanal. Rev.* 35, 229–252.
Seeman, M. V. (1976). Time in schizophrenia. *Psychiatry* 39, 189–195.
Simonson, E. (1928). Über das Verhältnis von Raum und Zeit zur Traumarbeit. *Imago* 14, 469–485.
Spencer, S. (1971). *Space, Time and Structure in the Modern Novel.* New York: New York Univ. Press.
Stern, W. (1924). *Psychology of Early Childhood.* New York: Henry Holt.
Uexküll, J. von (1909). *Umwelt und Innenwelt der Tiere.* Berlin: Springer-Verlag. Alsou *Streifzüge durch die Umwelten von Tieren und Menschen.* Frankfurt a.M: Fischer, 1970.

Available as 'A stroll through the worlds of animals and man' in C. H. Schiller (tr. & ed.), *Instinctive Behaviour*. New York: Int. Univ. Press, 1957.
Wallerstein, R. S. (1973). Psychoanalytic perspectives on the problem of reality. *J. Am. Psychoanal. Assn.* 21, 5–33.
Yaker, H. *et al.* (1971). (eds.) *The Future of Time*. New York: Doubleday.
Yates, S. (1935). Some aspects of time difficulties and their relation to music. *Int. J. Psycho-Anal.* 16, 341–354.

12. UNBOUNDING SOCIETY

It took over two millennia of social dialectic in the world of Western ideas before an intellectual movement arose that attempted to put into practice the Heraclitean vision of correlating social practices with notions of order, reason, measure, nature and man. To which was added the Christian idea of God. This happened in the 17th and 18th century European movement called the Enlightenment.

Among its many broadly known expressions is a palpably tangible yet also elusive idea, the American Dream. The emotive and cognitive content of the Dream may be represented by images of open spaces, wide skies, immense mountains, together with a matching state of mind, one that places no limits on what a person should be able to achieve with hard work and fair cunning. The unbounding of society so implied is the communal version of the view of the cosmos, recognized in the strikingly appropriate title of Alexandre Koyrè's work, *From the Closed World to the Infinite Universe*, cited in the essay that follows.

The intellectual ambience of the Enlightenment was informed of the idea of the infinite perfectibility of humans and of a demand to shape society so as to accommodate that belief. It retained salvation history in the garb of open-ended progress and identified such progress with the summum bonum of man on earth. From this philosophy—one may even call it a theology—followed both the strength and the naiveté of the Enlightenment.

TIME, INFINITY, AND THE WORLD IN ENLIGHTENMENT THOUGHT*

1. The Astronomical Backdrop Expands

For Johannes Kepler, as for the classical Greeks, an infinite world was of necessity formless and static, and hence, from Kepler's point of view, incompatible with the Christian confession of human destiny. In 1606 he admitted that the thought of an infinite world filled him with horror, and argued in favor of a finite universe of fixed stars.[1] But a century later, Christian Huygens considered Kepler's arguments, rejected them, and spoke of a universe so vast that the Copernican system in it is just a point. He speculated that the sun is an ordinary star and other such stars fill the infinity of space.[2]

By the last third of the seventeenth century, as a set of heavenly objects, known today as extragalactic nebulae, came to demand the attention of observers, the closed medieval universe began to grow.[3] In 1656 Huygens found such an object in the constellation Orion; and in 1714 Halley gave "An Account of Several Nebulae or Lucid Spots like Clouds."[4] What understandably puzzled later observers was that even their telescopes failed to reveal any structure to these objects. In 1750 Maupertuis speculated that the realm of stars may be without boundaries, and that these curious bodies may be much larger than ordinary stars.[5]

In 1734 Emanuel Swedenborg (1688–1772) speculated that just as the earth goes around the sun, so the sun revolved around a universal center of attraction, and added that nature "extends her force" and her sway into infinity, in which there are neither boundaries nor ends.[6] But it was Thomas Wright of Durham who extended the limits of Newtonian physics to the "innumerability of the worlds" by suggesting a "regular and rational theory of the universe before unattempted by any."[7]

Wright began his reasoning by evaluating "the moral probability proper for conjecture," then argued systematically about the distances to, and the multiplicity of, stars. He calculated that there are 10 million suns, 100 million satellites, in all, some 170 million objects. The Milky Way is finite, but not so the world of all stars: "What less than infinity can circumscribe them, less than an eternity comprehend them." Perceiving the Divine presence in the very center of that vast universe, he described God as "a Being, whose center is everywhere and circumference nowhere," sounding very much like Nicholas of Cusa, three centuries earlier.

From observational astronomy let me turn to some of the philosophical aspects of astronomy. The universe of stars—whatever its size—was placed by the Creator into a preexistent matrix of time and space, or is it the case that the universe itself created these matrices through some properties of matter? This question is the central issue of the famous Leibniz-Clarke correspondence of 1715–1716.[8]

Samuel Clarke defended Newton's theory of absolute space and time by asserting that both are infinite; space came to be filled with matter and time with events when God created the world. For Leibniz such conditions appeared whimsical, because he could not find sufficient reason why God would create the world at any particular instant of time. Also, space devoid of things was as unimaginable to him as time devoid of events. He insisted, therefore, that space is the order of coexistence, time the order of succession, both in the perception of simple, indivisible, and indestructible atoms of reality he called monads.

The universe of Leibniz may be described as an immense clock shop wherein all the clocks tick simultaneously, not because they are synchronized through signals but because they were set in motion by a single master plan called preestablished harmony.[9] Regardless, however, of the primacy of space to matter and of time to events, God was, for both Leibniz and Clarke, an observer outside the system, uninfluenced by his observation of the system, be the universe finite or infinite. But by the end of the eighteenth century God, or at least His representatives, were underattack, and the ideas of order imposed from without through preestablished harmony came to be transmuted into the idea of self-sufficiency in nature, imposed entirely from within.

The Newtonian concept of time and space found strong support in the work of Leonhard Euler (1707–1783). He held that since Newton's laws were true, the metaphysical conclusions to which they lead are to be preferred to *a priori* arguments, such as those of Leibniz.[10] He suggested that metaphysics ought to be built on physics and observation, a belief in which we recognize the philosophy that underlies mathematized science: the world functions by certain preexisting laws which diligent search may discover. It then follows, for instance, that answers to such questions as what is time, or whether the world is finite or infinite may be found if one but knows how to formulate the inquiry.

But there were other opinions. Two generations before Euler, John Locke (1632–1704) had argued that such a hallmark of time as the idea of succession derives from the experience of trains of ideas continuously passing through our minds.[11] The notion of time is thus referred back to man, rather than assigned to a discoverable world that is independent of man. Only when we reflect upon the experience of passing ideas, argued Locke, do we acquire the notion of duration which, then, may be applied to sensory impressions, such as motion. But issues of the infinity of space, the nature of duration or that of number remain beyond the comprehension of the mind. The Lockean reference of temporality back to man the clock-watcher and clockmaker has a long ancestry. St. Augustine, almost thirteen centuries earlier, urged his readers not to ask questions about time in terms of metaphysical presuppositions, but to seek their answers in self-observation.

One should not leave the issue of infinity and the universe without consulting some of the relevant entries in the *Encyclopédie*. Unfortunately, they are rather prosaic. We learn, for instance, that *Eternité* should be compared with a line that has no beginning and no end and whose nature is inconceivable by the mind of man.

2. The Mathematical Backdrop

A simple way to reach infinity would be to keep on counting without stopping. But Aristotle already recognized that such a request has something odd about it.[12] He concluded that in one way infinity exists, in another way it does not. Since, in his view, the world was finite, infinity could only be something potential, for instance something that permits us in principle to go on counting, but it could not be anything actual. Besides, since "nothing is complete which has no end," infinity remained an imperfect condition. Although his ideas on infinity informed Western thought almost to our own day, they did not stop even the Greeks from using in their geometry the method of exhaustion, the ancient equivalent of infinitesimal calculus, involving actual infinities. This method consists of consecutive subtractions of certain fractions of a remaining magnitude and approaching by this means a certain limit as the number of steps approaches actual infinity.

Beginning in 1669, Newton composed several accounts of his infinite analysis. In them he looked upon mathematical variables as flowing quantities, or fluents, of which certain other quantities, the fluxions, were instantaneous rates of change. Both he and Leibniz developed, independently, a mathematical method based on the assumption that adding together an infinite number of infinitely small quantities and obtaining a finite number makes sense. Euler's *Introductio in Analysis Infinitorium* (1748) is about infinite processes, infinitely continued fractions and infinite series. Six of the eight mathematical Bernoullis contributed to the mathematics of the infinitely small, and one of them, Jacques, wrote a poem about it:

> Just as the infinite series by the finite is bounded
> And the unlimited is delimited,
> So traces of boundless power in a small body reside
> And the delimited is unlimited.
> What joy it is to discern the small in the infinite!
> The immense in the small to perceive, how divine![13]

The mathematical idea of the infinitely large seems to have been acceptable to Kepler, notwithstanding his opposition to an infinite universe, for, he noted in 1609, that as an ellipse changes into a hyperbola via a parabola, one of its foci passes on to infinity.[14] Some fifty years later Gerard Desargue (1591–1661) extended this idea in his famous work on conics.[15] And in 1655 John Wallis (1616–1703), founder of the Royal Society, proposed a streamlined version of the Roman numeral for 100 million as the now familiar sign for mathematical infinity.[16]

During the eighteenth century mathematical knowledge grew relatively slowly. It was not until the first half of the nineteenth century that Bernard Bolzano (1781–1848) broached the question of actual infinity in mathematics, not as a tacit assumption useful in infinitesimal calculus but as an explicit concept. He asserted that "there exists an infinite even in the realm of the actual, and not merely among things which make no claim to actuality."[17] His arguments derive from theodicy (defense of God's goodness) but are treated in terms of infinite sets. In 1874 George Cantor published his revolutionary paper on transfinite numbers. Through his work actual infinity was explicitly introduced and assimilated into mathematics, though only by widening the notion of number.

3. Distancing Creation

Infinite or otherwise in size, when did the world come about?[18] After the middle decades of the seventeenth century, the universe came to be regarded as much larger than had been believed. A similar fate befell the estimated age of the earth: the day of Creation began to recede into an increasingly distant past.

Although Xenophanes of Colophon, at the turn of the sixth century B.C., had already speculated about the gradual emergence of living forms, it was the many implications of what has been called the fossil dilemma that eventually forced the view of an evolutionary process upon philosophers and others.[19] Fossils, discovered in every corner of the earth, could not be accounted for except by assuming that they had been deposited over periods of time immensely longer than the biblical chronology would have permitted.[20]

The increase in the estimate of the age of the earth, and with it, that of the world, was gradual but intellectually irresistible. It was a transformation from a divine calendar to a geological and eventually to a cosmic one. This development was paralleled by the transformation of the sacredness of the divine laws into the immutability of the laws of nature.

The Cartesian Benoit de Maillet (1659–1738) maintained that the universe was eternal, matter in constant motion, and dedicated his clever discourse on the oceans, the earth and the origins of life (1748) to Cyrano de Bergerac, "Author of the Imaginary Travels thro' the Sun and the Moon." The universe was filled with seeds of all species, deposited at Creation and "prepared for life, in the waters of the sea, as in their first uterus."[21] Afterwards, these seeds found a diversity of dispositions, such as being deposited in slime and other substances, as in their second uterus. All this, asserts one of his protagonists, agrees with the Sacred Book of Christianity. After development, the species migrated to the land and left their marine ancestors behind. It is thus that we have sea calves, sea dogs, men with tails, men without hearts, men with one leg, and the like. Beneath this fanciful story, de Maillet recognized in the succession of geological strata the progression of life; if the strata were old, so did life have to be. He even stated the unmentionable:

> let us be here content not to fix a beginning to that which perhaps never had one. Let us not measure the past duration of the world, by that of our own years. Let us carefully consider what presents itself to our view in their universe, this immensity of the firmament, where we see so many other stars like our own sparkling, and which without doubt, only appear so little on account of their distance.... Let us inquire after the manner in which this universal system perpetuates itself.... A knowledge of this will teach us how it has been formed, better than our conjectures.[22]

Thirty years later Buffon, in his *Epochs of Nature* (1778), appealed for an increase in the estimated time interval since Creation, so that the coming about of the world could be explained.[23] Through hypotheses, geological and astronomical data and laboratory tests and by making certain assumptions about the initial condition of the earth, he estimated the age of the world, at the end of the eighteenth century, to be about 75,000 years. To reconcile this figure with the biblical count he suggested that the "days" of creation were in fact epochs in Nature. Descartes and Leibniz, among others, also proposed ideas of cosmic evolution, but Buffon is probably the first serious scholar to have put numbers into the expanded history of the world.

But even 75,000 was not long enough. A decade after the publication of the *Epochs of Nature* the Scottish geologist, James Hutton, argued for a succession of worlds, and a system of nature in which "we find no vestige of a beginning, no prospect of an end."[24]

4. Clocks and the Clock Universe

Whatever its actual or potential age, the world ran its course and, for those who knew how to experiment, it offered a laboratory. Into this laboratory God must surely have placed such qualities as enlightened minds would judge appropriate for stones, animals, and people. Thus, for instance, Maupertuis perceived in nature the deliberate thrift of the Divinity. In 1744 he formulated his Principle of Least Action: "la Nature, dans la production de ses effets, agit toujours par les moyêns les plus simple."[25] A sociologist of science might speculate that his idea about an economy in nature was an expression of the dedication to proper order and decorum admired in France during the Enlightenment. Whatever its inspiration, combining thrift and praiseworthy conduct with emphasis on quantitative knowledge, Maupertuis turned to Leibniz' monads from the exploration of consciousness, and devised a calculus of pleasure and pain so that one might quantify happiness.[26]

Thus, the earth carried its cargo of people looking for happiness; nature minded its own economy of action and followed the laws of gravitation as prescribed by Isaac Newton. Maupertuis, who was the first among leading French intellectuals to appreciate Newton's laws of gravitation, led an expedition to Lapland in 1736 to check by measurement the flattening of the earth towards the pole, such a flattening being a demonstration of the validity of the principles of gravitation. Surely, then, one might reason, either God had had to create the earth already flattened, so as to conform to Newton's law, or else the earth needed a very long history with an initial state when it was not yet solid. The history of nature could then account for the change of the earth's shape from a sphere of Platonic perfection to the likeness of a slightly squashed apple. (It remained for our epoch to change that apple shape into that of a pear.)

But whatever the age of the earth, whatever its shape, how did the earth and the universe function? The answer was found in a metaphysical device: the clock.

During the seventeenth century the metaphor of the clock became the metaphor of the new mechanical philosophy of Galileo, Descartes, Boyle, and Newton.[27] By the end of the seventeenth century the clock universe was a widely accepted model. It fitted the discoveries of science, the revelations of the history of nature and the new brands of philosophies.

The explanatory and aesthetic power of the clock universe is great. The clock connects the temporal organization of the individual with that of society and connects both with the cyclic and divinely reliable functions of the heavens. The clock lends the rhythm of the stars to the activities of man and projects the periodic needs of man onto the heavens. To Enlightenment eyes the world itself was a clock, albeit a very complex one. Pope tells us in his *Essay on Man* (I.267–68) that its gears praise their Creator:

> All are but parts of one stupendous whole,
> Whose body nature is, and God the soul.

The clock was only one product of technical progress. In 1707 the first physician's pulse watch was constructed. Able to measure seconds, it was a distant offspring of Gallileo's pulsilogium of the 1580s. The year 1711 witnessed the first sewing machine, 1714 the typewriter, 1727 an improved version of Father Kircher's magic lantern of 1646. And in 1778, just a decade before the French revolution, the water closet was invented. By the turn of the eighteenth century the already impressive gadgetry of the *Encyclopédie* was transformed into working reality in the industrial revolution of England. But 200 years earlier, at the beginning of the seventeenth century, the clock was still the most complex scientific machine, a small universe in motion. Astronomical clocks that showed the positions of the heavenly bodies came to be admired for their ingenuity and praised for their symbolic significance.[28] The talent of the clockmaker who fitted together the parts of a clock was surpassed only by the infinite ingenuity of God who put the world together so that it might function smoothly, efficiently, and purposefully. And just as God had built His universe as though it were a clock, so man built his clocks as though they were universes.

During the seventeenth and eighteenth centuries, as the precision of timekeeping improved, there arose an increased need to correlate the numerous astronomical cycles that were used to tell time. This was an ancient enterprise: the history of timekeepers demonstrates the reduction of successively more refined ideas about natural processes to working devices that confirm those ideas.[29] But it was not until the eighteenth century that a vogue of owning astronomical clocks had spread across Europe. I wish to represent this cultural hallmark by an outstanding example.

Father Francesco Borghesi, inventor and designer of astronomical clocks, born in 1723, lived out his life in the region of the Tyrolean border between what are now Austria and Italy. Sometime in the 1750s he encountered a good clockmaker, Bartolomeo Antonio Bertolla. They seem to have become fast friends, representing, between them, the religious-cosmic and the mechanical know-how of their epoch.

Father Borghesi designed and Signor Bertolla built two fine clocks. Borghesi then wrote two books about them: *The Most Recent Perpetual, Astronomical Calendar* (1763) and another in 1764, giving the design of a more sophisticated instrument.[30] The latter has a chapter, "An Exposition of the Latest Theory of the Universe," describing what today we would call a planetarium. The author exhorts the reader to examine the device:

> Now you, reader and lover of the stars, turn it, and revolve it as long as it pleases you, and compare it astronomically and physically with the Copernican and the Tychonian system...and judge which one seems more consonant with nature when all things are examined.[31]

Borghesi also gives us a glimpse of the mood of his epoch:

> All this work I had performed eagerly, so that, while in my room, I might contemplate leisurely, both during the day and in the night, the true face of the heavens and of the seas unobscured by clouds, even though I had no astronomical equipment.[32]

Father Borghesi's second clock brought him substantial monetary rewards from Empress Maria Theresa. But as he aged he came to feel increasingly guilty for having paid too much attention to things mechanical and not enough to things spiritual.

5. Epistemic Problems of the Clockwork Universe

Father Borghesi shared his personal dilemma with the clockwork universe itself. What structural part of the world is mechanical, predictable, and stable, and what part is spiritual, unpredictable, and creative? That is, how are the Eleatic categories of being and becoming divided in the nature of time?

The clockwork universe was in motion and changing, but its structure was as permanent as divine laws. In our own epoch this division is tested in laboratories, as our scientists separate the permanence represented by equations from the unpredictable details of boundary conditions. In the eighteenth century, nature at large was thought of as the most superb laboratory in which man could learn to reconcile the laws of nature, and of nature's God, with the emerging sense of an unpredictable history of nature. The reconciliation between being and becoming, as the classical Greeks already knew, is not easy. It is interesting to consider some of the details of the intellectual cross-currents that bore upon this issue.[33]

One way to solve the epistemic problem of change and permanence in the nature of time is to concentrate on permanence. Voltaire held to that approach. In 1785 he suggested that nothing which vegetates or which is animated changes: all species have remained invariably the same, humans have never been fish, all things are what they are by immutable laws.[34] This stance is a distant intellectual offspring of the Parmenidean philosophy of being. Condillac, in his celebrated *Traité de Sensations* (1754), built his *Homo sapiens* adding sense channel to sense channel, not taking note of the unpredictable and creative functions of the mind. He failed to see that a person so constructed would be a Frankenstein's monster, exactly as it as was born in the mind of Mary Wollstonecraft Shelley in 1818. Julien Offray de La Mettrie took the ideas of unchanging systems seriously, asserting that there were no essential distinctions between the conscious and the simply vital manifestations of behavior.[35]

Another way to solve the epistemic problem of being and becoming in the nature of time is to concentrate on change, growth, and the unpredictable. In David Hume's *Dialogues Concerning Natural Religion*, the sceptic Philo asks, "does not a plant or an animal which springs from vegetation or generation, bear a stronger resemblance to the world, than does any artificial machine, which arises from reason and design?"[36] In botany and zoology there were ramblings that presaged the Darwinian revolution of historical time.[37] Charles Bonnet (1720–1793) argued for an evolutionary theory where the ultimate element was the primordial germ, continuously present since Creation.[38] Theories of preformation and encasement were firmly established, even though debated by epigeneticists. Some thinkers, such as Immanuel Kant, embraced the idea that man evolved from four-footed creatures, but did not make the generalization to biological evolution.[39]

Because of the mutually exclusive character of being and becoming, the more impressive the arguments came to be in favor of the clockwork universe, the less it was possible to find a place for man, life, and the history of life and nature. There existed, of course, a worldview that could accommodate man, God, and the universe; this was the linear salvation history of the Christian West. But under the pressure of rationalism, salvation history gave way to a multitude of deistic, secular and atheistic ideas. As the medieval model of the universe became less convincing, new ideas about the nature of time were born, but salvation history remained part of the clockwork universe. This association gave charm to many eighteenth-century ideas about nature but is also responsible for the naive panaceas for the world's ills which we have learned to associate with Enlightenment thought.

Consider, on a comparative basis, the following statements:

> "As every individual... intends only his own gain, he is in this as in many other cases led by an invisible hand to promote an end which was no part of his intention" (Adam Smith, 1776).[40]
>
> "Man will do right because it *is* right, and the new eternal Gospel will surely come." Let me not despair that Providence is inscrutable, for "the vast slow wheel, which brings mankind nearer to [its final] perfection is only put in motion by smaller, swifter wheels, each of which contributes its own individual unit thereto." Mankind's progress is endless and, to accomplish it "is not a whole eternity mine?" (Lessing, 1780).[41]
>
> The history of the species is "the unraveling of a hidden plan of nature for accomplishing a perfect State of Civil Constitution for Society" (Kant, 1784).[42]

Let me now turn to Kant for his synthesis of time, infinity, and the world.

6. The Kantian Synthesis

The cosmologies of the early Enlightenment tended to be traditional. The entries in the *Encyclopédie* under *Monde, Universe, Cosmogonie* and *Cosmologie* are undistinguished. Eighteenth-century cosmological texts were Aristotelian, the *Physica* and *De Coelo* being the major texts, together with Johannes Sacrobosco's *De Sphaera.*

Out of this background arose Kant's vista of a *Universal Natural History and Theory of the Heavens* (1755).[43] He went not only beyond Newton (who considered the solar system to be isolated from the stellar universe) but even beyond Thomas Wright, for, as Whitrow has stressed, he assumed that the nebulous objects of Huygens and Maupertuis were aggregates of stars that resembled our Milky Way.[44]

In his cosmogony Kant argued that just as the zodiac with its planets forms a solar system, so the host of stars that make up the Milky Way constitutes a single system. The Milky Way, he said, is the zodiac of the higher world. He added that there were innumerable such higher worlds, each immensely greater than our solar system, though each arose the same way: from originally diffused primitive stuff. They all revolve, they all keep their distances. In a remarkable guess, he argued that if light can reach us from these very distant objects, so must also attraction, for it "can be stopped by no obstacles, because it penetrates into the inmost recesses of matter."[45] He speculated that the universe began from a center and expanded gradually so as to fill infinite space in the progress of eternity. In our own region of the world nature had already evolved out of chaos, but if we could go elsewhere, we would find forms still "buried in a silent night."

Because of the infinity of God the universe must have no limits. It cannot simply be vast; it is necessary for God to have created an infinite universe. "Eternity is not sufficient to embrace the manifestations of the Supreme Being, if it is not combined with the infinitude of space." Creation is not the work of a moment. "Millions and myriads of millions of centuries will flow on" while new worlds are being created. "If we could embrace eternity, we would see infinite space filled with systems of the world."

In an adumbration of his famous antimonies, Kant remarked that infinite space and eternity are undemonstrable, concealed from human understanding. One can almost hear his inaugural dissertation of 1770: "The idea of time does not originate in the senses but is presupposed by them." In the eternal and infinite universe God continuously creates matter and living forms. While certain parts of the world die, others maintain themselves as new ones come about. The whole universe is held together by a systematic constitution because

> a constitution of the world which did not maintain itself without a miracle had not the character of that stability which is the mark of the choice of God.... If the presentation of all this perfection moves the imagination, the understanding is seized by another kind of rapture when, from another point of view, it considers how such magnificence and such greatness can flow from a single law, with an eternal and perfect order.

We can almost hear the famous concluding words of the *Critique of Practical Reason* (1788):

> Two things fill the mind with ever new and increasing admiration and awe, the oftener and the more steadily we reflect upon them: heavens above and the moral law within.

This was one man's superb synthesis of Enlightenment thought on the world, infinity, and time.

7. Infinity and Timelessness on Earth

The size of the universe revealed by the telescope, ever larger than theretofore imagined, was already acknowledged in Milton's *Paradise Regained* (1671). Christ beholds the vastness of earth and history, as Satan shows Him the world from the mountaintop through the "strange parallax or optic skill, / Of vision multiplied through the air, or glass / of telescope."[46] Both Milton and Henry More (1614–1687) responded to the vastness of the universe as did Kant after them. They all sensed the infinity of God in the infinity of space.

Marjorie Hope Nicolson found this aspect of Enlightenment thought sufficiently impressive to write at length about the aesthetic of the infinite.[47] She identified it as the transfer of vastness from God to interstellar space and thence to the terrestrial mountains. At the turn of the seventeenth century Joseph Addison, in his *Spectator* papers on the pleasures of the imagination, was also impressed by "the great." He was inspired by the beauty of the Alps, those glorious, heavenly, and frightening mountains that seventeenth- and eighteenth-century men of letters discovered and rediscovered, somewhat as one rediscovers the rites of spring.

In 1749 Christian von Kleist in his poem *Der Frühling* gave articulate expression to a type of romantic nature poetry which remained a hallmark of the Alpine atmosphere to our own day. Earlier, in 1729, Albrecht von Haller, a Swiss psychologist and poet (1708–1777), expressed sentiments which Kant found sufficiently sympathetic to include in his *Universal Natural History*:

> Infinity! What measures thee?
> Before thee worlds as days, and men as moments flee!
> Mayhap the thousandth sun is rounding now...
> And when its work is done, again in heaven
> Another shines. But thou remain'st! To thee all numbers bow.[48]

In these sentiments, from the point of view of our theme, we find a clockwork universe of continuous creation, together with actual infinity.

If all this earthly immensity can be put into a single Law of God, discoverable by reason, is not reason the highest good on Earth? Perhaps so, but two people holding the same opinion may often come to opposing conclusions. So the German reaction against rationalism was inspired by the same immensity of the mountains which, to others, suggested the rule of reason.

After the first decades of the eighteenth century, German Enlightenment thought tried to place rationalism in perspective by complementing it with an emphasis on feeling. The *Naturphilosophie* of Schelling (1775–1854) sought the essence of man not in reason but in the capacity of symbolic transformation of experience, such as myth-making or poetry. This *Naturphilosophie* is quite distinct from English natural philosophy, for it has its roots in mysticism and not in experimental philosophy. Having been filtered through the German *Aufklärung*, nature philosophy came to be of great interest to twentieth-century existentialism. What appealed to Schelling, and what appeals to modern existentialists, is the conviction implicit in nature philosophy: the world is essentially incomprehensible and hence unspecifiable by rational methods.

This attitude was expressed in the literary movement called *Sturm und Drang*, a revolt against the French version of the Enlightenment. The followers of this movement admired Shakespeare and Rousseau more than they did the classical French taste. Goethe's poem "Dauer im Wechsel" (1801), one example of many, is a melancholy view of the impossibility of a rational reconciliation of the two, a cry of the homelessness of the romantic soul. Its conclusion is a call for the final negation of the possible infinity of time:

> Lass den Anfang mit dem Ende
> Sich in *eins* zusammenziehn!

Shakespeare, two centuries earlier, had little interest in the starry universe except as a backdrop to the affairs of men and women. Not so for the Romantic poets of the Enlightenment whose feelings were saturated with the love and horror of infinity. To them the heavens, and heaven's earthly images, such as the Alps, were projections of the infinity of creation. Goethe's "Mailied" of 1771 is again an example of the romantic adventure of feeling the immense:

> Wie Herrlich leuchtet
> Mir die Natur!
> Wie glänzt die Sonne!
> Wie Lacht die Flur!

Blaise Pascal, in the mid-seventeenth century, hoped to overcome the indifference of the sceptics by recording his own feelings about the eternal silences of the infinite spaces of the universe which, he said, terrified him. He surely would have accused the deists of the Enlightenment of over-intellectualization and hence, dehumanization of the deity. In England John Wesley (1702–1791) preached man's concern with salvation, while in the land of the *philosophes*, Jean Jacques Rousseau (1717–1778) preached his own anti-rationalist revolution. No supernatural grace for him: instead, reach for the innermost self, arrest the passing time, stress the enduring moment and satisfy the burning human desire to attain a continuous ecstasy of timelessness, a yearning for a state of mind in which time stands still.[49]

He identified such conditions with the affects of pure selfhood and held that it was achievable by the pure, thinking self. He knew intuitively, though he did not use these words, that the ecstasies of timelessness are precious metastable conditions and that once experienced, a person will either perish as an individual or return to the fullness of his self, placed in the irreversible flow of historical time. In his social philosophy he bemoaned that civilizations have lost such experience of timelessness which, he held, was that of the child. In modern terms, he called for a regression of the self from the full temporality of the mature self to lower temporalities which correspond to the worlds of the child and the higher animals.[50]

The pure man-beast, in Rousseau's discourse on inequality (1754) lives wholly within himself, and there is no more urgent task for man than to reestablish this pure and early mental state.[51] The method should be one of *depuillement*, the removing of the foreign layers from the pure self, as one removes the layers of the onion. His intellect failed to note, though his failing mind demonstrated, that if all layers of the onion were removed, there was no onion left. The purely thinking self sought by Rousseau is not the human self any more but a state that resembles that of the pre-human animal.

During Rousseau's last years his persecution mania seems to have lost some of its intensity. In his last work, *Les Rêveries du Promeneur Solitaire*, he felt and reported on the freshness and beauty of nature, while still reaching for the timeless moment as a permanent state. Thirty years later, in the *Studienzimmer* scene, Goethe's Faust sought the same from Mephistopheles:

> Werd' ich zum Augenblicke sagen:
> Verweile doch! Du bist so schön!
> Dann magst du mich in Fesseln schlagen,
> Dann will ich gern zu grunde gehen.

Rousseau, and this particular Faust, were not alone in looking for the timeless moment. Toward the end of the eighteenth century there appeared a large number of Faust stories with the authors divided into the traditionalists and salvationists.[52] They represented two distinct views of time. The traditionalists condemned Faust to hell because he made a bargain with the Devil: such will be the fate of all who do so, when the Final Judgment comes—at the end of time. The salvationists, such as Lessing and Goethe, saw the striving of humanity finally redeemed in a heaven of completeness. Theirs was an eighteenth-century form of salvation history, a linear and progressive story with an open end of unlimited accomplishment:

Wer immer strebend sich bemüht
Den Können wir erlösen.

“Those who ceaselessly strive may be saved,” according to the Angels bearing to heaven the immortal in mortal Faust, at the end of the drama. Heinrich met his Gretchen, albeit in a disembodied form, and henceforth time was no more.

9. The Technical Aftermath

From mathematics, philosophy, and literature, let me turn briefly to some more mundane aspects of our theme.

In 1790 the Académie des Sciences set up a committee to propose a reform of weights and measures so as to make the unlimited progress of Enlightened minds possible. One of its suggestions was the calendar reform adopted in 1793. In retrospect, this effort proved to have been an experiment in trying to manipulate society by influencing the rhythm of collective activities. But less than twelve years after its adoption, France went back to using the Gregorian calendar.

Opinions about the reasons for this failure differ.[53] But it seems that the arbitrariness of all calendars, not ordinarily evident in daily life, did become all too evident to the French. Through the emphasis on secularization, the sanctity of time-reckoning was removed both in the old and the new schemes. Since it was impossible to eradicate the old scheme by decree, double dating remained a fact of daily life, though eventually the pull of the old proved to be too powerful.

What the *académiciens* failed to accomplish in their hurry, modern technocracy is in the process of completing. Our calendars are mainly secular, industrial, and economy-based schemes for counting the passing instants. The concentration of our epoch on the “now” and the impersonal time-madness of the age are all too well known. Judging from the present state of the enlightened Faustian man, the traditionalist writers of the eighteenth-century Faust stories were better prophets. Deism has died but the eschaton of progress survived in deed and, as Gregor Sebba has pointed out, even in style.[54] “Man was born free,” wrote Rousseau, “but is everywhere in bondage.” “The proletariat has nothing to lose,” wrote Marx and Engels in 1848, “but its chains.”

In 1854 George Friedrich Bernard Riemann, then 28 years old, laid the foundations of a formalism in which our universe appears as finite but unbounded. The Encyclopedists would have surely discarded such claims as nonsense. Five years after Riemann's paper, the *Origin of the Species* was published, laying the groundwork for a view of life whose time is open-ended and, in that sense, unlimited.

Voltaire once quoted an unidentified Count C. commenting on the *Encyclopédie*: "Everything is here," said the Count, "from the infinitely small to the infinitely great; the men who made this work served the entire world." Perhaps this anecdote sums up our theme.

The Enlightenment, especially in its eighteenth-century form, was interested in the world because it was interested in man and society. Through the Romantics, both in France and in Germany, it was concerned with infinity because it was concerned with whatever might lift the soul from its painful historical time into a state where time is no more. On the continent and in England, the Enlightenment retained salvation history in the idea of progress, rejected its supernatural trimmings, yet helped identify it with the *summum bonum* of man on earth.

13. THAT AWESOME GIFT

By the 21st century, progress, as the secular version of salvation history, has been accepted in many, though not in all regions of the world. The American Declaration of Independence even asserts that certain truths, related to progress, are self-evident, such as the unalienable right of each person for "Life, Liberty and the pursuit of Happiness." With such philosophical, religious and legal underpinning, one should hold with Robert Browning that, "Ah, but a man's reach should exceed his grasp / Or what's a heaven for?" People must be free to imagine and desire conditions better than their here-and-now and must have an opportunity, guaranteed by the community, to bring those conditions about. Opportunities ought to be as endless as the open sky appears to be.

But, can they be?

During the Christian centuries, questions pertaining to man's grasp and reach, together with questions about what are and what are not within human grasp and human reach, have been discussed in great detail and at great length. The perspectives were those of theology, philosophy, ethics and civil law. The most frequently used heading and theme under which grasp vs. reach were considered, an idea upon which much of Western law and morality is founded, was that of free will. Sometimes it was called free choice, sometimes free decision. The essay that follows examines the contemporary umbrella concept of human freedom, from the point of view of the natural sciences. It identifies human freedom as the most evolved, most advanced form of the open-ended evolution of freedoms and constraints. The inquiry is not directed toward the identification of the rules of right conduct or thought but toward an improvement of our understanding of the nature of human freedom. This is carried out by locating human freedom within the broad scheme of causations, temporalities and conflicts known to the integrated study of time. The essay concludes that as we learn how to use human freedom in ways appropriate for the time-compact globe, the conflicts among the archaic and the newer forms of realities simultaneously present in the mind, as discussed in Chapter 11, are going on with in creasing ferocity.

HUMAN FREEDOM

Abstract

With the help of the hierarchical theory of time, this essay (i) attempts to locate the sources of desires which, in their qualities, are peculiar to humans, (ii) explores the nature of freedoms available for the satisfaction of those desires and (iii) identifies the dynamics of conflicts that arise from possessing unbounded imagination in a bounded universe.

In an orchard near ancient Troy, imagined in Shakespeare's *Troilus and Cressida*, two lovers are discussing their shared concerns. The background to their exchange is the heroic misery of a cruel war. Troilus, son of the Trojan king, holds forth. "This is the monstruosity in love, lady, that the will is infinite, and the execution confined, that the desire is boundless, and the act slave to limit."

If we replace "love" with "human life," three questions pertaining to human freedom, emerge. What are the sources of the boundless desires? In what ways are the acts, driven by those desires, slaves to limit? What is the nature of the resulting conflicts? I will attempt to answer these questions with the help of the hierarchical theory of time, a theory in the natural philosophy time.[1]

1. A Backward Glance

This section places the issue of human freedom into its historical context.

The power of human desires and the limitations to human actions have surely been of concern to our species since we separated from our closest relatives. "I'd love that apple, but can I reach it?" It seems likely that the abstract notion corresponding to such a predicament—to the idea of a freedom of choice with limitations—originated in the recognition that the paths of action one could have taken in the past were, usually, more numerous than the single path one did actually take.

This recognition became articulated in the language of philosophical reasoning invented by the ancient Greeks. Later, beginning with early patristic thought, freedom of choice was seen as a rational faculty of the human soul. Under the name, freedom of the will or free will, it was understood to mean the capacity of humans to turn away from conduct that would threaten their eternal salvation. In that form, the idea remained fundamental to Western thought throughout the Christian centuries. During the Reformation it assumed a central position in the debates between Humanism and the emergent Protestantism. An example of those debates was the dialogue between Erasmus of Rotterdam (1466–1536) and Martin Luther (1483–1546). While Erasmus held that the freedom of the will enabled humans to select and follow paths to salvation, Luther believed that no person can do so by him- or herself but only with the grace of God.[2]

Since the Reformation the locus of discourse on the nature of man has changed. It left the Platonic-Christian tradition as its sole and necessary guide and shifted closer to the views of the first century Latin poet, Lucretius. His belief in the gods of his epoch did not prevent him from maintaining that "nature is free and uncontrolled by proud masters and runs the universe by herself without the aid of gods."[3]

The idea of human freedom, the theme of this paper, is the contemporary version of the idea of the freedom of the will. As did free will, human freedom also pertains to the privileges and burdens of choice among alternate paths of conduct. But its metaphysical foundations are not those avowed by Erasmus and Luther but those endorsed by our contemporary understanding of nature. Also, to be acceptable to critical thought, my approach is formulated as a synthesis that remains open to relevant ideas from all forms of knowledge: science, philosophy, history, ethics, aesthetics and religion. The inquiry is not directed toward the identification of the rules of right conduct or thought. Its purpose is to improve our understanding of the nature of human freedom.

2. The Physical Boundaries of Nature

This section explores the physical boundaries of our cosmic home, our "cosmic cocoon," to use Lawrence Fagg's expressive term.[4] These boundaries are the outermost reaches of human freedom of action, both in principle and in practice. At first sight, they seem to be far from, and irrelevant to a discussion on free will. But, as we shall see, as a family of boundaries they share certain qualities that will be useful to know when, in the following section, we identify our position as thinking and feeling beings in a universe indifferent to human desires.

We begin with considering the natural upper limit of speed. It is the velocity of light, reachable only by particles with zero rest mass, of which photons are the only certain examples. We may imagine infinite speeds of information transfer and with it, instant connections across the cosmos, but there is nothing in nature to which this imagined condition would correspond. Furthermore, "To a hypothetical observer, moving with a photon the whole span of our time would pass in a flash.... The usual interpretation of this curious result is that we cannot associate a 'clock,' i.e. a time-keeping system...with anything that is moving with the critical speed *c*."[5] The photon's world is atemporal. In its reality the temporal world of living organism, and the human experience of time's passage, have no meaning.[6]

To keep company to an upper limit of speed, one may want to seek a boundary of zero speed or complete (local) rest. Particles do move slower and slower as absolute zero temperature is approached, but near zero degrees Kelvin they turn into a quantum fluid in which there is motion even at the theoretical limit of zero temperature.

Our cosmic cocoon also has some cosmic or world horizons. These are information boundaries set by combinations of the expansion of the universe and the finite velocity of light. Their intricate details depend on the model of the universe employed.[7] We may imagine worlds beyond these horizons but there is no way we could learn anything about them.

We may seek boundaries to density such as an upper limit of infinite density in the singularity of the Big Bang. But, we "cannot trace the history of the universe back to infinite density...because the beginning consists perhaps of a primordial chaos of space and time."[8] The lower limit of density would be totally empty space. But vacuum is now preferably thought of as conditions of minimum, energy states and not as the much debated idea of nothingness.[9] Nothing is better understood as boundary to meaning.

There are boundaries to temperature. There is the absolute zero temperature mentioned above but, by the Third Law of Thermodynamics, although it may be approached, it cannot be attained. Reasons have also been given for the existence of a highest temperature, reachable in theory, but not in practice.

There are boundaries to duration. No object may be older than the age of the universe and no meaning may be assigned to durations shorter than the chronon, a span of about 10^{-24} sec. One may think of periods longer than the age of the universe or shorter than the chronon, but there is nothing in nature to which such imagined durations would correspond.

All these boundaries of our cosmic cocoon share two features.

One is that each may be approached gradually along a continuous path. The other is that the conditions found there are unpredictable from the phenomenal world of human dimensions and experience. There is no deductive reasoning that could extend our knowledge of galloping horses or rushing planets to light as a limiting speed; extend our familiarity with the starry sky to world horizons; our seeing small insects under a microscope to quanta;

our experience of the cold of the Arctic to absolute zero. Yet, once the boundary conditions are identified and its features understood—for instance, the limiting speed of light and its consequences, such as time dilation—and we know what to look for, we can identify those peculiarities in the rest of nature far from equilibrium. For instance, the total time saved over a 36 hour airplane trip, attributable to time dilation, is calculated to be around 10^{-8} second. It is there, but rather difficult to notice.

As we insert our living, feeling and thinking species into the immensity of an inanimate, unfeeling and unthinking cosmos, we shall reencounter the peculiar properties of nature's boundaries.

3. The Human Boundary of Nature

This section suggests that the complexity of the human brain represents a boundary of nature, joining the other boundaries so far considered.

What makes the organization of the human brain so remarkable is not the immensity of its component parts—10^{10}–10^{12} neurons—but the much larger immensity of the different ways in which those parts may be interconnected so as to form distinct brain states. The number of those distinct states has been calculated as being of the order of $10^{1,000,000}$. Based on such "hyper-astronomical numbers," Gerald Edelman does not hesitate to assert that the human brain is "the most complicated object in the universe."[10]

Before one can declare that the human brain is not simply complicated but is, in fact, the most complex known object of the universe, it is necessary to develop a universal definition and measure of complexity. Using that measure, a scale may then be constructed, along which the complexities of other natural objects and processes may be placed, for comparison. This is not an easy task. It was only recently that, with the help of algorithmic information theory, it became possible to construct such a measure and, with its help, create a scale of complexities found in nature.[11]

Supported by these calculations, I submit that the human brain constitutes a natural boundary to complexity possible to achieve by biological means.[12]

As humans, we live at prohibitive distances from the physical limits of speed, density, temperature, size and duration, and also from the cosmic horizons. But because we possess human brains, we are at the complexity boundary of nature. More precisely: our brains are.

Let me recall from the preceding section that all of nature's physical boundaries (i) may be approached gradually and also (ii) that the processes found along those boundaries are unpredictable from conditions far from them. I assume that these properties of nature's physical boundaries also hold for its boundary of complexity.

Certainly, that boundary was approached gradually, through organic evolution. But: what is the unpredictable property we find along the human—the complexity—boundary of nature? It is, I believe, the seemingly infinite imaginative powers of the mind and all the consequent desires which, in their qualities, are peculiar to humans. Namely, the ever-present, unsatisfiable need to expand our stores of knowledge felt and knowledge understood. Both biological and noetic knowledge.

I noted earlier that once a boundary to nature and the peculiarities of that boundary were identified and we know what processes or conditions to look for, then we can demonstrate their presence far from the boundaries. This also holds for the imaginative powers of the brain. They may be identified as creativity, already present in the world at large. My suggestion is hardly new: it has been voiced in philosophical and religious terms as the claim that humans share in the creativity of God or nature.

It is disciplined imagination that makes possible the definition of human reality, including the perception of the world in terms of noetic time. It is also imagination that creates, shapes, and maintains the symbolic continuity we call personal identity. Human freedom is the freedom of that symbolic continuity, functioning in noetic time.

4. Freedoms and their Limits

The sources of the boundless desires that concerned Troilus are the imaginative powers of the human mind. In turn, as I reasoned, these powers are hallmarks of the complexity of the humans brain as a boundary condition to nature. With this thought in mind, we are ready to explore the nature of freedoms available for the satisfaction of those desires and the limits to which those freedoms are slave.

We start that exploration by tracing the evolutionary emergence of (i) increasingly richer domains of possible chance combinations for inanimate and living matter and (ii) increasingly broader freedoms of choice for living matter. As we shall see, all the potentialities and limitations so identified are simultaneously present in human freedom. To make our entry into the survey easier, we begin along a devolutionary path. Then we will turn around and climb back up.

When a society collapses, those forms of collective conduct that demand instant by instant coordination among its members—the creation of a social present necessary for collective planning and action—become impossible.

Individual members of the society may well remain around but they will not be able to organize hunts, build bridges or make and enforce laws. A family of freedoms of choice and action have been lost.

When a person loses his mental powers, those forms of his conduct that demand instant by instant coordination of a large diversity of the neural population of his brain—the creation of an a mental present necessary for personhood—become unavailable. He may well remain alive but he will not be able to conduct business, run a household, much less compose a symphony. Another family of freedoms of choice and action have been lost.

When a living organism dies, whether tree, mouse or man, the coordination of its biochemical processes from instant to instant—the creation of an organic present necessary for maintaining life and for goal-directed behavior—ceases to be available. The body may well remain where it died but dead moles do not molehills make. Yet another family of freedoms of choice are lost.

When inanimate matter, whether a dead body or a cold moon, changes into the gases of its elements, the potentialities of matter for forming seas, mountains and galaxies by chance combinations, are lost. No freedom of choice is lost because inanimate matter had no such freedom, to begin with.

When a gas changes into electromagnetic radiation, as in an atomic blast on earth or in an exploding supernova, the possibilities available for gases—such as getting separated by molecular weights—are lost.

In each of these collapses certain degrees of freedom vanished. Obviously, these freedoms had to come into being during the course of cosmic and organic evolution. We will now trace that course along the path of emerging, qualitatively different modes of causations.

Causation is a relationship between two events, called cause and effect. A cause is anything that may be interpreted as responsible for change (its effect). When the form of connectedness between causes and their effects appears stable, the relationship is said to be lawful. These, the laws of nature—different types of causations—determine what possibilities are available to a living system. in the pursuit of its goals. They determine what is and what is not possible to achieve through conduct available for humans.

We begin with the most primitive level of nature, that of *absolute chaos* or *pure becoming*. Absolute chaos—pure becoming—is believed to have made up the primeval fireball and makes up black holes. In such a world "[t]here is no 'now' and 'then' and no 'here' and 'there,' for everywhere is torn into discontinuities."[13] The absence of "now" and "then" from time and the "torn into discontinuities" of space, signify atemporality. It is a form of reality to which future, past and present—categories of time appropriate for living organisms—do not apply. Here, atemporality, just as in the case of photons mentioned earlier, is not to be mistaken for non-existence nor should it be confused with "timelessness."[14]

The pure becoming of the primeval singularity is continued in the cosmic vacuum with its seething sea of creation and annihilation, revealed by quantum physics. Thus, beneath all natural phenomena lurks atemporal chaos wherein no meaning may be assigned to the idea of causation, lawfulness or time, into which all processes and structures may collapse and out of which, under certain conditions, different permanent structures may arise. Because of the nested hierarchical character of nature, pure becoming remains a constituent of all higher forms of causation.

If we are to remain with the evolutionary model of the universe, and employ in our reasoning what we know about nature, then we must conclude that what at the instant of creation arose from the primeval world of pure becoming could not have been the "evening... morning, one day" of Genesis, nor Monday morning and time to get up. It had to be the simplest integrative level of nature, that of particle-waves and prototemporality. In that universe the separation of space from time is incomplete, there are no well-defined points in space nor well-defined instants in time. In the peculiar logic of that integrative level, its denizens are "particles-and-waves" as well as "particles-or-waves." Its level-specific mode of causation is probability.

With the coming about of the prototemporal world *the freedom of probabilistic relationships* emerged from pure becoming. Probability is the most primitive form of lawfulness, it is the simplest form of being that arose from a universe of pure becoming. It is with probabilistic laws that the universe began, then continued along stages of increasingly more sophisticated forms of being or lawfulness. Probabilistic relationships, in the form of statistical laws, remained constituents of all higher order causations. They are present in the behavior of solid matter, of life, of human conduct and of historical change.

As the universe cooled, some of the particle-waves jelled—froze out—to form the ponderable masses of the galaxiesand, with solid matter, a new form of causation came about. It was determinism. To the freedoms of chaos and probability, determinism added a new degree of freedom, the *freedom of stability*, the possibility of identifiable macroscopic structures and processes. Because of the nested hierarchical character of nature, determinism remained a constituent of all higher order causations. All organic, mental and social processes display deterministic rules. The time of a deterministic world, eotemporality, is still undirected but, unlike the time of particle-waves with its probabilistic instants, it is continuous.

The relationship between the two lowest forms of freedoms—the probabilistic and the deterministic—is the substance of Heisenberg's Principle of Indeterminacy. Let me represent that Principle by one of its forms as $\Delta\mathbf{p}\,\Delta\mathbf{x} \geq \mathbf{h}$. The left hand side is the product of the probabilistic spreads of the momenta and of the location of a particle, appropriate in their forms of causation to the prototemporal world. The right hand side is a constant of nature, an unchanging, non-probabilistic quantity. The equation identifies a numerical link between the probabilistic and deterministic worlds of inanimate matter.

The left hand side of the formula informs us of the intrinsically probabilistic nature of the prototemporal world; it represents a principle of indeterminacy. The right hand side tells us that certain combinations of probabilities in the world of particle-waves follow deterministic relations. If the relationship had been named according to the nature of its right hand side, it would have been called the principle of determinacy. It is of course, both.[15]

But note that the constancy (the continuity of magnitude) of the quantity **h** could be so recognized only in a world—and expressed in the language of a world—that is above the prototemporal. As David Park reminds us, "Quantum mechanics is probabilistic in content and classical mechanics is not: no amount of mathematical fiddling will cross the gap, but it is often more conceptual than practical."[16] The conceptual difference becomes a pragmatic difference if we follow the cosmic history of the universe from its initial absolute chaos, and realize that determinism is a more advanced form of causation than is probability.

Neither side of the Heisenberg equation refers to the direction of time. The reason is that the directedness of time, the metaphorical flow involving past and future, requires a "now" as its anchor or reference. But what is known about the inorganic world through physics, cannot be used to define a "now."[17] Yet, the nowlessness of the physical world (eotemporality, in the hierarchical theory of time) does not signify that that world is only a figment of imagination, that it is unreal.

Nowness was introduced into the nowless universe of nonliving matter by the necessary inner organization of life. These assertions need elaboration.

What sets life apart from inanimate matter is its programmed internal coordination whose purpose is the maintenance of the integrity of the organism. In the language of thermodynamics, life is programmed to keep itself an open system. That program consists of the instant by instant coordination among the cyclic processes of systems of oscillators, a coordination whose phenomenal manifestation is the "now." Every living creature creates and maintains its organic present, with respect to which future and past may acquire meaning.[18] There is no similar condition in the inanimate world: no inanimate system is in danger of losing its thermodynamically open condition because it is not so open. The distinction between nowless, physical systems and living systems has its corollaries in their different types of freedoms. Sticks, stones and galaxies cannot select behavior; their fates depend on chance encounters.

Selection among alternate paths of action requires a distinction between futurity and pastness which demands, as I stressed, a present as its reference. Causation native to and characteristic of the world of living systems is organic intentionality. This, *the possibility of choice*, is a form of freedom that life added to the probabilistic and deterministic freedoms of nonliving matter.

Can organic intentionality be identified in the physical world? This has been discussed since Aristotle. In theology it appeared as the teleological argument for the existence of God, also known as argument by design. Since the 16th century, it has also been discussed under the name, physicotheology. In recent times Max Planck maintained that some of the laws of physics, such as Hamilton's principle of least action in dynamics and Fermat's principle of least time in optics, are witnesses to the presence of teleology in the physical world. Whitrow, having considered teleology in the physical world, concluded that (i) goal-directedness cannot be deduced from these principles but also that (ii) one cannot disprove the existence of teleology in the physical world as a consequence of that failure.[19] In sum: organic intentionality and its corollary, freedom of choice, came about with biogenesis. It cannot be identified in the inanimate world, as that world is understood through physical science.

One species, that of humans, acquired an ability to create sophisticated symbolic transformations of experience. It is with its assistance that men and women learned to serve distant future goals, guided in their conduct by individual and collective memories of the past. This advanced form of causation is *noetic intentionality*.

Can noetic intentionality be identified in organic evolution? Teleological arguments, when informed of theology, do sometimes claim that the universe was designed with the intent of creating humans. This is known as the strong version of the anthropic principle. That the behavior of each living organism is goal-directed—and hence could be called teleological—is obvious. Ernst Mayr proposed to describe such behavior as "teleonomic" rather than teleological and noted that teleonomic processes owe their goal-directedness to the operation of programs.[20] Is there a program in organic evolution? One that directs long-term processes towards the creation of humans? No such a program is evident in an understanding of organic evolution through biology.

The modes of causations we have considered—chaos, probability, determinism, organic and noetic intentionality—are the *canonical forms of causation.* They are paradigmatic modes of connectedness among events, they constitute the grammar of the languages in terms of which the level-specific laws of nature are written. All of these causations and all of nature's laws apply to humans. "Biology is destiny"—a broadly cited misattribution to Freud—is too narrow a specification.[21]

Pure becoming, probabilistic and deterministic connections, organic, noetic and social intentionality: all of these, together constitute the freedoms available for the satisfaction of human desires.

5. Multitasking

When Luther and Erasmus debated the nature of free will, the nested hierarchy of causations has already been in place for a few billion years. But its 21st century perspectives were still in the future. The debaters saw the world in terms of salvation history. The universe and man were understood to have been created a week apart, with the physical world serving as no more than a backdrop to the drama of man: fall, incarnation, redemption and henceforth, a path to salvation. This divine teleology was a vast and impressive drama.

The view espoused by contemporary natural philosophy is no less vast nor less impressive, though the roles of man and the universe have been reversed. Human history is now a part of the history of the cosmos of which people are a negligibly small fragment. Salvation history was replaced by the evolutionary story, summed up beautifully by the 13th century Sufi poet Jalal al-Din al-Rumi.

> I died from mineral and plant became,
> Died from the plant, and took a sentient frame;
> Died from the beast, and donned a human dress;
> When by my dying did I e'er grow less?[22]

Having donned human dress, we learned to combine the image of the self with images of future and past. This skill proved useful in the struggle for survival because it enabled its practitioners to seek guidance for their conduct in scenes that lived in their memories. With this help, they could subject possible future scenarios to present tests, then pick the most promising course of action.

The same imaginative powers also informed us of the inevitable demise of the self in death. That reminder conferred upon our species a form of insecurity unknown in other species. Although all advanced forms of life necessarily die, whether by accident or by aging, only in humans has the knowledge of an inevitable end to the self become a generalized malaise, an ever-present tension. That tension makes us seek reassurance against decay, such as in the form of ideas, processes, and things that seem to defy the passage of time. Foremost among such ideas are those of religions. The hymn of Henry Francis Lyte, Scottish born churchman and poet of the early 19th century, sums it up beautifully.

> Abide with me! Fast falls the eventide....
> Oh Thou, who changes not, abide with me!

S. G. F. Brandon, scholar of comparative religions, wrote that it was in the urge "to seek security from the effacing flux of time that I see the origin of religion. I would define my conclusion as follows: religion is the expression of man's instinctive quest for security, which results from the sense of insecurity caused by his consciousness of Time."[23] I suggest that the drive identified by Professor Brandon is also responsible for the creation of philosophies, the arts, the letters, the sciences that is, for the totality of human cultures.

It is within cultures that humans consider their own births and deaths, those of other humans and those of other forms of life. It is also within the furnishings of cultures that, by extension, men and women ask questions about the possible birth and death of the universe, including the possible beginning and/or ending of time. In Book IX of *The City of God*, Augustine gave reasons for his belief that "the world and time had one beginning, and one did not anticipate the other."[24] By "time" he could only have meant noetic time. He did not anticipate quantum theory with its probabilistic instants, general relativity theory with its temporal horizons or special relativity theory with its atemporal photons. If Augustine was correct then Bishop Ussher, placing the day of Creation on October 23, 4004 B.C. might have been wrong in his calculations but not in his principles.[25] The same holds for the "evening... morning, one day" of the Book of Genesis, already cited. Although the biblical assessments of the date of the origin of time and the world may not be acceptable to contemporary thought, it remains, nevertheless, the almost universally held view. It holds—to use contemporary idiom—that when time was born, it hit the ground running.

In the conclusion of his classic *The Natural Philosophy of Time*, Whitrow contrasted the hierarchical theory of time with the view he himself maintained, one that corresponds to traditional understanding. He wrote that "at all levels time is essentially the same, although certain aspects of it become increasingly significant the more complex the nature of the particular object or system studied."[26] The millennia old discussion about free will has maintained that, since Creation, time remained an aspect of the world independent of the furnishings of the world.

At variance with the view that when time was born it hit the ground running, the hierarchical theory of time maintains, in the words of Alexander Argyros, that time "is no longer[to be] understood as a background for reality, nor simply as the human experience of flow, but as an evolving palimpsest of emergent temporal levels constitutive of reality."[27] Such an understanding amounts to a radical change in some broadly held views about time. The change is not the much-heralded replacement of Newtonian absolute time with the time of relativity theory, because both Newtonian and Einsteinian times are Platonic. There is "no suggestion in either theory" Argyros continues, "that, over and above its plasticity and relativity, time is anything but a Platonic essence whose basic features have remained constant since creation." Both theories assign qualitatively identical temporalities to the probabilistic world of particles, the deterministic world of ponderable matter and to the worlds of organic, noetic and social intentionality.

The radical change in perspective is the recognition that, in the reality of the most complex known system, that of the human brain, the nature of time is unique, unshared and unsharable with other species. Noetic time, a necessary reference of human freedom, is an evolved form of temporality that could not have been present at Creation any more than human brains could.

The change, from Platonic (Newtonian and relativistic) to the hierarchical theory of time, makes an interpretation of human freedom more difficult. For, in this new view, we are not only the actors in, and audience of the human drama. We are also the makers of the stage—that of nootemporality and biotemporality—upon which we see our selves move and our lives unfold. It is the human mind that formulates attitudes toward future and past, then surveys and interprets history in their terms. It is within that interpretation that the mind situates the identity of the body of which it is a function, defines our goals, and tries to govern our conduct. This is multitasking of a weighty kind. The security of a God-given Newtonian and relativistic time is removed. Now we are in the world of Lucretius, cited in the opening section of this essay, where "nature is free and uncontrolled by proud masters and runs the universe by herself, " with ourselves as the observers, the observed and the stage designers.

6. The Permanent Revolution

We left Troilus telling Cressida about infinite desires being slaves to limit. He did not have to wait long for an example. Cressida fell for a handsome Greek no sooner than she swore eternal love to Troilus; she even gave the Greek an arm band, a token of commitment, one she received from Troilus. Whose was it, the Greek wanted to know? "Twas one's that loved me better than you will, But, now you have it, take it." In the background, the heroic war turned to be an endless slaughter and suffering, a "clapper-clawing," a life in "vile, abominable tents." The great heroes of Homer prove themselves to be course, ignoble opportunists. All characters progressively diminish in a storm of folly and weakness. Cressida's uncle, who served as the go-between the Trojan hero and the Greek beauty, seems to have acquired venereal disease. The play ends with his lonely plans to "sweat and seek about for eases" and eventually, "bequeathing you my diseases." Ecce homo.

The reasoning in the preceding sections had its inspiration and sources in a humanistic interpretation of the natural sciences. How may these, sometimes abstract, arguments be linked to the concrete men and women, such as those of the Bard of Avon, people who copulate and murder with gusto, who pray and curse, cry and laugh? Trough human values, through that portable kit whose tools we use to manage the images of the $10^{1,000,000}$ different potential brain states. Here are four examples of those values, summed up in Joe Darion's lyric in the play, *Man of La Mancha*.

> To dream the impossible dream,
> To fight the unbeatable foe,
> To bear with unbearable sorrow,
> To run where the brave dare not go...

Learning to live with impossible dreams and acting upon them could never have been easy. Guidelines on how to use human freedom were surely always sought. Since Greek antiquity, those guidelines have been classed under the categories of the true, the good, and the beautiful. All three categories of values need the human sense of time.

Truth may be defined as the recognition of permanence in reality; it serves as a guide to what to believe. Goodness may be defined as an assertion that certain conduct, intent or character trait will promote lasting harmony in the minds and affairs of a person and in the dynamics of society; it serves as a guide to conduct. The beautiful and the ugly are guidelines to the sorting out and managing of emotions. If the quality of feelings mobilized by external or internal events or conditions is such as to make one desire their perpetuation, then whatever is believed to be responsible for those feeling is judged beautiful. If the quality of feelings is such as to make one desire them to be absent, then whatever is believed to be responsible for that feeling, is judged ugly.

Details of what specifically are thought to be true, good and beautiful vary immensely. Still, as a group of guidelines, they have been regarded as conservative forces that bring balance into the personal and collective affairs of humans. Yet, by the very nature of human knowledge, testing for truth cannot ever be complete and hence it can never come to a halt. Likewise, there are no examples of instructions for right conduct that remained valid, unmodified, for historically significant periods of time, and no aesthetic judgments that remained stable for historically significant lengths of time. It follows that in matters pertaining to the beautiful, as with issues of truth and goodness, the search must go on.

When this reasoning is critically examined, it becomes obvious that, contrary to broadly held opinions, human values are not conservative but revolutionary forces. Their long-term roles have been the creation of changes in the very teachings which they are said to uphold.

We are now in a position to return to the question asked in the very beginning of this paper. What is the nature of the conflicts that arise when the desires, created by an unbounded imagination, are set against the boundaries of human freedom?

The answer is: they drive human conduct to ceaselessly new adventures, because satisfying them can offer only temporary relief. They create and maintain an ever-present yearning for more lasting truth, more stable and satisfying goodness, more complete beauty. These needs, these peculiarly human yearnings play the same role in the noetic world as hunger does in the organic world. In living organisms, recurrent hunger is maintained by the conflicts between the self-organizing and decay processes that are always simultaneously present. These conflicts, "increasing store with loss, and loss with store" remain continuous as long as an organism is alive. The hunger for human values as guidelines to conduct is due to the conflicts that define mental life and that continue as long as a person maintains his identity. Both the organic and mental conflicts are but specific examples of the chronic incompleteness and incompletability of nature.[28] It is thus that human freedom is responsible for a permanent revolution.

Indeed. Summed up in the words of Prof. Paul Harris, more concisely than the author of the idea could have done, "contentions over values create conflicts that drive history."[29]

A visual representation of the conflicts perpetuated by human values may be found in a 1924 engraving of the German artist, Käthe Kollwitz. Its title, in free translation, is "No more wars!" But the female figure is so hell-bent on achieving her goal of stopping wars that she is ready to go to war to achieve it. It is no wonder, then, that the true home of the mind, seeking rest, is not the ever-changing present but the non-present categories of time: the yet to come, and the long ago. And the non-present category of space: the far away.

The pursuit of inner peace in a world of conflict is helped by a paradoxical privilege: human freedom includes the possibility of giving up that freedom. I would like to represent that capacity by a musical example or, failing that because of technical difficulties, by words that speak for the musical example.

Käthe Kollwitz, Nie Wieder Krieg. *(1924). Lithograph. Gift of Richard A. Simms. Reproduction, courtesy National Gallery of Art, Washington., D.C.*

In Verdi's *La Forza del Destino*, as in "Troilus and Cressida," whatever can go wrong, does. Leonora, abused and haunted for deeds others—or no one—has done, is ready to exchange her human freedom for the security of heaven. *Pace, pace, mio Dio*! Soon thereafter she is murdered by her brother, essentially because no one understood anyone else. Alvaro, her equally star-crossed lover, is ready to deliver a voluble curse upon fate but is interrupted in that search for relief by the Father Superior. *Non imprecare, umiliati....* "Do not curse: humble yourself before Him who is just and Holy.... You see this angel [Leonora] is winging her way to the throne of God."

Decisions in matters of value spread between Kollwitz' s "No more wars!" and Verdi's "humble yourself." Between taking up arms against a sea of troubles, and suffering the slings and arrows of outrageous fortune.

If dealing with human freedom by means of human values includes the perpetuation of conflicts and their accompanying misery, would it not be advisable to function without them? That is not a possible alternative. Mental powers are defined by, they are identical with the conflicts of human freedom, just as simultaneous self-organization and decay are identical with the life process. Living without the conflicts of human values would not be human life. Robert Browning had it right when he wrote that "a man's reach should exceed his grasp, Or what's a heaven for?"

7. A Forward Glance

In Section One I wrote about Erasmus and Luther. Theirs was an age of great socio-economic changes as well as great ideological differences pertaining to the interpretation of the message of Christianity. In retrospect, though long and bloody, those disagreements appear to have been but dress rehearsals for the conditions and disagreements of our days. The Reformation was a problem only for Christian Europe, whereas today's conflicts involve all religions and all ideologies. Our epoch involves economic imbalances on a scale unimaginable in the sixteenth century as well as cultural upheavals across a restless, globalized earth.

In this hubris, the sense of the holy is being relocated.

W. Norris Clarke, S.J. noted that the "fascinating task-challenge-opportunity for natural theology today is to speculate imaginatively as to what the 'personality' or 'character' must be like of a Creator in whose image this astounding universe of ours is made, with its prodigal abundance of energy, its mind-boggling complexity, its fecundity of creative spontaneity, its ever-present fluid mixture of law and chance, etc."[30]

To find the contemporary holy, remarked the Greek novelist and poet Nikos Kazantzakis some time ago, we ought to revise our perspectives. "It is not God who will save us—it is we who will save God by battling, by creating, and by transmuting matter into spirit."[31]

Our collective experiments in the use of human freedom, in search for a workable balance, vacillate between militant piety and moral nihilism. The stunning success of the technological tooling of human functions employed, as they are, without guiding principles, demands that the natural environment adapt to humanity's bootless dreams. But humans can change faster than non-humans, we are outrunning nature, our environment is becoming fragile.

There is no world war but there are many local wars. Some are between nations, others among ethnic groups, some between religions and there are many local skirmishes among mercenaries of commercial interests. While technology, communication and economic needs are pushing the globe toward homogenization, the ceaseless wars create and maintain fragmentation. These conditions permit the destructiveness of our species to become efficient and keep up, as different but equal, with the many remarkable examples of the human readiness for sacrifice and good will.

Upon our high-tech, tightly wired planet, nightmare-like fantasies are played out and, by those acts, become acceptable forms of conduct in individual, national and international affairs. In the global supermarket food, water, women, men, children, infants, software and drugs flow in all directions, in support of commercial and military interests. So does stolen scientific information, investment and medication. There is a trade in body parts. Men sell their kidneys, women rent their vaginas and their wombs to get money for the daily needs of survival of their families and themselves. Every nation makes and transports arms with no regard to their origins, destination or intended use. In this global wheeling and dealing the difference between the haves and have-nots increases while Marx mumbles in his grave, "It is very different and much worse than I thought."

As we learn how to use human freedom in ways appropriate for the time-compact globe, the conflict between the noetic and the organic is going on with increased ferocity. Almost exactly two-hundred years ago, William Blake demonstrated his sensitivity to the human predicament as he penned his *Songs of Innocence and Experience, shewing the Two Contrary States of the Human Soul.*

> In every cry of every Man,
> In every Infant's cry of fear,
> In every voice, in every ban,
> The mind forg'd manacles I hear.

14. OPIATES THAT CIVILIZE

The essays of this book, so far, addressed time-related issues in the physical universe, in the life process, also those present in the human mind and—in a single context—in society. The preceding essay went so far as to place human freedom in the general scheme of other forms of freedom. Since fools rush in where angels fear to tread, it is now appropriate to ask: how can we live with human freedom, "that awesome gift"?

In 1844 Karl Marx completed his *Critique of the Hegelian Philosophy of Right*. In its introduction he wrote that "religion is the general theory of this world, its encyclopedic compendium, its logic in popular form, its spiritual *point d'honneur*, its enthusiasm, its moral sanction, its solemn complement, its universal basis for consolation and justification." Then follows an epigram that is often cited but hardly ever in context. "Religious suffering is at the same time an expression of real suffering and a protest against real suffering. Religion is the sigh of the oppressed creature, the feeling of a heartless world and the soul of soulless circumstances. It is the opium of the people."[1]

Marx failed to give a reason or even speculate about the likely reasons that drive people to seek opiates, to begin with. That reason was identified in the writings of S.G.F. Brandon, a professor of comparative religions. The history of religions demonstrates, he wrote in 1972, "that the basic motivation in religion itself is the quest for security from death or the consequences of death."[2] This is also the central theme of his *History, Time and Deity*.[3] In the lyric words of the 17th century poet, Robert Herrick,

> Gather ye rosebuds while ye may,
> Old time is still a-flying,
> And this same flower that smiles today
> Tomorrow will be dying.[4]

In terms of the hierarchical theory of time—and stated in its shorthand—what people seek relief from, through the consolation of religions, is the tension between time felt and time understood. That tension is the subject of the essay that follows. The formulation that sees the tension between time felt and time understood as in need of "opiates of the people" permits the enlargement of the family of medications. It suggest that we recognize other potentially civilizing opiates among which, the essay maintains, are the arts, the letters and the sciences. For all these continuously mediate the conflicts between time felt and time understood.

TIME FELT, TIME UNDERSTOOD[5]

Abstract

The experience of time's passage is intimately familiar, the idea of time is strangely elusive. Mature, healthy humans find it is easy to act consistently with the notions of tomorrow, yesterday and today. Yet, explaining what is meant by future, past, and present, without assuming prior familiarity with time, seems impossible. The asymmetry between the obviousness of the experience of time, and the unobviousness of the idea of time, has been a source of perplexity to reflective thought for at least fourteen centuries.

This essay sees the sources of that asymmetry in the evolutionary structuring of the cognitive capacities of the human brain. It draws attention to certain conflicts rooted in the differences between those cognitive capacities—copresent in the mind—and notes the importance of the humanities in the management of the conflicts.

The experience of time's passage is intimately familiar. Present, future, and past and the flow of time appear to be simple, obvious aspects of the natural world. Yet, trying to explain to someone, who does not already know what is meant by the passage of time is not simply difficult, but it seems impossible. Why so?

This essay proposes an answer to that old question, formulated in terms of a contemporary understanding of man and the world.

First, it searches for the field of learning where the inquiry ought to be anchored, and launches the inquiry. Based on what is learned, it distinguishes between knowledge felt and knowledge understood in general and time felt and time understood, in particular. Next, it recognizes certain unresolvable conflicts between these two forms of familiarity with the nature of time. It concludes by identifying the arts, the letters and the sciences as collectively developed means for dealing with—being able to live with—those conflicts.

1. Situating the Inquiry

In what domain of learning should we anchor our inquiry into the reasons that make experienced time so obvious and the idea of time so unobvious?

Eastern and Western philosophies have a history of at least twenty-four centuries of disciplined speculation concerning the nature of time. But, since the Renaissance, philosophical reasoning came to depend on non-philosophical forms of learning to such an extent as to challenge the validity of purely speculative thought. Therefore, let us listen to philosophy but not anchor our inquiry there.

The notion of time has long been associated with such other concepts as those of God, death, fate, faith and eternity. Should we anchor our inquiry in religion and theology? I do not believe so, because these forms of learning concentrate on purpose, history, hope, the supernatural, life after death and rules of conduct and cannot accommodate what we know of the lawfulness of the physical, the organic, the mental and the social worlds. Nor are they sympathetic to the kind of critical testing and analysis that contemporary explanations demand. Let us, therefore, listen to religion and theology but not anchor our exploration in their domain.

For interpretation of all things knowable, contemporary judgment prefers to appeal to the sciences. They have been placed on a pedestal that, in past ages, was reserved for high priesthood and witchcraft and served, as do the sciences today, as starting points of exploration. It is, indeed, to the sciences, especially to physics, that the reading public turns for information about the nature of time. But physics has no epistemology that could accommodate its indigenous teachings on equal footing with the human experience and feelings for time's passage.

We might look to any and all other fields of learning: to history, sociology, music theory, psychiatry and end with the clockmaker. She will send us back to the philosopher and the theologian.

This predicament reveals the awesome capacities of the human mind, functioning in a universe of infinite wealth. Two millennia ago that wealth was recognized by the apostle Paul, in a context appropriate to his epoch and place. He wrote that "There are... so many kinds of voices in the world, and none of them is without significance. Therefore, if I know not the meaning of the voice, I shall be unto him that speaketh a barbarian, and he that speaketh a barbarian to me."[6] So as not to remain a barbarian when it comes to considering the obviousness of time as experienced, and the unobviousness of time as understood, one will be well advised to be pragmatic: acknowledge the many-sidedness of the problems and have confidence in the possibility of appealing to any and all of the different domains of knowledge, as needed.

An attempt to make that kind of inquiry possible was the founding of the International Society for the Study of Time in 1966, of which *KronoScope* is a publication. Within a few years it became clear that the creation of an intellectual framework that permits contributions to the study of time from all fields of knowing had certain systematic problems. Namely: each of the fields upon which one must draw, has its jargon. Each has its peculiar rules as to what constitutes necessary and sufficient proof for truth or fact. Also, the different modes of knowledge have, what may be described as, different personalities. The different demands and personalities become built into the ways of thinking preferred by those in each field and are often incompatible.[7]

It also became clear that the differences in methods and languages are not arbitrary. They are demanded by the fact that Nature is divided into distinct orders of organization and complexity. It follows that the pluralities of jargon, proof and personality should not impede the development of an integrated study of time but serve as guides to it.

Because of the nested hierarchical organization of nature, each of us belongs to the world on several levels. We are made of matter; as living beings we belong with all other living organisms; as thinking beings, with all people; as social beings to many kinds of social groups. These different ties are the domains of different sciences: physics, biology, psychology and sociology. The humanities are the glue, the method, the means, the protector and the enhancer of the sciences. They cover all of them as the hen covers her eggs.

What is—what was—needed was a natural philosophy that allowed for the legitimacy of different jargons, proofs and personalities and permitted their interactions. Such a conceptual framework has been developed, its foundations in natural philosophy, and its propositions have been elaborated.[8] Here I wish to single out only one of those propositions because it is relevant to our theme. Namely, that "time," as time is thought of to signify the human experience of passage, comprises a nested hierarchy of qualitatively different temporalities. Not that time may be associated with different concepts but—it is important to repeat—that it comprises of qualitatively different temporalities.

Suggested by the recognition in nature of a nested hierarchy of qualitatively different temporalities, and for reasons that will become clear as we progress, I now turn to evolutionary biology and neurology as my primary, but not exclusive guides to distinguishing between, what will be called, time felt and time understood. Hopefully, at the end of my reasoning, the reader will carry the conclusions back to his or her special field of knowledge.

2. Knowledge Felt and Knowledge Understood

Humans have an anatomical structure that is sensitive to the dynamics of the total organism. It is not itself a receptor but the correlator of all receptors. It is this structure—the central nervous system—or rather, its coordinating organ, the brain, that I now propose to consider.

The vertebrate brain evolved by growing new structures around and about old ones. In its most advanced form—that of the human brain—it became a system of immense complexity, comprising some 10^{10} neurons with some

10^{14} synapses.[9] This population is organized into three layers. They correspond to stages in the evolution of the anatomy of the brain and in the associated stages in the evolution of its cognitive sophistication.

The oldest, in sequence of appearance on the stage of history, are the brainstem and midbrain. Together, they are sometimes called the reptilian brain. The reptilian brain is responsible for the working of reflex systems and for inframammalian cognition. The next layer, sometimes called the mammalian brain, comprises the hippocampus and the limbic structure. It controls instinctual drives and motives. The outermost and uniquely human structure is the cortex, possessing the speech and associative areas of the brain. These layers retained the functions that correspond to their origins in the human lineage. Each does for the human body what it did for its archaic owner before the later layer(s) evolved. Each level retained its peculiar way of assessing reality and hence formulating—gathering and processing—knowledge.[10] The older parts of the brain may be said to command knowledge felt, the newer parts knowledge understood. It is this architecture that makes possible the telling of the pain of a toothache from the pain of personal loss, or the ecstasy of the bower from the sublimity of metamathematics. Distinguishing between the older and newer layers of the brain—upon which my reasoning is built—is amply justified. But one must keep in mind that they share the same skull, that they continuously interact and have been evolving together. What emerges into consciousness is a fluid summation of both the old type and new type assessments of reality. Their collective behavior is governed by the rules of psychodynamics, appropriate to individual motives and drives.

3. Time Felt, Time Understood, and their Conflicts

The different judgments of reality made by the different layers of the human brain extend to the assessment of the nature of time.[11] To help introduce the notion of qualitatively different temporalities, let us think of walking back along the evolutionary stages of our species. I would like to use a formula used by St. Paul, but in a manner he would have found unacceptable. He wrote that "When I was a child... I understood as a child, I thought as a child...."[12]

When, four million years ago, I was a mature, fully grown Australopithecus, a Southern Ape-man, my awareness of time was limited to a capacity of dealing with immediate challenges. I could also employ a hazy awareness of a narrow horizon of future and past, superimposed upon a biologically programmed behavior.

When 3 billion years ago my life was vested in the protists—in fungi, algae, and protozoans—I did have a biologically patterned temporal behavior but I had no sensibilities that could have helped me interpret the world in terms of the passage of time.

When, before the hazy origins of life some three and a half billion years ago, my ancestors were oscillating crystals, the reality of time was that of the physical world, without a definable "now."[13]

To our backward walk along the path of organic evolution let me add a forward walk along the evolutionary steps of the brain.

The temporal reality of the oldest, the reptilian part of the human brain, may be imagined as a world of cyclic variations. Predictability, demanded by need satisfaction ("the next breath cannot wait very long") extends only to an immediate future. Individual memory, related to satisfaction or danger, extends only to an immediate past.

The temporal world of the mammalian brain may be imagined as having narrow horizons extending to "soon" and "gone". For the meeting of contingencies the oldest levels of the brain have only such skills available as natural selection has distilled in the biological endowments of the body.

Only for the uniquely human level of the brain does time possess open horizons of past and future, populated by long-term individual intentions and memories.

One of W. H. Auden's poems captures the dynamics of this process, if not with scientific precision, then certainly with powerful poetic sweep.

> Sessile, unseeing
> The Plant is wholly content
> With the Adjacent.
>
> Mobilised, sighted,
> The beast can tell Here from There
> And Now from Not-yet.
>
> Talkative, anxious,
> Man can picture the Absent
> And Non-existent.[14]

When we react to change or permanence in the environment or contemplate the world in terms of time, we enlist the evolutionarily older as well as the more recent levels of our assessments of the nature reality. They differ, as I sketched, in their temporal horizons and cognitive sophistications. Now we are in a position to ask about the relationship between time felt and time understood.

The tasks of the deeper regions of the brain are the semi-autonomous controls of biological processes. They are unavailable for cognitive examination. I cannot explore the timing of my heart rate or the programming of my digestion except indirectly, through the use of instruments and theorems. This unavailability makes it difficult or even impossible to give a complete description of experiences we call feelings. Sixteen centuries ago the Bishop of Hippo could not explain "to someone who asketh" what time was, because what he would have had to explain was a form of knowledge felt. We can write volumes about pain, love making or the experience of time's passage, but the experiences themselves remain of the character of personal knowledge. Only once so known, may then we refer to them, using our skills for the symbolic transformations of experience. This answers the question, "Why so?" asked in the first paragraph of this paper.

The tasks of the newer regions of the brain are those of speech and associative functions. By time understood I mean the temporal dimensions of the world as perceived and hence, defined, by these newer regions. It includes our awareness of our aging and eventual death, as well as an appreciation of the temporal organization of societies and cultures. The logics or laws of these, the noetic functions are alien to the older layers and are, therefore, inexpressible in their language. It is impossible to explain to a reptile what we mean by tomorrow. I cannot lessen my present hunger by being told that next month I shall have food. For the primitive layers of our brains, time's open-ended passage does not exist.

The relation between the newer and older regions of the brain are hierarchical: we can feel without thinking but we cannot think without feeling. Time understood subsumes time felt. This asymmetry often finds expression in poetry, as in these lines of Dylan Thomas.

> Time held me green and dying
> Though I sang in my chains like the sea.[15]

Our organic selves are always green and are always dying. In the words of another poet, "from hour to hour we ripe and ripe / And then from hour to hour we rot and rot...."[16] While all this simultaneous greening and dying happens to our organic selves, our noetic selves, functioning in the reality of time understood, are trying to free themselves from the limitations of the ripening and rotting of their bodies. They "sing in their chains like the sea." They reach for an imagined, timeless, indestructible life in peace, health and well-being. And thereby hangs a tale.

Namely, there are conflicts in human life that are successfully resolved, day after day, minute after minute. For instance, the left- and right-eye-views of the visible world are different. Their differences are continuously mediated by the mind; their conflicts (contradictions) are resolved through the perception of a three-dimensional space. But time felt and time understood are not two different views of the same, temporal dimension of the world. They are reports about two qualitatively different temporalities appropriate, respectively, to a more primitive and to a more sophisticated evolutionary state of the brain.

Let me illustrate the two different, copresent, hierarchically related temporalities, displaying their autonomies, by two realistic examples.

The body of a man with metastasized cancer, functioning in its reality of time felt, is ready to die, while the symbolic construct known as his self, wants to keep on living in its reality of time understood. The soma has no choice, the person dies.

At the other extreme: the body of a suicide bomber is ready to keep on living in its biotemporal world, while the mental construct of the noetic self wants to die. The person has a choice, difficult as it is, because of the presence in his mind of a collective assessment of time, known as a historical or sociotemporal perspective.

There is a universal conflict, an ever-present one that is surely coeval with the human mind. It is that between the knowledge of the certainty of death (a noetic piece of information about our bodies) pitted against the dream, in time understood, of a fulfilled, complete life without its earthy problems. Either here on earth or, if that is not possible, then somewhere else. This conflict vanishes only with the death of the body or that of the mind. It cannot be solved in the sense of making it disappear. It may only be maintained as a perpetual malaise, an ever-present tension that makes us seek signs of permanence—of deathlessness—in the hope of decreasing that tension. This search is a trait characteristic of our species and is better known as conscience. Hamlet's claim that conscience makes cowards of us all is correct but incomplete. Conscience makes cowards, or heroes, or any of their infinitely many possible amalgams of us all.

Practical assistance in exploring those infinitely many ways comes from the great cultural continuities of mankind, from experimental tests of reality performed in imagination, known as…

4. ... THE ARTS, THE LETTERS, AND THE SCIENCES

Time felt and time understood are, so I reasoned, evolutionary stages in the assessments of the nature of time. They are simultaneously present in the mind in a nested hierarchical fashion. If explored independently, they show mutual exclusiveness. Thus, if I increase the keenness of my understanding of the nature of time I end up with imposing intellectual structures, such as that of General Relativity Theory, from which all hallmarks of the human experience of passage are absent. That experience must be brought to the theory, it cannot be derived from it.[17] If, instead, I attend more and more to my feelings of passing time, I arrive at experiences best described as mystical. I will be hearing unrepeatable voices and experience indescribable ecstasies, from which all hallmarks of my noetic identity as a person are absent.

The human experience of time, in its everyday sense, is a balancing act between these two extremes. It is an inner selection pressure. It is a permanent call for watchfulness to protect our integrity as individuals, such as by maintaining the conflict between time felt and time understood, without becoming unthinking zombies or abstract heads unfit to survive. The arts and the letters are collectively created and controlled experiments offering guidance for that balancing act. They explore all imaginable scenarios, while involving both knowledge felt and knowledge understood. Then, it is up to individual men and women to internalize those explorations—to learn from them—and select paths of behavior based on their judgments of human values. Let me turn to details.

The tapestry of shapes, motions, utterances, colors, sounds, rhythms and tempos found in the arts and letters, carried by an uncountably numerous host of narratives, delivered in tongues alive and dead, and kept coherent by meaning, is overwhelming. Their oldest and most general form, the mother lode of the arts and letters, is dance. Music is the audible dimension of the dance; poetry is dance mapped into words; prose is poetry changed to shuffle, walk, or run; painting, sculpture and architecture are snapshots of dance steadied by meaning; film is dance, music, literature and painting in virtual reality. Each and every encounter with any lode of the mother lode—whether or not the cognoscenti would judge it an example poor or outstanding—entices us to scan our store of thoughts and feelings. These inner journeys join our memories and, together with them, form a network of paths or maps. From them we may creatively select patterns of balance between time felt and time understood and, combining them with our native ideas, plot our plans for conduct both short and long-term. We may even select ways in which we wish to modify our memories. Such plots, though they may be chiseled in marble, are never final. They must be continuously redrawn in terms of new aspirations, values, and interpretations.

The sciences may be regarded as forms of the humanities, following certain rules of their own—as do all the arts and letters. Although early science was a mixture of inquiries into the nature of Nature and interest in (what later came to be called) technology, modern science originated in the desire of finding out about God's plans and logic. This is from an ode by the mathematician Edmond Halley, "prefixed" to the *Principia.*

> Lo, for your gaze, the patterns of the skies!
> What balance of the mass, what reckonings
> Divine! Here ponder too the Laws which God,
> Framing the universe, set not aside
> But made the fixed foundations of his work.[18]

To the store of humanities let me, therefore, add the forms of knowledge known as the sciences. They differ from the arts and letters only in their shared and necessary assumption that the universe is lawful, that these laws may be identified by humans, and that the cosmos follows a certain stable logic. They share with the humanities their spirit of exploration, their search for coherence by plot that is, by meaning (known as hypotheses) and their readiness to mine their imagination. To the immense store of imaginary scenarios offered by the arts and letters, the sciences add their tests—with their validities forever challenged and subject to change—as to what is possible, difficult or impossible for humans made of matter, possessing life and possessed by ideas.

In 1804 William Blake concluded the introduction to his *Jerusalem* by remarking that "Nations are Destroy'd, or Flourish, in proportion as Their Poetry Painting and Music, are Destroy'd or Flourish! The Primeval State of Man, was Wisdom, Art, and Science."[19]

Since in the mind of man there can be no closure to the conflicts between time felt and time understood, no civilization can be maintained without the benefits of its arts, letters, and sciences continuously mediating those conflicts.

15. HOW TO PERPETUATE CONFLICTS

As already and amply stressed, the Enlightenment retained the perspective of salvation history but secularized it into the notion of endless progress, a belief now randomly rampant around the globe. A parallel fate befell philosophical and theological precepts of faith, conduct and feeling. A narrow sampling of them became the guiding principles of fundamentalist movements. An extended sampling came to be subsumed under concerns with the classical notions of the true, the good and the beautiful.

According to received teachings and representable by an image dear to this writer, believing what is judged to be true, conducting one's self by what is thought to be right and guiding one's emotions by standards that favor the beautiful will lead the practitioner to a country of green pastures and shining rivers. To a land where sheep may safely graze, where the music played is that of Johann Sebastian Bach and where the company one keeps is that of all those one ever loved, and where explanations of passion and knowledge are not needed because everything is obvious, necessary and free.

So far the hope. But a prosaic examination of the roles of human values in the ecology of survival comes to a very different conclusion. Namely, that living by the principles of the true, the good and the beautiful may, indeed, help one navigate successfully among the gifts and dangers of human freedom—for the short term. But the long-term effects of commitments to human values are the opposite. To wit, they generate and maintain conflicts which give rise to, and perpetuate, both the stunning creativity and frightening destructiveness of our species.

CHANGE, PERMANENCE, AND HUMAN VALUES *

Abstract

The nature of the true, the good, and the beautiful have been of continued interest to reflective thought at least since the birth of dialectical reasoning in ancient Greece. In contemporary scholarship, the study of what constitutes truth is of concern mainly to epistemology, the scope and nature of the good is dealt with in ethics, the ramifications of what is judged beautiful is a subject of aesthetics. Although these three domains of critical thought embrace a substantial portion of the world of the intellect, it is difficult to find within their boundaries a recognition of the significance of time awareness in the generation of human values.

This paper examines the role that the experience and idea of time have played in the creation, maintenance, and revision of values. The framework for the inquiry is the hierarchical theory of time.

The need for human values is seen as arising from the necessity to control the unbounded imaginative powers of the mind. That control takes the form of collectively approved guidelines for beliefs, formulated in terms of truths and untruths; of guidelines for conduct, stated in terms of right and wrong; and guidelines for the management of emotions, in terms of the beautiful and the ugly.

To relate human values to time and remain consistent with the general usage of the terms, truth is defined as a predicate which asserts that a belief is judged permanent; good is defined as a predicate which asserts that certain conduct, intent, or character promotes stability in the affairs of man and society; beautiful is defined as a qualitative judgment upon feelings which we would like to remain lasting. When so understood, human values are seen to join the perennial attempts of our species to oppose the passage of time.

Normative teachings about values do so in many ways, among which their bid for permanence is the most obvious. They generally maintain that believing in what they say is true, doing what they claim is right, and appreciating what they judge beautiful will help achieve stability in the life of the individual and society here on earth, or at least for the person post mortem, elsewhere. But the witness of history suggests that value judgments as guides to thought and action—such as the pursuit of truth through science, justice through law, or beauty through the arts—have never secured lasting balance among the passions of mind, body, and society.

The purpose of the paper is to defend the thesis that the role of human values in the development of civilizations has not been one of leveling. On the contrary, human values have helped generate and keep alive those unresolvable conflicts of the mind and of society which have been responsible for the stunning creativity and frightening destructiveness of our species.

This paper interprets the nature of human values in terms of the hierarchical theory of time. It concludes that although value systems often claim that abiding by their judgments would secure stability in the individual and communal affairs of people, the evolutionary role of value judgments has been one of creating and maintaining instability and thereby promoting social change and individual development.

The Need for Value Judgments

The need to select responses appropriate to the immediate challenge of food, friend, or foe must be coeval with life itself. Such responses, useful for the perpetuation of the life of the species or the survival of one of its members, may be said to be of value for that species or for its member. This meaning of value also applies to men and women insofar as they are biological organisms. Protecting a child, for instance, involves patterns of behavior that humans share with all other species that protect their offspring.

But making sacrifices for an ideal, a belief, or in behalf of a person long dead is a human trait not shared with any other species. Protecting an idea, for instance, and trying to assure the continuity of its life requires patterns of conduct that could not have arisen before the human brain crossed a certain threshold of complexity, beyond which it could function in ways we call mental. The set of peculiarly human motivations that are said to justify such conduct are usually associated with the ideas of the true, the good, and the beautiful. As a group, these ideas constitute the family of human values.

To understand the origins of human values in the mind and, through that understanding, gain insight into the roles played by them in the lives of persons and civilizations, it is useful to take a historical and developmental approach. Accordingly, we begin by inquiring into the evolutionary origins of mental functions. By mental functions I mean an ability to perform symbolic transformations of experience and, with those symbols, draw a map of a world richer and freer in important ways than the material and living world from which it arose.

An example of such a transformation would be the changing of sense impressions about tigers into the line, "Tiger! Tiger! burning bright," writing down the line, and by that two-step process—a creative and a communicative one—adding one's self-expression to the myriad variety of cultural goods held by a civilization.

The evolutionary origins of mental functions may be closely identified with the coming about of our species. Guided by what is known about the emergence of our species from its hominid ancestors, I would describe that process as one which involved several simultaneous and mutually reinforcing changes. I would also imagine that in their collective effects they resembled positive feedback in that they amplified change, rather than negative feedback which would have maintained stability.

Those steps surely included an increasing division of labor together with refinements in cognitive powers that evolved toward the recognition of a highly differentiated environment. One may also safely assume the appearance of increasingly articulated methods of communication that helped enlarge a store of signals that stood for themselves, by adding a store of signs that stood for events and things other than themselves.

I imagine that these changes conspired to expand the temporal horizons of the early hominids. At one time, the future-related behavior of these hominids must have been almost entirely instinctive, that is, genetically based. Yet, at some later date a nongenetic type of conduct came to allow planning as a complement to instinctual action. During the eons that witnessed the coming about of nongenetic, goal-directed behavior the rudiments of the human sense of time, with its horizons expanding beyond the immediate future and past, must have been born.

Sooner or later, each member of the species must also have noted the existence of a curious object. It shared the external world with sticks and stones and creatures that moved, but it could also feel hungry, experience anger, and desire a mate. This object was the distant forerunner of the self. Somewhere along our ancestral history, the inevitable demise of that object must have come within the scope of a thinkable future. That discovery, I suspect, extended mortal terror—which is an episodic response to immediate threats to life—by a type of concern unknown in the animal world. It was a lingering anxiety related to the necessary passing and the eventual demise of the self. Nonpassing—permanence—expressed or expressible in structures, whether physical or mental, then came to acquire a special appeal to the creature who knew that sooner or later, he had to die.

Besides suggesting to the mortal self that death may only be a mirage, being able to perceive the world in terms of structures and processes that appeared unchanging had its advantages in the strategy of survival. For that reason, so I speculate, it was favored by selection. Specifically, images—thoughts—of permanent objects and patterns of behavior made it possible to play out scenarios of future behavior and, with their assistance, place present conduct in the service of long-term planning.

Let me now make the deliberate mistake of representing human time by a straight line, with one of its ends standing for the conception of a person, the other for his death. On this geometrical metaphor the beginning and ending of life are symmetrical events. But in our minds the relation between the beginnings and endings of our lives are anything but symmetrical because the self, which will die with the body, did not originate with conception but was mentally constructed during the first years of life. It is "I" who will die, but the fertilized egg in my mother's womb was never "I the egg" but "the egg from which I came."

Where a person was before he or she was conceived is a question of limited interest, at least to the Western mind, whereas the question of where a person will be after death, if anywhere, mobilizes the whole spectrum of human emotions as well as the intellect. A longing for immortality in the form of life after death, made possible by symbols which stood for the self and could be thought of as surviving the body, has informed the development of all civilizations.

When our distant ancestors began burying food and weapons with their dead, they attested to their belief that death was not final. To borrow a line from Dylan Thomas, they expressed their "rage against the passing of the light." To be able to live with this rage, people created a variety of opiates—objects, acts, and ideas—that could make the passage of time appear Unreal or, at least, insignificant. In his novel *Two on a Tower*, Thomas Hardy (1905, p. 36) makes his protagonist remark that if "you are restless and anxious about the future, study astronomy at once. Your troubles will be reduced amazingly. But your study will reduce them in a singular way, by reducing the importance of everything" of ordinary human concern.

These opiates of all people, these bids for conquering time, when taken together are the stuff of human knowledge. Among the many forms of theoretical and practical knowledge, we have to include the ability to construct scales of values pertaining to beliefs, to conduct, and to the management of emotions. Human values then join the other efforts of our species directed toward lessening the threat of passage. They disclose that role in their self-descriptions: I know of no assertion of truth, or teaching about right and wrong, or the claim of beauty for a piece of art or a poem that would not make a claim for a measure of permanence.

But, if human time and human values are part and parcel of the same constellation of feelings and thoughts, then a natural philosophy of time should be an ideal location for an epistemology of human values. For, through the many fields of knowing that contribute to the interdisciplinary study of time, the architect of such an epistemology would have direct access to insights relevant to the origins, scope, and nature of values.

In what follows, I shall attempt to sketch the outlines of an epistemology built on a particular natural philosophy of time.

The Hierarchical Theory of Time

The hierarchical theory of time, also called the theory of time as conflict, is a conceptual system in the natural philosophy of time. As is often the case with theories that are built on assumptions alien to established views, it is difficult to understand some of the details of the theory before it is understood as a whole, and it is difficult to understand the theory as a whole before its details are understood.

Yet, I cannot elaborate here the content of that theory in any detail but only sketch those of its teachings that are relevant to the theme of change, permanence, and human values. Unfortunately, such a sketch tends to leave the impression that the ideas outlined were born ready and complete, as was Athene from the head of Zeus. Of course, they were not so born. Those who would want to learn more about the hierarchical theory of time may wish to consult some of my other writings.[1]

The theory adopted, and extended, the idea of stable integrative levels of nature, argued in the works of Joseph Needham and others, who speak about those levels as being arranged along a scale of increasing complexity. Let three works represent the extensive literature: Needham (1944, pp. 233–272), Needham (1968, pp. 106–169), and Pattee (1970). Specifically, the theory recognizes five distinct integrative levels, each possessing a different complexity, each governed by different kinds of laws and logics, and each characterized by a different temporality. The integrative levels form a nested hierarchy of increasing richness of content as we proceed from the world of electromagnetic radiation to that of particles, to massive physical systems, to that of life, and of the mind of man. As we rise along the scale of complexities, each temporality is seen as subsuming in it that or those beneath it, and each allows for qualitatively new forms of creative freedom.

Let me define and describe these temporalities, starting with the immensely rich yet most familiar kind of time: that of the human experience of time's passage.[2]

Nootemporality or noetic time. This is the temporal reality of the mature human mind. It is characterized by a clear distinction among future, past, and present, by unlimited horizons of futurity and pastness, and by the *mental present* with its vast temporal horizons.

Biotemporality, or biological time. This is the temporal reality of living organisms, including man as far as his biological functions are concerned. It is characterized by a distinction among future, past, and present. But the horizons of futurity and pastness are very limited when compared with those of noetic time. The boundaries of the *organic present* are species-specific and appear to be fixed.

It is with respect to the organic and mental presents that future and past, and with them the metaphor of a flow of time, acquire meaning.

Eotemporality is the time of the physicist's *t*. So named after Eos, the Greek goddess of dawn, this is the temporal reality of the astronomical universe of massive matter. It is continuous but is without a definable now and therefore necessarily undirected. It is a kind of time to which our ideas of a present, future, or past cannot be applied and hence it cannot be said to "flow."

The time of the world of elementary objects bears the name *prototemporality* for proto-, the first in a series. It is an undirected time and also one in which events have only probabilistic locations.

The world of electromagnetic radiation is termed *atemporal.* Atemporality does not signify nothingness but a condition to which none of our ordinary notions associated with time apply. It is akin to Plato's idea of complete chaos (quite unlike the chaos of chaos theory) and to eternity, if thought of as a state without time.

From the atemporal, the most primitive level of time, let me pass back "up" to a very complex kind of time. *Sociotemporality* is the name given to time on the organizational level of societies. Although much has been written about it, it is extremely difficult to identify its hallmarks with the same authority with which we can describe those of the lower order temporalities. The reason for the difficulty is, I believe, that we are both the makers and the subjects of sociotemporality and hence an observational platform, a stable perspective, is not, or not yet, available.

The five, or perhaps six temporalities constitute the *canonical forms of time.*

The stable integrative levels of nature, as already mentioned, are distinct in their complexities. Complexity is defined with the assistance of the relationship between minimal programs and complexity, an extension of Gödel's incompleteness theorem proposed by Chaitin (1975). This relationship is then so generalized that it could be applied to the sets of members that comprise an integrative level of nature. This generalization made it possible to obtain numerical values for level-specific complexities. The numbers so obtained fall into five distinct ranges in apparent support of the qualitative differences among the integrative levels, a fact which is intuitively obvious.[3]

The distinction among integrative levels in terms of their complexities recommended that level-specific languages, causations, and with them increasing regions of undeterminacy be also recognized.

At this point the theory made one of those leaps of imagination that are not at all arbitrary, yet may be shown as having been justified only ex post facto. For it is not possible to arrive at them through deductions from propositions admitted as correct by earlier reasoning.

That leap of imagination was the postulate that the canonical forms of time are not different aspects of time which become noticeable as we rise to systems of increasing complexity, but are stations along the open-ended evolutionary development of nature.

Although speaking about the evolution of time seems to be a contradiction in terms, there should be no difficulty—on the basis of logic—in admitting the possibility of a genesis and evolution of time. It is generally held that the expanding universe does not expand to fill preexisting space but expands space itself in a manner that might be called creative. All that the hierarchical theory of time claims in this respect is that time does likewise. It does not evolve within a preexistent expanse of time but creates new kinds of temporalities. But whereas the evolution of space is one of quantitative increase, the evolution of time involves quantitative as well as qualitative steps both in the physical world and upon the integrative levels of life, mind, and society. Such qualitative changes include increasing degrees of freedom or unpredictability.

Before turning to the issue of human values, it is necessary to speak about one other teaching of the theory: that of the nature and origins of existential or constitutive tension. The best way to introduce that idea is by describing what the theory holds about the origins of life.

Imagine that in the Precambrian Era certain molecular aggregates began to display a collective rhythm, such as for instance, being able to absorb or emit energy in synchrony with the tides. If such an internalized and expressed rhythm offered any advantages for survival then it was selected for, and organic evolution was on its way.[4]

From one point of view, having internalized and expressed the tidal rhythm was a step in adaptation: it made the behavior of these molecular clock shops fit the appropriate niches of the cyclic spectrum of their environment. But the same improvements in this speculative scenario also opened the way for finer tuning to cycles that were unrecognizable earlier, such as for instance to the spring and neap tides or the semidiurnal inequality of the tides. After these and other periodicities were internalized, they had to be coordinated with each other and with yet other cycles, such as with the daily and yearly motion of the sun.

Each step of improved adaptation refined and enlarged those domains of organic functions where the expected and prepared-for conditions could differ from the conditions actually encountered. These ever-present differences amounted to a form of stress, to a chronic condition of life, to needs whose satisfaction had to be regularly delayed. I called this need existential tension. (In Fraser [1990] see index entries under "Existential tension.")

With the appearance of man 3½ billion years after the first forms of life, the scope of existential tension became greatly enlarged. For, to the biological needs that had to be satisfied there came to be added new needs for possible, as well as impossible things, events, and conditions. Existential tension in humans became manifest as an all-informing hunger for food, for mate, for knowledge and power, for life, and even for death.

The idea of existential tension, when interpreted in terms appropriate to the different integrative levels, is the basis of the metaphysics of the hierarchical theory of time, giving it its alternate name: the theory of time as conflict.

That metaphysics sees life, mind, and society as processes defined by their unresolvable, creative conflicts.

Here process means change that follows a stable pattern for a long enough time for us to recognize continuity, transient as the continuity itself may be.

Conflict means the coexistence of two opposing trends, regularities, or groups of laws in terms of which those processes may become intelligible.

Unresolvable means that if those conflicts cease, so does the integrity of the process.

Creative means that under certain conditions the conflicts may give rise to the processes and laws of a higher integrative level—itself subject to the principles of unresolvable conflicts.

Let me now illustrate the meaning of what was just said.

The life process is often thought of as one of self-organization or growth. This view is valid but it is only half of the story. To make it the complete story, one must think of life as a conflict between growth and decay, maintained and coordinated from instant to instant in the organic present. This conflict is unresolvable because if it ceases, if there is no more growth that may be opposed by decay, then the animal or man is either dying or dead. The existential tensions of life may thus be seen as functional corollaries of biotemporality.

Likewise, the process of minding—the use of one's mind—certainly involves the brain's labor of self-organization. Again, this view is only half of the story. To make it the whole story, it is necessary to think of minding as a conflict between those processes of organization and decay that go on in the brain when we think. This conflict is also unresolvable because if it ceases, such as in senility, if there is no more selfhood that must continuously struggle against the loss of identity, then there may still be a living body but not a mind. The existential tensions of the mind may thus be seen as the functional corollaries of nootemporality.

Analogous arguments hold for society. It is true enough that society is a group of people organized for a common purpose, but once again it is only one half of the story. It is better to think of society as a process, a conflict between the forces of growth and decay. If that conflict vanishes, such as when social organization collapses, there may still be persons around, but not a society.[5]

Now, with the hierarchical theory of time as an aid in the inquiry, we may turn to an epistemology of human values.

Time and Human Values

Let human values be represented by the ideas of the true, the good, and the beautiful. I do not mean Platonic Ideas lodged in eternity, but constellations of thoughts and feelings ordinarily designated by those notions.

I intend to argue that truth and untruth are socially approved guidelines to the selection of beliefs; good and evil are guidelines to conduct; the beautiful and the ugly are judgments upon the quality of feelings, useful in the management of emotions. I will consider each one of these categories of values in terms of the antinomies of change and permanence, in the context of the hierarchical theory of time.

Guidelines to Beliefs: Truths and Untruths

As a working definition, I propose that we think of truth as an assertion that a belief is judged permanent.

How may one test whether a belief deserves to be called permanent and therefore represents truth? Perhaps by making predictions based on the belief and test their correctness in the course of time. Examples may be: "Ten years from today Mars will be at its closest point to earth because its orbit is elliptical"; "by the turn of the twentieth century AIDS will have killed over 60 million people"; "Christ will come to judge the living and the dead."

Many readers will recognize the proposed prescription for testing for truth as having been borrowed from the practice of the exact sciences, where predictability is judged a necessary test for demonstrating verity. But the kind of testing for permanence that works well for physics cannot be uncritically extended to the testing for truth in the complex processes of life, human behavior, and society, for reasons that need to be explained.

The notion of permanence or changelessness, implicit in physical theory and experiment, first entered Greek thought some twenty-four centuries ago. It was then, as it is today, an idealization based on a knowledge of nature limited by whatever is available to sense impressions. The limitation of the received meaning of permanence is discovered, however, as soon as those domains of nature that can only be explored through scientific instruments and abstract, mathematical reasoning are included in the survey. Specifically, one can find only permanence-analogues; that is, aspects of the world that remind one of the ordinary notion of permanence but do not correspond to it. The character of the permanence-analogues are seen to depend on the integrative level one considers. As they vary, so does the relation of truth to time.

To illustrate the meaning of these remarks, I will first discuss what corresponds to the idea of permanence in the integrative levels of directed temporalities (those of life, mind, and society), then consider the physical world of undirected temporalities, then return again to the domains of directed temporalities so as to gain a perspective upon the distinctions among physical, biological, mental, and social forms of truth.

In the biological domain, organic evolution separates those aspects of the umwelt of a species that appear permanent from those that do not.[6] Behavioral traits of food, friend, and foe which regularly repeat themselves are retained by natural selection and are deposited as cumulative residues in the programs of biological inheritance. Genetic material constitutes biological memory of what has been learned about permanence through the evolutionary history of a phylum.

This biological memory (the permanence-analogue of biology) is quite unlike computer memory (which is an example of physical permanence) because of the vastly different complexities of organic and inorganic systems. Whereas a computer printout is a one-to-one mapping of what has been stored in the computer and is, therefore, predictable, genetic memory retains only regions of possibilities. This is the reason why phenotypic development and behavior—unlike planetary motion—is predictable only in very broad terms. The structures and functions of the phenotype are the results of responses, permitted by the genotype, to the unpredictable challenges of the developmental environment.

In man the slow biological learning process of sorting out the permanent (in the sense explained) and the contingent is enlarged by mental learning. Also, the horizons of time for which predictions could be made were opened up through a recognition of increasingly more refined patterns in the environment. While the ability of the mind to make predictions based on recognizable stable patterns of the environment has thus reached its present level of great sophistication, human conduct acquired an equally impressive level of unpredictability. The freedom so gained has been sufficiently broad to have permitted the rise of civilizations. The myriad details of those civilizations, though governed by the lower order regularities of physics and biology, are unpredictable from those lower order principles.

As I already mentioned, permanence, expressed in structures and processes that seem to be unaltered by time, has a special appeal for humans. There is even support for the belief that we instinctively search for whatever appears permanent (see the section on "Resolution of Perceptual Conflicts" in Fraser [1990, pp. 91–93, 312 and 374]). The search begins in infancy and is developed under social control. The universe of the infant is a chaos of sensory impressions out of which he constructs an increasingly patterned world. Those aspects of the world that change very slowly are judged permanent and are reified in symbols, such as in the nouns of natural languages. Processes, that is, aspects of the sensible world which involve changes that follow enduring patterns, are described by words called verbs. Those aspects of the world which involve change without a recognizable pattern cannot and therefore do not have names.

How does the child learn to identify processes? He learns to separate himself from his mother, from other objects, and eventually establish his identity. Selfhood (the "I") so established is a mental construct. It is precisely because it is a mental construct that it may be assigned a status of unchanging continuity. Although "I change," there is always an imaginary object represented by "I" that does not. Once established, the feeling of identity serves as a paradigm for permanence or oneness, and oneness is the basis of counting. In its turn, counting is the basis of quantifying experience and of studying the laws of quantities, such as those of mathematics, universally judged as permanent.

Let us, therefore, consider number—and quantities in general—so that by interpreting quantitative relationships revealed through physics, we may identify the permanence-analogues of the physical world. Our findings will then become a part of the inquiry about truth-as-permanence.

If one is unwilling to believe with Plato that mathematics is a direct link to a preexisting store of truths lodged in eternity, then one must ask this: Through what kind of evolutionary process did mathematical knowledge arise?

A case can be made for the belief that mathematical thinking arose through a selection process that favored quantitative thought. Mathematics may then be seen as a mental ability for creating symbolic transformations of experience, a skill that became programmed into our genetic makeup as was, for instance, three-dimensional vision. And, like three-dimensional vision and its innumerable expressions in art, mathematical thinking remains subject to individual talent and social conditioning.

Our animal ancestors might have counted their offspring, our distant human ancestors the bodies of their dead enemies; none tried to count electrons or light years. When we learned how to use mathematical science for the exploration of worlds not directly available to the senses, we began to receive counterintuitive reports about time, space, and causation in the physical world.

This, then, is what can be said about permanence-analogues appropriate to the different integrative levels.

In the atemporal world nothing can correspond to predictability, to permanence, to causation, or to rest, which is the spatial analogue of permanence.

In the prototemporal world, events in time and locations in space are of a probabilistic, statistical nature. Causation is also probabilistic and with it, so is truth itself. Since probabilistic truth means only likelihood, the truth of the quantum world does not correspond to our ordinary ideas of truth.[7]

As elementary particles combine into the solid matter of the eotemporal world, the probabilistic causation of the prototemporal world gets smoothed out to yield the deterministic causation of the eotemporal world. It is in the Newtonian (eotemporal) universe of massive matter that we first find permanence and predictability of the kind that is close to the received, idealized versions of these concepts.

This is not surprising, for the whole structure of classical philosophy as well as early modern science is built on those features of the Newtonian universe that are within the range of human senses, unaided by instruments.

As we enter the higher integrative levels, the nature of the way in which events are connected once again changes. Thus, the characteristic connectivity, or causation, of the bio-, noo- and sociotemporal worlds relates to final goals, defined by the self-interests of living organisms, persons, and societies. The deterministic predictability of the macroscopic physical world remains applicable, but in the case of life, mind, and society it is complemented by increasing degrees of freedom.

What emerges from these reflections is the following. Change and motion are primeval states of the universe and for that reason they must be taken as ontologically prior to permanence and rest.[8] Out of the totally impermanent world arose in the course of evolution different ways whereby events may be connected, to wit, in their hierarchically nested order: not at all, probabilistically, deterministically, and by final causation.

Western ideas of what constitutes truth comprise a large set of propositions that have varied with the philosophical, religious, and scientific fashions of the centuries.

Two millennia after Plato, the notion of truth as discoverable permanence reached its apotheosis in the magnificent structure of the natural sciences.

The ideas of scientific law and scientific truth were born when the Platonic ideal of timeless forms assumed the shape of mathematical images, known as formulas. In creating formulas, the early purpose of Western science had more to do with demonstrating the wisdom of God than with serving the practical needs of passing life. Bit by bit, however, the religious trimmings gave way to a metaphysics of science and technology. This metaphysics endowed scientific pronouncements with the same kind of uncritical admiration that, earlier, was reserved for religious pronouncements.

The idea of truth in exact science remained the image of a world divided into time and the timeless, a division to which nothing in nature corresponds. No wonder that this dichotomy has given rise to certain problems.

For instance, the many ways of knowing came to be separated into those which demanded predictability and precision as a test for truths and those that could not and, therefore, did not so demand. These two groups may be associated with C. P. Snow's two cultures. But if the hierarchical nature of time is taken into account and with it, we recognize the evolutionary character of predictability, and hence the dependence of the quality of truth upon the complexity of the systems considered, then we find that knowledge does not gather into two distinct domains. Instead, it forms a continuous spectrum of interpretations along a scale of permanence-analogues.

This reasoning places even mathematical truth into the perspective of an evolving world. It suggests that mathematics, appropriate to the handling of level-specific issues of society, of the mind, and of life is likely to develop in a direction that would make some of its branches approach the kind of qualitative reasoning which biologists, psychologists and social scientists have already found necessary for their metier.[9]

In the spectrum of different kinds of truths the human paradigm of truth remains "this above all: to thine own self be true," which says nothing about precision or predictability. Truth, appropriate to nootemporality and sociotemporality, is closer to the beatitudes than it is to any claim based on quantitatively measurable comparisons, no matter how significant the statistical handling of biological, psychological, and social data may be.

Thomas Hardy has sensed the mischief resident in mistaking such mathematical truth for truth as a human value. His *Two on a Tower*, is the story of a young astronomer who returns to claim his lady love after a lengthy and unfortunate separation. She sends him away with words that say, but do not mean, "Go!" But the young man leaves just the same because, as Hardy put it, "He was a scientist and took words literally. There is something in the inexorably simple logic of such men which partakes of the cruelty of natural laws that are their study" (Hardy, 1905, p. 331).

There is something inexorably simpleminded in the mad rush of our age for gathering raw data, thought of as knowledge, a kind of cruelty that comes from mistaking quantitative truth, uninformed of human passion, for truth as a human value. The confusion between the two helped make our age minute wise and century foolish.

Guidelines to Conduct: The Good and the Evil

As far back as it can be traced, people of all cultures have demonstrated their readiness to formulate rules of preferred conduct in terms of what is right and wrong, an equal readiness to uphold those rules against challenge, and again an equal readiness to challenge them. The cruelty of our species in meting out punishments to those who break the rules, and the suffering that people have been ready to endure to do just that, is nothing short of incredible.

Opinions as to what specifically constitutes right and wrong are tremendously varied. But there is a fair agreement among contending teachings that by doing what is right, a state of harmony and perfection may be approached. Descriptions of those states of fulfillment are also quite varied but, once again, they do have an element in common. They tend to claim that the state one ought to strive for should be a permanent one.

The goal of a permanent heavenly abode is clear in Christian salvation history. Plato, working within the Greek presumption of the cyclic rise and fall of all arrangements on earth, had hoped that after his soul had separated from his body upon death, it would gain eternal wisdom through the purification process it followed while still in his body. The great Mid-Eastern and Eastern religions tend to see the destiny of man in reaching a final state where in the words of Brandon's conclusion (1962, p. 384) "he will be secure from the everlasting menace of time's destructive logic." Confucianism and Taoism, philosophies and ways of life rather than codified beliefs, seek lasting harmony within the person, among individuals, and between society and nature.

The shared views of these many great traditions about the desirability of a lasting, final, and balanced state suggest a working definition of the good in terms of permanence.

Let good be defined, therefore, as a predicate which asserts that a certain conduct, intent, or character will promote permanent stability in the affairs of man and society, whether here on earth or elsewhere. The variations among moral teachings then shift from a shared desire for final, permanent fulfillment to the different specifications on (1) what form such permanent states would take; (2) the reasons why one should get there; and (3) the map of the road to be taken.

But neither moral courage in defense of approved conduct or intent, nor beastliness of attacks upon conduct or character judged wrong, have ever succeeded in securing anything but transient stability in human affairs. Whether in the face of unfavorable socioeconomic conditions or with the aid of favorable ones, human life has always been one of armed and unarmed conflicts.

I see history as a ceaseless turmoil of unresolvable creative conflicts, with an occasional decrease in the intensity of these conflicts, and not as an ascent of man with frequent troubles from the bad guys.[10] One may well object that this is too pessimistic a view, it is the seeing of a glass half empty when it should be seen as half full. My answer to this objection is that perceiving man as a creature in continuous conflict helps make sense of history, whereas an image of men and women as good by nature does not. This is an old debate, relevant to the theme of this section, but not one that can be discussed here in detail.

Is it, perhaps, that moral instructions of all kinds have failed to provide a steady balance among the passions and needs of man because they were poorly informed or their enforcement was incompetent? Would it be desirable to have across the globe only a single set of ideas about right and wrong? Is it the case that, given time, money, and more research, we should be able to identify the kind of moral tenets and modes of enforcement that would, finally, secure permanent social balance? Or is it that we misinterpreted the role of ethics and morality in the development of civilizations? The case, I believe, is this last alternative: a misjudgment of what ethical rules do, in fact, accomplish.

Let me defend this thesis by first noting an analogy between the history of life and that of morals. Organic evolution is made possible by the existence of genetic variations, for it is the availability of choice that allows for selection pressures to favor some and disfavor other phenotypes and through them, select genotypes.

In the case of man, the selective forces of biology are joined by forces resident in symbolic causes that are shared or opposed by members of a society. Considering that symbolic causes themselves change, it is justifiable to assume that there has existed all along a degree of choice in conduct between following or rejecting collectively held ideas of right and wrong. The choice I am postulating has been called free will or human freedom. With the assistance of the hierarchical theory of time, let me explore the dimensions of human freedom as a degree of unpredictability, specific to the nootemporal integrative level.

In the physical world causation is probabilistic or deterministic but in either case, unfree. One cannot speak of choice in behavior. There exist no autonomous systems in terms of whose needs choices could be made, nor is there a present, an anchorage in time to which the choices could be referred.

It is only in the world of living organisms that one may begin to speak of choices in behavior. The spectrum of life's freedom stretches from the meager alternatives available to prokaryotes to choices in rudimental planning available for the most advanced nonhuman species. Biological freedom, as far as it goes, also applies to man: the bodies of people in coma keep on making their choices in support of life.

The degree of freedom of choice takes a large step when one considers humans, because the mind is able to imagine both possible and impossible ways of behavior, and generate a practically unlimited store of recommendations for conduct. The mind can play with symbols on its inner stage, move them around in imagined space, and back and forth in imagined time. These gifts of imagination have been the most characteristically human driving forces in the story of man. An animal will fight if abused, but only humans can maintain that an injustice has been done and direct their long-term conduct accordingly. It is here that private conscience has been fighting with public authority, private needs with public demands, where social guidance for what is right and wrong is most clearly called for, and is very often uncalled for.

Socially endorsed values concerning conduct are built into language, art, artifact, and even into the rules of acceptable reasoning. These modes of communication tie individual and collective visions together and provide for unity as well as conflict among the symbolic continuities known as individual and social identities. The subtle ubiquity of ethical teachings make people, submerged in a single tradition, take the moral norms of that tradition as obvious and, therefore, undebatable truths. This pattern of behavior makes all societies conservative. As the Declaration of Independence observes, "all experiences hath shown that mankind are more disposed to suffer, while evils are sufferable, than to right themselves by abolishing the forms of which they are accustomed."

Still, symbolic continuities may be changed much more rapidly than genetic inheritance, a fact that leads to the paradoxical role of moral values in the development of societies. On the one hand, as I just maintained, they make all societies conservative. Yet on the other hand, it is the existence of moral values that provide the usual and most efficient means for changing societies.

But, effecting substantial changes in moral values is never easy. Socrates had to violate the laws of Athens, then drink the hemlock to make a point about freedom of speech. Christ had to violate the laws of Rome and drive himself to crucifixion to replace a thundering Jehova by the Son of Man. It seems that any substantial change in prevailing ideas of good and evil demands sacrifice. A nation, being a symbolic entity in the minds of its citizens, cannot experience joy nor can it make sacrifices. Nations may perish but only men and women can suffer and die. To examine the nature of sacrifice that is necessary for changing a prevailing set of moral tenets and relate that effort to time, let me turn to the form of drama known as tragedy.

The unfolding of tragedy demands uninterrupted reflections upon past and future and the continuous making of choices among alternative courses of conduct. The struggle of the hero involves decisions that pertain to a hierarchy of obligations (Fraser, 1990, p. 394). The tragic hero constantly weighs his duties and responsibilities by the standards of prevailing judgments, as well as in terms of new values which by the process of his struggle he, in fact, creates. Because of his decisions, the domain of possible alternative choices of conduct available to him systematically narrows, until he has no further choices left but must suffer the denouement of death.

The story of the tragic hero is a literary projection of the predicament of the human knowledge of time. On the biological level the body continuously makes its own decisions, manifest in a process called aging. Because of those decisions the domain of alternative biological choices available to men and women keeps on narrowing until there are no further choices left. Against the biological decisions that necessarily lead to death, the mind holds up the image of the self as a player on a stage of eternity, both in the sense of endless time and a world without time.

What for the tragic hero are external forces challenging his ideas of right and wrong is, in the life of every person, an awareness of time's passage. It is an internal challenge, a selection pressure by an idea called time, a power more implacable than any external one, more absolute than any social convention or political tyranny. The unavoidable conflict between the fantasy or image of eternity on the one hand and, on the other hand, the biological facts of passing lends human life its irreducible aspects of the tragic. This is a cross-cultural trait, independent of whether or not the literary form of tragedy is known to a culture, because all men are dreamers and all men are mortals.

Speaking about conflicts between imagination and reality in the context of conduct is a way of describing the existential tensions appropriate to the nootemporal integrative level. These tensions, expressed in the ideas of good and evil and in the guidelines of right and wrong in moral obligations, may then be understood as aids that help people manage to live with those conflicts.

If moral laws create that much tension, if we must force ourselves to live by them, would a society without moral laws be more stable and hence more desirable? I would think that in the absence of moral tenets there would be no human society at all. Communities are created when their peculiar conflicts are created.

With the emergence of the time-compact globe, individual ????? increasingly replaced by institutional memory; individual desires ????? communal needs. As a new global sociotemporality is being fashioned, ????? of obligations is also changing and with it the specifications of what is right and wrong, good and evil.

Organic evolution by natural selection could not, and never did, lead to lasting balance but rather to the creation of new structures and functions. Likewise, so I conclude, the role of morality has been one of maintaining those unresolvable conflicts of the mind that nurture the tragic dimensions of human life and, thereby, promote changes in the rules of preferred conduct.

Guidelines to the Management of Emotions: The Beautiful and the Ugly

The purpose of this section is to place aesthetic values, represented by the idea of the beautiful, into the perspectives of the hierarchical theory of time.

It is assumed that anything and everything that humans can perceive, experience, or think may become aesthetically relevant, provided it succeeds in mobilizing human feelings. Let beauty and ugliness be regarded as predicates upon the quality of feelings so mobilized. Beauty may be thought of as a feeling one would like to remain lasting, ugliness a quality of emotion one would prefer to be transient.

Debates about emotions in respect to ideas of beauty span the centuries. In the *Republic* (X-606d) Plato took poets to task for watering and fostering such feelings as pain and pleasure whereas, he wrote, "what we ought to do is to dry them up." In our own epoch philosophers, Gestalt psychologists, information theorists, semioticians, behaviorists, psychoanalysts, existential psychologists, sociobiologists, and theorists of art assign broadly different importance to feelings in the creation and appreciation of the beautiful. The reasoning of this section appeals to an understanding of the origins and evolutionary development of aesthetic judgments.

King Lear, as he approaches the end of his tragic life, wishes for a world wherein he could pray and sing and "laugh at gilded butterflies" (V.iii.12). Do butterflies like pretty patterns? Biologists tell us that those designs are mostly for the birds.

Butterflies, writes E. O. Wilson (1975), "tended to develop poisonous and distasteful substances to repel vertebrates [their predators] while simultaneously developing audacious color patterns to provide warnings about their unpalatable conditions" (p. 241).

In the world of insects whatever is beautiful (for us) cannot be separated from its utilitarian use (for them). Nor would my definition of beauty as a predicate upon the quality of feelings be valid because I would not know what to mean by Lepidopteric emotions. Let us take leave, therefore, of our beautiful winged cousins and turn to higher animals that give ample demonstration of having feelings.

Darwin maintained that expressions of emotions both in animals and man were remnants of previously useful adaptive behavior. Writing about hatred and anger, he spoke about uncovering the teeth as an expression of defiance. Sneering in humans, for instance, involves the retraction of the lips and the exposing of the teeth on one side, possibly the remnant of the behavior of a toothy ancestor, ready to bite. "The expression of half-playful sneer graduates into one of great ferocity when, together with a heavily frowning brow and fierce eye, the canine tooth is exposed" (Darwin, 1872, p. 248). He added that a trace of the same expression is present in a derisive or sardonic smile.

From the utilitarian designs on butterfly wings to the sneer of a dog or a person, to a sardonic smile, to such a smile acted out, written up, or drawn on paper one may trace a path to illustrate the continuity of behavioral patterns from animals to man and, perhaps, draw some conclusions regarding evolutionary path of emotions. But when it comes to trying to understand expressions of those emotions in the making of objects, or in the creation of conditions that might be called beautiful, our outsider status with respect to other species poses an unsurmountable difficulty.

Namely, familiarity with the inner experience associated with aesthetic judgment is an essential ingredient in understanding the ideas of the beautiful and the ugly. I do not listen to music in the privacy of my home or in a bistro because others—dancing men and women or artistic monkeys—do or would behave in certain ways. I listen because music fills *me* with joy, or with a sense of sadness that dissolves the feeler in the felt, or because it speaks about *my* memories and hopes. Sooner or later, all inquiries about the nature of the beautiful come upon the issue of personal identity. For it is in the selfhood of a person's mind and body that the feelings we associate with the beautiful are born and manifest.

In animals, even in the more advanced species, we can do no more than observe and ask questions by means of experiments. When it comes to humans the situation radically changes. There I can insert myself in the study as an expert witness in matters of emotional experience known from within, serve as an object of my own observation partly from without, and use the immense flexibility of human language to help explore the role that feelings play in aesthetic experience.

An aesthetic judgment made by someone attending to his feelings—those of a person different from all other persons—is a very intricate process because identity is an intricate symbolic construct (Fraser, 1987, pp. 157–159). Understanding the nature of such judgments becomes even more intricate when one considers the social dimensions of the exercise. Although I am the one who registers his feelings and attributes their causes to the outside world (by calling this object or happening beautiful and that ugly), what I so register is conditioned by collective standards. And, I am both the cocreator and the subject of those standards.

People express their emotions in a myriad of different ways. By those very expressions and by giving the feelings names (this is love, this is anger, this is sorrow) they standardize and socialize the qualities of affects. The decisions made by a writer or an artist, or the artist in all of us, concerning which thoughts and emotions ought to be recorded and for what aesthetic reasons, involves a bargaining within each person. It is between his or her judgment as an individual, and his or her judgment as a representative of the standardized and socialized qualities of affects.

Socially preferred views on the beautiful and the ugly serve him or her as guides to the management of those very emotions which help formulate the collective judgments in the first place. Aesthetic values, no less than those of truth and morality, are messages passed back and forth between a person and a group. They define certain feedback circuits as it were, conditions that are continuously close to resonance: they selectively amplify and mute but are never passive or quiescent.

Opinions as to what specific objects, conditions, or thoughts are beautiful and which ones are ugly differ from age to age and place to place, and what is ugly for one person may be beautiful for another. What humankind shares are not specific tastes but the desire to reach for whatever is judged beautiful and reject whatever is thought of as ugly. The readiness to make such distinctions joins the readiness to offer judgments upon beliefs and conduct, as the third great cross-cultural universal of human values.

Whenever there is even a modicum of agreement among members of a society about the nature of feelings engendered by aesthetically relevant aspects of the world, judgments of beauty or ugliness are promptly attributed to those objects or events. These attributes are then maintained through the tendency of language—and the total human communication network including art and artifact—to resist changes in the established direction of the cultural process (Fraser, 1978b, p. 350). Because of this inertia, certain images, actions, thoughts, will be judged intrinsically beautiful or ugly. These judgments tend to remain valid for the group until its collective ideas on beauty once again change through the dialectic of aesthetic evaluation.

What is known about the evolutionary origins of human feelings, the primary sources of aesthetic value judgments?

Jean-Jacques Rousseau, a keen observer of the inner landscapes of men and women, remarks at the very beginning of his *Confessions* that we suffer before we think. Be it suffering or pleasure, we certainly feel before we think. And there are sound biological and psychological reasons supporting Rousseau's example of common wisdom.

Emotions, such as fear, rage, sexual arousal, aggressiveness, and motivation, are regulated by the limbic system of the brain. The limbic system includes the hypothalamus, a group of cells that governs the internal environment of the body. Control of the internal milieu, no less than the control of the primitive emotions, is an archaic function of the nervous system. The neural center for phylogenetically more recent functions—speech, memory, thought—is located in the cerebrum, the part of the brain that surrounds the limbic system and evolved later.

Feelings are not only older than is reasoned thought, they also appear manifest in the development of the individual long before reasoning does. A child can tell what to him or her appears beautiful or ugly and express those feelings in words, drawing, music, or acting, long before he or she is able to offer reasoning regarded as mature. In the course of growth and development, these early judgments become integrated with other value judgments, on their way to becoming a part of the mental world of the grown person.

Although judgments based on emotions are more deeply seated functions of the mind than are those of reasoning, in a mature person they are not independent, neither are they on an equal footing. They are hierarchically related. It is possible to experience intense and rich feelings without the presence of reasoned thought, but it is not possible to reason without a world of underlying emotions. Scientific theories and even individual mathematical formulas are often described as beautiful. Calling them beautiful lends to reasoning the added dimension of emotions. But although qualities of feelings expressed in the arts, letters, or in daily discourse may profit from reasoned justification, they do not demand it.

It will be recalled that for scientific judgments of truth to be valid, agreement by others is essential; for aesthetic judgments, although the dialectic between the person and the group is always present, it may be irrelevant. What a person selects from among his feelings as those worth perpetuating and even sharing with others, is necessarily a privileged disclosure. The labor of the selection process is a part of the instinctive quest of all people for permanence. In its most reassuring version, beauty is an absolute and eternal Platonic form. In the theory of time as conflict it is neither eternal nor absolute. It is seen, instead, as both a guide and a strategy for creating order and permanence in the turmoil of the ill-defined and ever-changing shapes of feelings.

That strategy is formalized—expressed and revised and reexpressed—in the arts and letters. Each and every literary or artistic object is a record of what the writer or artist judged worth retaining. Every sculpture, painting, or flower arrangement represents whatever its maker judged as worth preserving. Each and every composition, choreography, or narrative writing expresses what its composer or writer regarded as an unfolding process which should not by that unfolding, pass.

Is longevity in public appreciation, so sought after by artists and artisans alike, a test of beauty? From a collective point of view it is a part of the test. Clocks and calendars are more or less the socially accepted scales used to demonstrate aesthetic value. The Venus de Milo, so the argument might run, is still admired after more than two millennia: it must be beautiful.

Individuals, however, have access to a method more subtle than measuring time by a clock. It is one that recommends itself as a check on whether or not a created work has been successful in offering a haven against the flux of time. What I have in mind is the experience of timelessness occasionally engendered by something judged beautiful, whether or not man made, whether actually present or existing only in expectation or memory.

Imagine that while reading a poem you were overcome by a feeling of timelessness. Was it due to a passing glimpse of the inexpressible absolute script of Creation, of which the lines you read were but poor, earthly copies? This could be a poetic, Platonic, or religious interpretation.

Experiences of timelessness have been recognized in all civilizations. They have often been regarded as privileged states, direct links to an almighty and everlasting divinity or to the Evil One. The explanation in contemporary terms is much more earthy and more intriguing. It maintains that the feeling of timelessness is the result of what may be described as a mental journey to the archaic ????? of one's own mind, with their temporal unwelts different from the nootemporal one. I have dealt with this phenomenon of "time travel" in various contexts, elsewhere (Fraser, 1981, pp. 5–7; 1987, pp. 290–309; 1990, pp. 74–75, 305–311). Here are the details relevant to the current argument.

Consider first that reports on feelings of timelessness never mean the complete absence of time. They mean the phrase, "an odd kind of time," one that lacks some of the features of noetic time. It may lack a sense of continuity, a sense of temporal ordering with respect to a present, it may lack a present itself, or a directedness, or open horizons. Such adventures, true time travels, are made possible by the fact that the human brain retains some of its history. Structurally, it comprises three evolutionary forms, with the later ones surrounding the earlier ones, as already mentioned in connection with the cerebrum that surrounds the limbic system. The innermost part of the limbic system has sometimes been called the reptilian brain.

The three layers—the reptilian, the paleomammalian (or limbic system), and the neomammalian (cerebrum)—share the cranial cavity and are in constant communication. Still, each does for the human body what it would have done for its archaic owner before the later layers evolved. Each retains its peculiar way of assessing reality, including the nature of time. The different realities, each with its peculiar temporality, are continuously present among our mental functions and interpret sense impressions simultaneously. In the normal, waking state the most recent, the noetic control system predominates, although, with Freud, not without the influence of the older systems. But one or another of the older control systems can become dominant while the evolutionarily younger systems are short-circuited as in sleep, illness, danger, or feelings of elation.

Let us recall the hallmarks of the canonical forms of time. Having severely limited temporal horizons is the hallmark of biotemporality; the absence of a preferred direction is the hallmark of eotemporality; fragmentation of a directionless time describes prototemporality, complete temporal chaos characterizes atemporality.

The affective dimensions of these temporalities may be described as their "moods" (Fraser, 1978, pp. 284–293; 1987, pp. 291–309; 1989, pp. 11–13).

Thus, the atemporal mood is the suffering of the schizophrenic who feels the pull of chaos and panics because of his or her loss of nootemporal reality.

The prototemporal mood is a feeling of disjointedness with respect to time. A visual representation of this mood may be a Jackson Pollack painting with its locally coherent islands of form and color which, when taken together, have no coherence.

The eotemporal mood is a sense of directionless time, it is the "oceanic feeling" described by Freud.[11]

The biotemporal mood is the happiness or fright of a moment that has no distant future or past, while the nootemporal mood is the reasoned feeling of human reality with its vast temporal horizons and sense of possibilities.[12]

There are an infinite number of ways in which the various moods of the temporalities may mix and enter conscious awareness. It is not surprising, therefore, that people often report about strange experiences related to time and, having no other words available, describe them as those of timelessness.

The Greek word for such an experience was *ekstasis*, which meant "to cause to stand outside one's self." Do experiences described as those of timelessness help us stand outside ourselves in a world that is somehow superior to that of the mature, waking mind? Not at all. Instead, they take us to the perceptive worlds of the archaic portions of the brain, less evolved than the noetic.

May we regard experiences of timelessness, related to feelings of beauty (or ugliness), as tests of some kind? Yes, but not the kind of tests that the exact sciences could admit as valid, because they are unrepeatable. Each and every one of them is unique, different, and almost as fleeting as the lives of certain elementary particles. Yet they are tests, nevertheless. They are measurements.

It is a fundamental principle of measurement theory that one must compare like with like. One measures length with length, variation in biological structures with other, similar variations. One can compare feeling only with feeling.[13]

For instance, a report of "oceanic feeling" may indicate that the temporal mood of a present feeling, as compared with the mood of noetic time experience, lacks something, such as directionality or a well-defined present. That such an inner comparison was made, in regard to the effects of an object or condition, suggests that the experience has mobilized the emotions to a sufficient degree to make the experience in question aesthetically relevant.

Decisions as to whether a particular aesthetic experience should be described as beautiful remains a socially conditioned personal one. According to the definition of the beautiful given earlier, each decision will depend on whether or not a person would wish that the feelings generated by the experience remain lasting.

Since peoples' states of mind brought to any experience are infinitely varied and also change with time, it is impossible to develop a universal, normative theory of artistic and literary taste. One has to remain satisfied with interpreting the nature of aesthetic judgments, as I have been attempting to do in this section.

One of the central claims of this essay has been that all humans seek to escape from the insecurity conferred upon them by the awareness of passage and reach for what appears to be permanent. The most complete philosophy that asserts the superiority of the permanent and the inferiority of process is that of Plato. He maintained that reality consisted of eternal, timeless forms or ideas, while temporal objects were only appearances or copies of those eternal, unchanging forms. He saw temporality itself as an inferior condition of man the creature, compared to the permanence of his creator (*Timaeus* 37c–d).

In Platonic philosophy the theory of values is closely tied to this view of time. We read in the *Republic* (509d–e) that human values may be arranged along a scale, a line which may be imagined as connecting heaven with earth. The good, the true, and the beautiful are placed close to the heavenly terminal of the line. Plato would say that these qualities always are (remain permanent), they never become (change). They are timeless and for that reason of the highest value. In contrast, all things that pass are placed close to the lower terminal of the line, and because they are temporal, they have the least value.

May we conclude, with Plato, that our aesthetic faculties open our minds to the vistas of eternal, unchanging, and therefore, most valued forms? I do not believe so. I believe that Plato's image of the divided line is upside down. What matters to people the most are those things that pass, such as life, love, and beauty. People seek truths and hold on to them for dear life because they know how vulnerable all verities are. The good is valuable because we know that moral judgments are subject to reckless and sudden changes, making what is judged good today evil tomorrow. Beauty is precious because a whiff of the wind can collapse the delicate configuration of the conditions—external and internal—by which those feelings have been generated.

The feeling of timelessness induced by the contemplation of man-made or natural beauty may bestow solace to a person under the relentless burden of individuation. But it does not open to his admiring view the vistas of eternal, unchanging forms and secure, through them, a lasting haven in the mind against the certainty of time's passage. The theory of time as conflict sees the role of our aesthetic faculties as lying in the opposite direction.

Aesthetic exploration is the foremost and freest method available to our species for the creation of new images and desires, including many to which no object or event in the external world does or could correspond. By this means, the perennial human search for lasting beauty generates not satisfaction stably fulfilled, but an intensifying conflict between the world as we find it to be and that other imagined, timeless world to which art gives a transient form and a changing name.

A Backward Glance, to Serve as a Summary

To the extent that any extensive and intricate argument may be summed up, what this essay proposes is that human values are self-generated selective forces which work through symbols. As modes of knowing, they may sometimes help stabilize human affairs in the short run. But in the long run they refine and amplify human conflicts. Their factual role in the development of cultures has been one of generating and maintaining, through conflicts, the stunning creativity and frightening destructiveness that characterizes our race.

References

Argyros, A. (1991), *A Blessed Rage for Order: Deconstruction, Evolution, and Chaos.* Ann Arbor: University of Michigan Press.

Brandon, S. G. F. (1962), *Man and His Destiny in the Great Religions.* Manchester, U.K.: Manchester University Press.

Bronowski, J. (1973), *The Ascent of Man.* Boston: Little, Brown.

Chaitin, G. J. (1975), Randomness and mathematical proof. *Sci. Amer.*, May, pp. 47–52.

Darwin, C. (1872), *Expressions of the Emotions in Man and Animals.* Chicago: University of Chicago Press, 1965.

English, H. B., & English, A. C. (1958), *A Comprehensive Dictionary of Psychological and Psychoanalytic Terms.* New York: McKay.

Fraser, J. T. (1978a), *Time as Conflict.* Basel: Birkhäuser Verlag.

—— (1978b), Temporal levels: Sociobiological aspects of a fundamental synthesis. *J. Social Biol. Struct.*, 1:339–355.

—— (1981), Temporal levels and reality testing. *Internat. J. Psycho-Anal.*, 62:3–26.

—— (1982), *The Genesis and Evolution of Time.* Amherst: University of Massachusetts Press.

—— (1987), *Time, the Familiar Stranger.* Amherst: University of Massachusetts Press.

—— (1989), The many dimensions of time and mind. In: *Time and Mind*, ed. J. T. Fraser. Madison, CT: International Universities Press, pp. 1–14.

—— (1990), *Of Time, Passion, and Knowledge*, 2nd ed. Princeton, NJ: Princeton University Press.

Freud, S. (1930), Civilization and its discontents. *Standard Edition*, 21:57–145. London: Hogarth Press, 1961.

Hardy, T. (1905), *Two on a Tower.* New York: Harper & Brothers.

Needham, J. (1944), Integrative levels: A revaluation of the idea of progress. In: *Time, the Refreshing River*, ed. J. Needham. London: Allen & Unwin, pp. 233–272.

—— (1968), Hierarchical continuity of biological order. In: *Order and Life*, ed. J. Needham. Cambridge, MA: M.I.T. Press, pp. 106–169.

Pattee, H. H. (1970), The problem of biological hierarchy. In: *Towards a Theoretical Biology*, Vol. 3, ed. C. H. Waddington. Chicago: Aldine, p. 117.

Penrose, R. (1989), *The Emperor's New Mind.* Oxford: Oxford University Press.

Plato, *Timaeus*, trans. B. Jowett, in *The Collected Dialogues of Plato*, ed. E. Hamilton & H. Cairns. Princeton: Princeton University Press, 1973.

Wilson, E. O. (1975), *Sociobiology.* Cambridge, MA: Harvard University Press.

16. THE TRUE

Human value were identified earlier as individually derived but collectively recommended guidelines for navigating among the gifts and dangers of human freedom. This chapter attends to truth, the first family of those guidelines, and explores its relation to time. Here are three examples that appeal to the notion of truth.

(i) The American Declaration of Independence asserts—as already quoted—that "we hold these truths to be self-evident: that all men are created equal, that they are all endowed by their Creator with certain unalienable rights...." As recently as in 2005, in a highly visible case of wrongdoing by the USA Administration, the Special Counsel asserted that "truth is the engine of our [USA] judiciary system."[1]

(ii) *Pravda*, the Russian word for "truth," was the title of the leading newspaper of the Soviet Union from 1918 to 1991 and was the official organ of the Communist Party. The validity of its title was assumed to be undeniable and appropriate to the utterances of the Central Committee of the Party.

(iii) $2^{1/2} : 2^{-1/2} = 2$ will be claimed as true by everyone who knows arithmetic. The truth claim resides in maintaining that the numerical value to the left of the equation sign equals the numerical value to its right.

(iv) Jesus of Nazareth is recorded to have said that "if you continue in my word...you will know the truth and the truth will make you free."[2]

What all these examples claim is that, whatever they assert may again be asserted, unchanged, as time goes by. In the words of a Latin saying that first appeared in the writings of Aulus Gellius, a second century Roman philosopher and poet: *Veritas filia temporis*. "Truth is the daughter of time." English monarchs from Henry VIII to Elizabeth I used it as their motto. It is best translated as "time will prove the permanence of the claim." Aristotle might have opined that "Today it is snowing" will have to be changed to "yesterday it was snowing" but never to "it was not snowing on that day."

A truth claim is a large order. Its details may be intricate, its meanings may be many valued. In the sixteenth century, *veritas filia temporis* was cited in favor of the causes of Protestantism as well as Catholicism. There even appeared dramas involving personified Truth and personified Time, together with personified qualities of what constituted right in human conduct and what constituted approved and disapproved thought. Here are two opinions about truth and time, both from "a merye enterlude," a drama with a very convoluted plot, published in 1553.[3] This is from the Prologue.

> ...Veritee the daughter of sage Father Tyme
> Shewith all yt ys, be it Vertue or Cryme.

A second example, from the same play, is an exchange between Avarice and Veritee.

> Ver: I am dame Veritee.
> Avar: What, the daughter of Tyme?
> Ver: yea.
> Avar: I knowe my Mr. [master] your father well afyne,
> Welcome, faire ladie, sweete ladie, litle Ladie,
> Plaine ladie, smoothe ladie, sometime spittle ladye,
> Ladie longtong, ladye tell-all, ladie make-bate,
> & I beseche youe from whence are ye comme of Late?

In the age of the Internet the position of truth remains no less ambiguous. This does not stop the essay that follows from taking a hint from the father-daughter relationship and examine the idea of truth as a permanent belief. It concludes, however, that all claims of truth-as-permanence necessarily include acts of faith and hence, subject to challenge and revision. The search for truth is driven by the conservative quest for identifying whatever is permanent, whatever is unchanging. But because of the necessary, continuous revision of all truth claims, the evolutionary office of truth as a human value has been the creation of conflicts and, through them, the bringing about cultural, social and personal change.[4]

TRUTH AS A RECOGNITION OF PERMANENCE: AN INTERDISCIPLINARY CRITIQUE

A survey of the cognates of the word "truth" in several languages suggests that, when people speak about truth, they mean a recognition of permanence in the phenomenal world. Permanence, in its ordinary sense, signifies a condition of unchanging continuity. But when the idea of truth as a claim for unchanging continuity is critically examined with the aid of a natural philosophy of time, it is revealed that there is nothing in nature to which such a definition could correspond.

However, there are a number of approximations or correspondences to the idea of permanence, appropriate to the different complexities of physical, biological and social processes. These correspondences permit the identification of distinct and qualitatively different types of truths, but with none of them being eternal verities. The conclusion reached is that, whether in science, philosophy, religion or even in the evaluation of numerical data, belief in a statement of truth-as-permanence always includes an act of faith that can only be based on moral judgements of right and wrong.

1. Truth as a Recognition of Permanence

In the contemporary division of scholarly labour, questions about the nature of truth are most likely to be discussed in what English speaking people call epistemology, German speaking people *Erkenntnislehre*. By whatever name, epistemology explores the nature, validity and boundaries of the effectiveness of different types of knowledge.

Specialised studies about truth in the sciences, in religion, in philosophy and in human law abound; there is no need to review them here. Instead of compiling a list of who-said-what, then picking and choosing "this man's art and that man's scope", I propose a semantic sketch of the cognates of the English word "truth" in a number of languages and offer the results of the survey as the initial material for suggestions about the relationship between truth and time.

A word is a record of a captured idea or feeling; it is an audible or visible signal that carries a message. Words even in themselves narrate stories through the feelings, associations and anticipations they engender and through memories to which they appeal. All words have long histories, even recently coined ones, because they either derive from earlier words or resemble them, and because they combine familiar phonemes. Words carry shades of meaning which may or may not be consciously recognised by their users, yet mature speakers will respond to the many meanings of words used in different syntax. It is because they are depositories of a spectrum of meanings and emotions as well as of the changing evaluations of reality that words can acquire semantic depths.

My reasoning now will proceed in a direction opposite to those who would advocate increasingly more precise definition of concepts in natural philosophy. I side, instead, with John Stuart Mill who, writing in 1843, remarked that if one would attempt to rectify the use of vague terms by giving them fixed, narrowly defined meanings, one would endanger one of language's "inherent and most valuable properties, that of being the conservator of former ages which may be alien to the tendencies of passing time... Language is the depository of the accumulated body of experience to which all former ages have contributed their part and which is the inheritance of all (ages) yet to come".

Let me ask, therefore, what kind of accumulated body of experience does the word "truth" conserve in English?

The word suggests disposition to faithfulness, fidelity, loyalty, constancy, steadfast allegiance. It signifies solemn engagement or promise, a covenant. In the context of the relationship between humans and their social and natural environments, truth means an agreement with reality, a conformity with fact, where "fact" stands for that quality of an event, object or condition which one believes to be unchanging with time. The word truth may also refer to moral stances which are consistent with revelations believed to be everlasting.

In German, combinations that include the word *Wahr*—as in *Wahrheit, wahrhaft, Wahrnehmung, Wahrmachung* signifies concepts in which the ideas of reality and truth are intricately, even intractably, mixed.

Sanskrit, the classic language of India, is another Indo-European tongue; it is believed to date to the second millennium B.C. In Sanskrit of the Hindu tradition, arising from Vedic roots, the cognate of "truth" is *satya*, a word derived from the present participle of the verb "to be" to which the suffix *ya* is added. The combination may be translated into English as "is-ness", meaning that which really exists. In a philosophical-religious context *satya* signifies an act of truth, a calling upon facts to witness that a person enacted his dharma, that he remained true, that is, steady to duty and divine law, themselves judged to be permanent.

Hebrew belongs in the family of Hamito-Semitic languages, presumed to have been spoken as far back as the 7th to 8th millennia B.C. The Hebrew word which is usually translated as truth is *emeth*. It signifies certainty, honesty, faithfulness. In common use it means permanence, lastingness, absence of change. In Judaism truth is primarily an ethical notion associated with the nature of God. Among Jewish philosophers the word *emeth* was generally accepted to mean correspondence with reality, with the way things are.

Japanese and Chinese employ the same ideogram to signify truth; in Chinese it is called *zhen shi*, in Japanese *shi* or *makoto*. As far as I could ascertain, it does suggest conditions that abide, but not unchangeability in an absolute sense. Its primary meaning is authenticity: it signifies something that is not a substitute but the real thing, something not mingled with impure elements. It has no reference to anything that transcends this world, it suggests a human rather than a metaphysical world. The ideogram for *shi* depicts a pot with feet and a spoon inside the mouth. The sense is that of a state of satisfaction, fulfilment, arrival.

These linguistic traces of thought offer no guidance as to what kind of philosophical, political, religious, scientific or legal meanings the idea of truth ought to have, but they do make obvious what the hundreds of millions of contemporary and past speakers of these languages associate with the nature of truth.

They believe that truth is a claim of correspondence to reality. In addition, Indo-European languages stress that truth also strongly relates to permanence. These shades of meaning are consistent with the metaphysical convictions and ways of thinking characteristics of Indo-European and Sino-Japanese people and are reflected in their philosophies, religions, politics and modes of reasoning. Let me, therefore, take advantage of this majority vote by meaning and propose a working definition of truth, by concisely restating what has already been said.

Let truth be defined as an assertion that a belief corresponds to reality. To what extent and in what manner it also involves permanence depends, then, on the role that permanence assumes in the natural world.

This working definition determines the path that my inquiry into truth and time must take. First I shall have to outline what is to be meant by reality, then explore the nature of time in a manner appropriate to that understanding of reality.

2. Reality and its Changing Boundaries

In this section I propose an understanding of reality, appropriate to the natural philosophy of time.

Early in this century, the German theoretical biologist Jakob von Uexküll drew attention to the fact that an animal's receptors and effectors determine its world of possible stimuli and actions and through them, the nature and extent of the animal's universe. He called such a species-specific universe the umwelt of the species. Von Uexküll's idea of species-specific worlds is a contemporary form of the opinion of St. Thomas Aquinas that "whatever is known is known according to the manner of the knower".

What is not in the umwelt of a species must be taken as non-existent for the members of the species. For instance, ultraviolet patterns on certain butterflies exist for other butterflies but do not exist for vertebrates, because vertebrates have no sense organs through which they could read those patterns.

The word *umwelt* has been naturalised into English, as have been hundreds of other German words. It is defined as "the circumscribed portion of the environment which is meaningful and effective for a given animal species". Note that the "environment" of which the animal's umwelt is a part is our own human umwelt, carved out of a much larger universe which, we must assume, also includes structures, processes and displays, perhaps certain features of time and space of which, at the moment, we do not know. We may become aware of such features, however, as our knowledge of the world becomes richer and with it our species-specific umwelt expands. For instance, we discovered that butterfly wings have ultraviolet patterns because we learned to make photographic plates sensitive to ultraviolet rays and with their help we can translate those patterns into the visible spectrum of light. Note again the unreducible homocentric bias: "visible" means, visible to humans.

The concept of umwelt is easily refined and applied to the world of different human sensory systems. For instance, our auditory, tactile and visual realms determine distinct, though continuously interacting umwelts whose boundaries may be enlarged through the use of scientific instruments. In yet another step they may be even further enlarged by including in it processes and structures of nature that are knowable only through mathematical formulas.

What I have sketched here is a generalisation of von Uexküll's idea of species-specific universes into a view of reality with expandable boundaries. Taking advantage of this understanding of reality I propose the following:

If the careful exploration of the umwelt of a class of processes reveals that time in the umwelt possesses certain hallmarks, then we must regard those hallmarks of time as complete and sufficient in themselves, even if they appear strange and incomplete when compared with the ordinary human experience of time. This view of reality, called the extended or *generalised umwelt principle* is useful in discovering the nested hierarchical structuring of time and through it, the temporal structuring of truth.

The extended umwelt principle is a differentiated mapping in the natural philosophy of time of what Helga Nowotny calls "the human agency", that is, the ever-present and necessary reference to the formulator of the questions being asked. In the present context, the questions are about time and truth. Both Nowotny's "human agency" and the extended umwelt principle are distant and highly refined derivatives of the Protagorean doctrine of man as the measure of all things.

With these details in mind, we are now ready to learn about the way time enters human reality.

3. The Hierarchical Theory of Time

Specifically, I want to turn to a class of principles in the natural philosophy of time known as the theory of time as conflict or, by another name, the hierarchical theory of time. The same theory has also been serving as a framework for interdisciplinary studies. Its principles and its significance for the different sciences and the different domains of the humanities have been elaborated in several books and in some fifty papers. Here I want to sum up some of those aspects of the theory which are relevant to our concerns with truth as a recognition of permanence in reality.

3.1. *The general background*

The theory maintains that time is not a homogeneous single thrust in which all processes equally partake, but a hierarchically nested system of qualitatively different temporalities.

There are many ways to enter the theory. A good way for our purposes is through the structure of the mental present of human experience. That present combines cognitive and affective interpretations of sense impressions with the feeling of continued identity, and with ideas about future and past. The ceaseless classification and reclassification of events into future, past and present is the experience described as the flow of time. Humans see themselves as partaking in that flow and conduct themselves accordingly. Not only humans, however, but all life forms function by a temporal ordering of their actions. They all respond to their species-specific present needs which demand satisfaction in the future, and they do so according to the patterns of behaviour inherited from or learned in the past.

In stark contrast to the behaviour of living and thinking beings, the temporalities of the physical world, as revealed through the equations of physics, are without a definable present. That is, the laws of physics have no features to which the idea of a present could correspond. This fact was noted by and disturbed Albert Einstein. He maintained that there is indeed something essential about the "now" but identifying what that essential characteristic is, he wrote, was not within the task of science.

I disagree with Einstein's opinion and maintain that an understanding of how the "now" is defined in a nowless physical world is very much within the capacity of science. I shall even sketch how this is and has been done, later in this paper. But I do agree with Einstein that the consequences of the physical world *not* having a definable now are, indeed, far-reaching. Namely, since future and past make sense only with respect to a now, the flow of time metaphor cannot be applied to the time of the physical world. Even the assignment of the experiential arrow of time to the thermodynamics of closed systems is arbitrary and could not be so interpreted unless one assumed that we already knew what is to be meant by the direction of time. Details of this reasoning have been elaborated elsewhere.

The absence of directed time from all formal statements of physical change have often been cited by philosophically minded physicists as evidence that the foundations of the universe are timeless. This presumed timelessness, contrasted with the human certainty of passage, favours the idea of a Platonic division of the world into the eternal or unchanging on the one hand, and the temporal or passing on the other hand. But such a division is too crude to accommodate the nature of time as it is revealed through contemporary scientific knowledge and humanistic understanding.

A much richer epistemic framework is needed. The new framework must admit and correlate the qualitatively different temporalities that are found spread between, at the one end, the primeval chaos beneath the contemporary cosmos and, at the other end, the sophisticated organisation of cultures. The hierarchical theory of time offers such a framework. It respects the different assumptions, modes of reasoning and tests for truth that the different fields of knowledge demand and by so doing it reveals the dynamic structuring of what, by a single word, we call time.

3.2. *Temporalities and causations*

The hierarchical theory of recognises in nature a nested hierarchy of five organisational levels along a scale of increasing complexity. Beginning with the most complex but most intimately familiar one, they are: the mental processes and their universe of symbolic representations; the life process; the physical processes of the astronomical universe of massive matter (the worlds of Newton and of General Relativity Theory); the quantum processes of elementary objects; and finally, the most primitive level, which is the chaos of radiative energy. Each integrative level subsumes the functions and structures of the one or ones beneath it, adding to it or to them new degrees of freedom.

When the time-related teachings of the sciences and the humanities are systematically surveyed and arranged to correspond to the hierarchy of organisational levels, five distinct types of time emerge. By the generalised umwelt principle these temporalities are not to be regarded as different aspects of time which reveal themselves as we rise along a scale of increasing complexity, but are to be taken to correspond to the reality of qualitatively different temporalities which form a nested hierarchy. Each temporality is appropriate and complete for the organisational level where it is identified, even if it appears incomplete when compared with the human experience of time.

It is not possible to arrive at the idea of qualitatively different temporalities through deductions from propositions admitted as correct by earlier reasoning. The claim of the hierarchical theory of time is a leap of imagination. It is an inductive generalisation of the kind that is frequent in scientific reasoning.

For an appreciation of the meaning of qualitatively different temporalities, intuitive notions alone are insufficient. It is difficult enough to comprehend the temporal world of an infant, let alone that of a rabbit, a flea or a virus. And, we have no intuitions concerning the temporal worlds of galaxies, molecules or of objects that travel at the speed of light.

The description of the temporal levels that I will give proceeds from the most complex yet most familiar temporality, that of the human mind, to the most primitive and therefore least familiar temporality, that of the electromagnetic world.

Nootemporality is the time of the mature human mind in its waking state. Its hallmarks are: a distinction among future, past and present, unlimited temporal horizons, and the *mental present* with its continuously changing boundaries and cognitive content. Nootemporality may be represented by the picture of a long straight arrow: shaft, head and tail. The mental present is not imagined as located along the shaft, but as a family of processes for which the arrow itself is a visual metaphor. The characteristic connectivity among events of the nootemporal world is that of *symbolic causation*: humans live by, and often die for symbolic causes and are governed by truths of symbolic nature appropriate for the noetic and sociotemporal worlds.

Biotemporality is the time of living organisms, including man as far as his biological functions go. Its hallmarks are: a distinction among future, past, present, limited temporal horizons, and the *organic present* whose boundaries are species-specific. Biotemporality may be represented by the picture of a very short arrow. The organic present is not imagined as located along the shaft but as a family of processes for which the arrow itself is a visual metaphor. The characteristic connectivity among the events of the biotemporal world is *final causation*, that is, action in the service of a self-appointed goal.

Eotemporality is the time of "the physicist's *t*", that is, of the astronomical universe of massive matter. It was so named after Eos, the Greek goddess of dawn. It is a temporality without a present and hence one to which ideas of future, past and the flow of time do not apply. It may be represented by the picture of the shaft of an arrow, without a head. Its characteristic connectivity is *deterministic causation*.

Prototemporality is the time of elementary objects dealt with in quantum theory. It is not a continuous form of time but one in which instants may only be specified statistically. The visual metaphor appropriate for prototemporality is that of the fragmented shaft of an arrow. Its characteristic connectivity is *probabilistic causation*.

Finally, even the picture of the fragments may vanish and we are left with a blank sheet of paper, a symbol for the *atemporal* world of electromagnetic radiation. Atemporality does not stand for non-existence but for a world of complete chaos: the primeval universe at the instant of the big bang, or the inside of a black hole.

These temporalities are the *canonical forms of time*. As I already stressed, they are not different aspects of time which become noticeable as we rise to organisational levels of increasing complexity but stable aspects of reality along the open-ended evolutionary development of nature, including the genesis and evolution of time.

Although the idea that time itself evolves appears to be a contradiction in terms, it cannot be rejected on the basis of logical reasoning alone. For, just as the expanding universe does not fill preexisting space but expands space itself, so the qualitatively different temporalities do not evolve within a preexisting dimension of time. Rather, they emerge as correlates of the different functional and structural complexities of the distinct organisational levels of nature.

With an understanding of what is to be meant by reality and some rudimentary ideas about time as a nested hierarchy of temporalities, we may now turn to the recognition of permanence as an evolutionary aspect of time, and with that recognition learn to accommodate qualitatively different types of truth.

4. The Evolution of Permanence

The dictionary definition of permanence is that of a continuously enduring state without change. The question to ask is this: is there anything in nature that corresponds to this definition of permanence? In search for an answer, I want to consider different approximations to the idea of permanence upon the different integrative levels of nature: the physical, biological, mental and social worlds.

4.1. *Permanence in the physical world*

The macroscopic physical world as seen both in Newtonian and Einsteinian physics is deterministic. For that reason, at least in principle, one should be able to discover its governing laws and expect them to be permanent. The deterministic laws of physics are, indeed, remarkably appropriate to massive matter. But they are only approximations because all macroscopic change is under multiple control with atemporal and prototemporal governing principles always present beneath the deterministic one. Deterministic laws are neither absolutely precise nor are they eternal verities; they are only tentative proposals. That this is the case is already explicit in the method of scientific inquiry set forth by Newton. "In experimental philosophy we are to look upon propositions inferred by general induction from phenomena as accurately or very nearly true, notwithstanding any contrary hypotheses that may be imagined, till such time as other phenomena occur, by which they may be made either more accurate or liable to exceptions."

Let us leave the eotemporal world with its approximations of permanence and turn to the elementary constituents of matter in the prototemporal world. The immutability of elementary objects was abandoned long ago. As seen today, the objects which make up matter are wave-particles on one scale of distances and times, and pointlike geometrical properties of primitive space-time on another scale. There is nothing in their behaviour that would or could be familiar to us or to any living organism through direct experience. They are those parts of our umwelt which may be known only through mathematical models. They are the firstborns of chaos. They are, at one and the same time, the constituents of massive matter and the object-processes of a sea in which massive matter floats. Their probabilistic type of permanence is a step between the pure becoming of chaos and the approximately deterministic processes of the eotemporal world. In the prototemporal world an event is followed by another event only by a degree of likelihood, but never with certainty. It is as if God's commandments, originally written for the human-sized children of Israel had been revised as follows: "On the average, you shall not commit adultery. Generally, you should not kill. Most of the time, do not covet your neighbour's oxen."

The probabilistic character of the prototemporal world, as represented by formulas, may itself be claimed to be an unchanging and continuous and hence permanent and therefore true feature of that world, were it not for two problems. The more obvious problem is that our knowledge of that world keeps on changing and is anything but permanent. The much more subtle problem is that a judgement of permanence cannot be arrived at within the prototemporal umwelt. Judgements of permanence about probabilistic processes may only be delivered from the biotemporal or higher integrative levels where continuity and permanence do have meaning.

Based on what has been learned, we may now identify three types of truths which first appear in the physical world: the *chaotic or atemporal*, the *probabilistic or prototemporal* and the *deterministic or eotemporal*. Because of the nested hierarchical character of nature, deterministic truth always includes probabilistic and chaotic components.

4.2. *Permanence in the organic world*

What corresponds to the idea of permanence in the biotemporal world? To answer that question, it will first be necessary to sketch a model of life entirely in terms of processes, without reference to structures.

I start with the frequency spectrum of biological oscillations. This spectrum extends in species alive today across the incredibly broad range of 24 orders of magnitude; the human body possesses clocks across 22 orders of magnitude in frequency.

It has often been asserted that the role of biological clocks is to help organisms adapt to the cyclic changes in their environments and, thereby, assist them in their struggle for survival. This is certainly among its uses but, as a perspective upon the life process, it is a misleading one. It is analogous to saying that the sounds of the instruments of an orchestra assist the orchestra in making music. But those sounds do not *assist* an orchestra in its performance. They constitute, they make up the music. Likewise, the trillions of oscillators that comprise living organisms do not "help" the organisms to survive: they make up, they constitute the living organisms.

To continue the analogy, orchestral music demands not simply a collection of instrumental sounds but the synchronisation of those sounds according to the rules of harmony. Likewise, for an organism to remain alive it is necessary that the multitudes of its biological clocks be kept cycling according to the demands of their mutual dependence. Specifically, certain biochemical events that should happen simultaneously must happen simultaneously while others that should not, must not, or else the organism will come to harm.

The schematic representation I described is the *coordinated clock shop model* of life. The evolutionary continuity of life since its genesis has consisted in the passing on of the skills that are necessary to maintain the synchronisation of the biological clocks which make up that clock shop. These skills are in the form of instructions. The transition from inert to living matter was a transition from non-instructed to instructed synthesis of biological macromolecules. The phenomenal manifestation of the inner synchronisation is the organic present. Although there is an immense variety of life forms built on carbon chemistry, and although other life chemistries may certainly be imagined, all conceivable processes that could be called living have to be characterised by their capacity to establish inner coordination and with it by creating an organic present in the presentless world of non-living matter.

It is with respect to the organic present that futurity and pastness acquire meaning. And when they do, they do so in terms of the self-interest of the organism.

Let me use this model of life and this understanding of the creation of the organic present to sketch what, in the organic world, corresponds to the idea of permanence, as continuity without change.

I submit that what corresponds most closely to the idea of permanence in the biotemporal world are the slowly changing skills necessary for the maintenance of the organic present, and the more rapidly changing patterns of goal-directed, purposeful behaviour.

The ability to maintain the organic present, and to act in ways appropriate for a purpose, have been continuous for a 3.5 billion years. This is an impressive example of continuity but it is hardly one without change nor is it an eternal feature of the world, for it did not exist before life was born.

Based on what has been learned, we may now add a fourth type of truth to the earlier three. It is truth by self-directed purpose. It is the truth of a goal, defined by a living organism or group of organisms, and serving the eventual satisfaction of present needs. Because of the nested hierarchical character of nature, all such truths necessarily include deterministic, probabilistic and chaotic components.

4.3. *Permanence in the social process*

What approximations for permanence may be identified among the principles that govern man in societies? What I am searching for is Helga Nowotny's universal rules beyond the specifics. But these words are mine because among her extensive writings I was unable to relocate her statement in this matter. Therefore, faute de mieux, I will focus on *historicism*, a translation of the German *Historismus*. In the pragmatic English speaking world, with its dislike of anything that suggests metaphysics, historicism came to stand for a methodological belief concerning explanation. It is an assumption that human history, in the framework of cosmic history, may become interpretable from a single stable point of view. But so far no one has identified a pattern of covering laws which the opinion of most, even if not all historians, could be declared to be a principle of permanent, unchanging continuity and hence, by the definition I suggested, a statement of social truth.

Based on this thumbnail sketch, we may imagine the highest kind of truths as a so far unidentified family of principles that govern history. All one can say about them is that, because of the nested hierarchical organisation of nature, any and all such principles will necessarily have chaotic, probabilistic, deterministic and goal-seeking components.

5. Truth as an Act of Faith

The ideas I have proposed are unconventional. Let me, therefore, sum them up once again.

We have identified qualities that are akin, but never equivalent to the notion of permanence as unchanging continuity.

The atemporal world of chaos is one of pure becoming; nothing in it can correspond to permanence. In the prototemporal world truth is probabilistic, in the eotemporal world it is a changing approximation to determinism. In the biotemporal world permanence comprises a slowly changing instruction for maintaining the life process. In human history, covering laws and, with them, permanent governing principles have often been and are being continuously proposed but there is no single one so far upon which there has ever been general agreement.

What we observe here is the genesis and evolutionary emergence of qualitatively different forms of lawfulness, a process I described elsewhere as nomogenesis or the coming about of permanence-like features.

Since there are a number of qualitatively different kinds of permanence-like relationships in the world, there are also a number of qualitatively different kinds of truths, depending on the temporalities of the umwelts to which they pertain.

But, since none of them corresponds exactly or even closely to the notion of eternal verity or unchanging continuity, I conclude that all claims of truth-as-permanence necessarily include acts of faith subject to challenge and revision and all claims of truths are being continuously revised. A claim of truth definable as a permanent belief can exist only in the mind of a person who decided to maintain an opinion come what may; but even that kind of truth changes because the person ages and dies and others must come to maintain the continuity of their convictions.

Since all statements of truths involve acts of faith, claims of truths are never totally divorced from issues of morality that is, from prior judgements upon what is right and wrong. Such value judgements also show a nested hierarchical organisation reflecting the nature of time—but that is another lecture.

References

J. T. Fraser, *Of Time, Passion and Knowledge*. Princeton: Princeton University Press, 1990.

—— *Die Zeit—vertraut und fremd*. Basel: Birkhäuser, 1988; and München: dtv, 1990. Translation of *Time, the Familiar Stranger*.

—— *Il Tempo: una presenza sconosciuta*. Milano: Feltrinelli, 1991. Translation of *Time, the Familiar Stranger*.

—— *Genesis y evolution del tiempo*. Pamplona-Iruña: Pamiela, 1993. Translation of *The Genesis and Evolution of Time*.

17. MUSIC DO I HEAR? HOMER, BORGES AND THE PIED PIPER

It has been said that music is a universal language. This metaphor is defensible, provided one immediately adds that it is spoken, so to say, in many and different dialects. Also, that these dialects are often mutually unsharable in their emotive contents. Music travels easily across cultural and ethnic boundaries but it is unlikely to naturalize in foreign settings without readjustment to the new ambience. As a consequence, music across the globe presents humanity in its manifold, creative best. To illustrate this, the reader is urged to listen to Nigerian juju, Hawaiian chant, American country music, Cuban son, New York hip-hop, Cajun dance music, Beethoven's Choral Symphony, Caribbean steel band, Jewish klezmer, country fiddling and Navajo healing music. There are no cultures without some form of music, just as there are no cultures without some form of belief in a higher, universal power. Also, just as religions are numerous and differ substantially in their teachings, so do musics in their forms. What are the origins of the power and ever-presence of musics in the lives of men and women, both collectively and severally?

The task of exploring music-making all over the world and in all styles, as music is practiced today and was practiced in the past, is that of ethnomusicology. The term was coined by the Dutch musicologist Jaap Kunst in 1950 to replace the earlier term, "comparative musicology." I propose to seek the reasons for the universal power of music with the help of the integrated study of time.

Once, listening to Bach's Toccata and Fugue in D Minor, I was hypnotized by the running and rushing of its musical emotions, its cosmology of ambivalent and paradoxical feelings. It spoke about the elusive notion we call timelessness, also about the freedom of passion and about the pain of passing. The notes were gone before I had time to grasp that they went by. I was pummeled by the onslaught of ever new emotions, by the climbing of one breath upon another in a ceaseless descent into chaos and ascent out of chaos into a world of deeply felt but only vaguely understood timelessness—as considered in Essay 11. But music is more than a search for timelessness. Paradoxically, it is also a search for the temporal. One summer evening, in an Andalusian village, I watched the long, undulating skirt worn by a woman dancing the flamenco. It was a flamenco *jondo*, or profound flamenco. Intense. Brooding. Metaphysical. Celebratory. The skirt rose and sank, folded and unfolded. Set against the warm brown color of her skin, it was music praising the temporal, at its best.

Plotinus, the great 3rd century Hellenized Egyptian philosopher, wondered, "What man with music in his soul, beholding the harmony in the intelligible world, but must be moved by the harmony in sounds that are heard with the ear?"[1] Indeed. But why so? I will try to answer this question in three steps.

1. By visiting Homer and exploring the power of words spoken and heard.
2. By visiting the Pied Piper of Hameln and exploring the power of music played and heard.
3. Then, with the help of steps 1 and 2, interpreting music as a tool of individuation, a mode of defining the emotive self.

1. Homer and Borges: The Power of Words Spoken and Heard

It is possible for a person to be blind and, literally, lose sight of whatever is visible in space—and yet, possess a complete identity. But it is not possible to lose the inner sense of time and still have a functioning human mind. Let me illustrate this claimed by early poetry: poetry remembered, recited and heard.

During the seventh century B.C. the island of Delos, the smallest of the Cyclades, was a flourishing center of religious, political and commercial life. Citizens of ancient Greece gathered near a colossal statue of Apollo to celebrate the god with contests among poets who offered hymns in his praise. The extant record of such a contest, known as the Hymn to the Delian Apollo, ends with the singer bidding farewell to "all you maidens." Remember me, he tells them. If anyone asks you

> Whom think ye girls, is the sweetest singer that comes here, and in whom do you most delight? Then answer, each and all, with one voice: 'He is a blind man, and dwells in rocky Chios: his lays are evermore supreme.' As for me, I will carry your renown as far as I roam over the earth.....[2]

The singer, speaking these words, is believed to have been Homer. The passage is the earliest record attesting to his blindness. In 1955 H-J von Schumann, a German psychiatrist, examined the nature of dreams in the Homeric epics and noted that they were predominantly aural. Homeric characters tended to hear dreams rather than see them.[3] By way of comparison, dreams in classical Greek plays are primarily seen and only secondarily heard. Von Schumann and others, familiar with the dreams of the blind, considered this quality of Homeric dreams and concluded that they resembled the dreams of those who lost their sight in early life.[4] Their work added intriguing support to the twenty-seven centuries of tradition which maintained that Homer was blind, even though there is no explicit mention of this fact in the Iliad or the Odyssey.

What are the reasons for the remarkable endurance of the Homeric works? There are many. I shall concern myself only with two because they contribute to our understanding of the magic of music: the Homeric sensitivity to time in human life and the force of Homeric language.

A number of scholars observed that Homer was uncaring about what—today—we call "time," an abstract idea, something that flows independently of what happens in time. That idea did not yet exist in the epoch of Homer.[5] But Homer was very caring about the constellation of time, life, love, and death and also about the experience of passage. I believe that Homer's handling of the concrete temporal experience in the lives of his heroes is one of the great strengths of his poems.

Let me turn to his attitude to time and his use of language, by quoting from one of the beautiful narratives of the Argentinean poet Jorge Luis Borges, also blind. "Until then," wrote Borges, meaning until the time Homer lost his sight "he never dwelled on the pleasures of memory. Impressions have always washed over him, fleeting and vivid." He walked the "bustling marketplaces [where] he had heard entangled stories which he accepted... without attempting to find out whether they were true or imaginary." But "little by little, the beautiful world began to leave him; a persistent mist erased the lines of his hand, the night lost its multitudes of stars.... When he knew that he was going blind, he cried out.... But one morning he awoke... and unexpectedly felt... that all this had already happened to him and that he had faced it with fear, but also with joy and curiosity. Then he went deep into his past, which seemed bottomless, and managed to draw out of that dizzying descent" the memory of his first victory in combat and that of "a woman, the first given to him by the gods.... With slow amazement he understood. In the nighttime of his mortal eyes into which he was now descending, love and danger were also in wait for him... he already divined... a murmur of hexameters and glory... of black ships roaming the seas... [and the murmur] of the Odysseys and Iliads it was his destiny to sing and to leave resounding forever in mankind's hollow memory."[6]

In Borges' interpretation, Homer's blindness made him turn inward. It forced him to change the predominant area of his interest from that of seeing in space to that of speaking and hearing—in time. I imagine the spell-bound people of Delos, twenty-seven centuries before Borges, sitting and listening to Homer for days on end. The Iliad alone contains some 16,000 verses. He was an oral poet. a "sweet singer" in his own words, a member of the tradition of teller of tales whose stories were passed from generation to generation by word of mouth.

As did Homer, Borges also wrote of heroism, love and hatred. But, true son of our epoch. he continually played with the idea of time. In his wittings the past is always unfinished, it is always being created, not unlike the future. His cadences of intellect and emotion speak of a world in which all parts of reality hang together by the music of creation, composed by the Ordainer of Order. In the epilogue to a collection of his stories he wrote this:

> A man sets himself the task of portraying the world. Through the years he peoples space with images of provinces, kingdoms, mountains, ships, islands.... Shortly before his death he discovers that "the labyrinth of his lines traces the images of his face."[7] When humans of any epoch think of time, they trace, not the images of their face but the music of their minds. They hear and manipulate a narrative. For the mind of twenty-first century man—to coin a contemporary metaphor—time is a voice, narrating a story, heard on the cell phone, calling from eternity. The lasting voices of Homer and Borges illustrate the power of spoken language. The Homeric ideal of the noble hero, whether in battle or in the Olympic games, became the evaluational basis of Greek conduct and through it, remained influential in forming the Western scale of values, including taste in art and literature.

Obviously, then, the intellectual and moral foundations of a civilization may be laid with the assistance of and through the power of human voice spoken and heard. Is it also the case that the emotive identities of civilizations may be expressed, recognized and modulated with the assistance of and through the power of music played and heard?

2. The Pied Piper: The Power of Music Heard

Along the broad spectrum of different musics, spread between Bach's Toccata and Fugue and the flamenco jondo—both mentioned earlier—music explores, shares and expresses the human experience of being and becoming in the nature of time. It combines the predictable and the unpredictable, the permanent and the contingent.[8] I imagine a man who experienced a similar constellation of feelings, thirty-thousand years ago, as he watched a woman dance. Then, to stop the passage of her dance vanish into the past, he carved the Dancing Venus of Galgenberg.[9] This made it possible for him to carry the memory of the dancing woman with him, in a tangible form. That figure is an early form of a snapshot.

Let us join the long history of music and dance and visit the town of Hameln in northwestern Germany. In the year of 1284 the town was overrun by rats. They carried flees that were infected by *pasteurella pestis*, the bacteria of the bubonic plague. They spread the plague through their bites. That year a man appeared in Hameln. He said he was a ratcatcher and that, for a fee, he would rid the town of its rats. The fee was agreed upon. The man took a fife from his pocket and began to play it. From every nook and cranny rats began to gather around him. When he thought he had all of them he began to walk toward and into the river Weser. The army of rats followed him into the river, where they drowned. The man climbed back out and went to collect his fee. But the town fathers had second thoughts and refused to pay. By then, because of his many-colored coat, the man came to be called by the town folk, the Pied Piper. The piper took out his fife again. This time he played a different tune. He reached a different audience but produced similar results.

> There was a rustling, that seemed like a bustling
> Of merry crowds jostling at pitching and hustling
> Small feet were pattering, wooden shoes clattering,
> Little hands clapping, and little tongues chattering,
> And like fowls in the farmyard when barley is scattering
> Out came the children running.
> All the little boys and girls....[10]

These events, though not in these words, were inscribed in the wall of town hall and in a town gate of the village of Hameln. The inscriptions spoke about a magician who led one hundred thirty children away from the city. Other German and Austrian towns soon claimed to have been the locations of similar events. This is not surprising. It is quite interesting, however, that similar stories are found in England as well as in Syria and China.[11]

It has been suggested that the appearance of a Pied Piper relates to the Children's Crusade. Beginning in 1212, a French shepherd boy started gathering other children around himself for a crusade to the Holy Land, to take it back from the Muslims, not by force but by love. It is estimated that some 30,000 children crossed the Alps to Italy on their way to Jerusalem. The pain of the families, back home changed to horror when, later, many of the villages learned that during the year after the children left, a large number of children were sold on the slave markets of the Middle East.

This writer never heard Homer sing his lays nor did he watch the Pied Piper walk away with the children of Hameln. But he did witness, early in World War Two, the ecstasy of the crowds brought about by the oratorical triumphs of Adolf Hitler, routinely followed by choreographed marches.[12]

That theater of the virtual absolute was not unique. Similar dramas were surely witnessed by those who sang the Marseillaise or the Communist International or the Giovinezza of the Italian Fascists—each in its proper year and proper Hameln. All these were of small dimensions compared to a 1964 performance by the Beatles, heard by 73 million people.

Homer did not need eyesight to recite his epics and, through them, set the scale of Western values of conduct though, I would think, he had to have had his pre-blindness years to have access to the visual elements of those epics. Ludwig van Beethoven began to lose his hearing at the age of thirty, lost it totally by the age of forty-nine, died at the age of fifty-seven. A look at the list of his works documents no decline in the intensity of his work. To the contrary: his work matured in sophistication and complexity.

The examples of Homer and Beethoven illustrate the ability of humans to imagine the sharing of their thoughts and feelings through the sound of human voice and the sound of music. Also, that they can do so without recourse to the total functioning of the very senses that helped establish their personal identities. I submit that this unique, minding ability is at the foundations of the universal power of music.

3. The Art of the Audible "Now"

It has been said that music is the art of time. Indeed. But—why so?

An answer, from music theory, is likely to cite the harmonies of musical sounds and the creation of a dialectic that modifies the cognitive and emotive dimensions of our experience of passing time. But, as it has been demonstrated in the preceding essays, there is nothing in inanimate nature that corresponds to the experiential flow of biological and noetic time. To understand music as the art of (organic, noetic and social) time, one must employ the integrated study of time. Let us start with the coordinated clockshop model of life, introduced in Essay 9.

> Biological oscillations, observed across 24 orders of magnitude [in frequency] and synchronized from instant to instant, constitute the life process. In the phenomenal world, the inner synchronization is manifest as the organic present. It is thus that life creates a 'now' in the presentless world of nonliving matter. The organic present, so born and maintained, allows distinctions to be made between present and nonpresent conditions. In their turn, the nonpresent categories of time may be separated into futurity and pastness in terms of the present needs and available means of an organism.[13]

The spread of the frequency of biological oscillators across 24 orders of magnitude, when expressed in measures employed in music, is a range of 78 octaves. The spectrum of audible cycles spreads from about 21,000 Hz down to perhaps 15 Hz. This is only 12 ½ octaves, although the spectrum of music is broader because it includes repetitive units which may take up seconds, melodies which may take up many seconds, compositions that may take hours. It is interesting to note that as we leave the range of audible frequencies we enter the domains of the mental and social cycles and processes.

In complete and valid analogy to the organic present, created through instant by instant coordination of the 78 octaves of biological oscillations of living systems, the musical present is created when cacophony is replaced by polyphony, when the different musical parts are synchronized from instant to instant according to whatever stable rules composer, performer and listener find acceptable. While the organic present attests to—is demanded by—simultaneities of biochemical necessities, the musical present attests to—is demanded by—simultaneities of aesthetic needs. Once the musical present has been defined through sound, musical memories and expectations may be created, with reference to it, through combinations of pitches, rhythms, melodies, stases, variations in timber and texture. Musical expectations and memories constitute the experience of the flow of musical time.

The immense wealth of musical sound that leaves the listener with a sense of motion or passage is anchored in the musical now This is strikingly illustrated by the fact that the engraving on a phonograph record or the numerals of a digital recording—no matter how complex the sound—are but single lines. A composer of unusual gifts should be able to draw a single wiggly line or write a single, continuous series of 1-s and 0-s, directly on a paper, without employing the instruments that change the oscillations of air molecules to an engraved line or to a line of recorded binary digits. Once such a wiggly line or one made up of 1-s and 0-s are transformed, by electronic devices, back to audible sounds, a listener will be able to distinguish particular voices in a choral work of two-hundred voices and even identify a mistuned violin in an orchestra.

That magic of identification is accomplished by the listener's mind which, as a part of its task of maintaining the mental present, analyzes all impressions, then integrates them within the nested hierarchy of organic, mental and social presents. The same analysis and synthesis permits the longitudinal organization of music. A symphony or an opera may thus become the sonic cosmology of an individual life, of an ethnic group or of the universe, from its *Urklang* to the twilight of its gods.

It has been suggested that the origins of language involved the separation of the emotive from the intelligible content in voiced utterances. It is an interesting thought which may be spun further. The intelligible components became language, the emotive ones music, with no well-defined separation between them. They are profoundly interwoven. This permits each to address the domain of the other. Because it is created through the sense of hearing, as is spoken language, music shares with language the task of identifying the individual selves. Musical experience thus enters the labor of individuation and becomes as intimate and as important as are our ceaseless negotiations with life and death.

Analogous reasoning holds for collective identities. While the flag as a symbol of collective identity is in the external world, an anthem or a march is a part of the audio loop of self-definition. If sung together in the shared musical present, they help establish group identity. No comparable power may be attributed to, say, the Great Seal of the United States.

Music is created and is recognized through hearing, as is spoken language. Musical experience enters, as does language, directly into our faculties of articulated voice. It shares with language the task of self-definition and becomes as intimate as is the coming to terms with life and with passing. The universal power of music—just as the power of spoken and heard words—resides in this immediacy. Since human concerns continuously demand the definition and redefinition of personal and collective identities, musical practices and preferences may be seen as the temporal corollaries of self-images in space.

18. A DIFFERENT WONDER

To build his burial chamber within a pyramid, King Cheops of Egypt (26th c. B.C.) ordered people to work in gangs of one-hundred thousand, for periods of three months. "The pyramid itself took twenty years in the building. It is a square, each side is eight-hundred feet long and the same in height…." wrote Herodotus, in his *History*, early in the fifth century B.C.[1] So began the recorded history of the oldest and only surviving example of what came to be known as the Seven Wonders of the Ancient World. Whatever was to be classed among them was decided in the second century B.C. by the Antipater of Sidon, a Macedonian regent. Each of the Wonders, in its time, was viewed as the greatest structure on earth. Each combines great skills of architecture and engineering, with religious convictions appropriate to place and time. Each expresses the collective pride of its makers and hence each may serve as an icon of the people who built it. Here they are, listed in the sequence they were built.[2]

- The Great Pyramid of Giza is made of almost two and a half million blocks of stone, each weighing about two and a half tons. Built some time between 2575 and 2467 B.C., it served as the traditional tomb for the Pharaohs of Egypt,
- The Hanging Gardens of Babylon were terraced gardens. Their outer walls were said to have been 56 miles long. They were built early the 6th century B.C. by Nebuchadnezzar II, to please his wife who missed the greenery of her native Médéa, an Algerian town near Algiers that still bears the same name.
- The Temple of Artemis was a shrine to the goddess, built around 550 B.C. Constructed entirely from marble, it housed many statues, including those of Amazon women. A goddess of fertility, Artemis was pictured as draped with eggs and multiple breasts.
- The Statue of Zeus at Olympia was erected around 450 B.C. It served as the location of the Olympic games, celebrating Zeus, the king of Greek gods. His sandals and his robe were made of gold. His scepter, inlaid with metals, had an eagle perching on it.
- The Mausoleum at Halicarnassus, Turkey, was the burial place of King Maussolos. Built early in the 4th century B.C., it was decorated by the statues of people, horses, lions and other animals. Its top formed a 24 step pyramid. Visitors described it as magnificent.

- The Colossus of Rhodes was a bronze statue of the sun god Helios. It was said to have been 32 meters high. Erected around 300 B.C., it was toppled by an earthquake seventy years later. Nine centuries later it was broken to pieces by Arab invaders. Its bronze was sold for scrap.
- The Lighthouse of Alexandria, built around 280 B.C., was said to have been 130 meters high, with a spiral ramp leading to the top where a fire burned at night. It was a technological triumph of its age and a prototype of all later lighthouses. It was still standing in the 12th century but toward the end of the 15th century, it collapsed. Its material was used to build a fort.

What may the post-Renaissance centuries offer to match these Wonders? The Eiffel Tower? Saint Peter's in Rome? The Big Ben? Westminster Abbey? The sculptures of the Mount Rushmore National Monument in South Dakota? Memories of the Twin Towers of Manhattan? The Berlin Wall and its copies? These objects, though impressive, do not radiate that self-possessed immensity which the ancient Seven Wonders seem to have radiated for the people of their epochs and places. They did so through a hallmark they all shared. Namely, each was built to conquer the passage of time, one may say, by brute force. By immensity. They were intended not simply to last but to outlast all other structures.

To serve as the icon of the last few centuries, I propose to identify a very different type of Wonder. As did the ancient Wonders, this one also attempts to conquer time's passage, but not by being a structure that outlasts all others. It is not even an object. It is a way of understanding nature and placing that understanding in the service of opposing the ravages of time. It is a way of knowing that already brought about many changes. Such as an explosion of (unevenly) burgeoning familiarity with the universe and man. Such as an eruption of demands for food and general wellbeing. Such as cultural transformations that made the globalization of the earth not only possible but unavoidable.

This Wonder is called science. It may be represented by an aphorism. It is: an art of conversing with stones, with plants, with animals and with humans. The question-and-answer periods are called experiments. When it involves talking with stones, the language is mathematics. As the systems with which science converses become more complex, when stones are left behind for plants, animals and humans, the language spoken must accommodate the increasingly freer forms of causations.[3]

Instructions for these conversations with nature were born in ancient Greek philosophy, but they grew to maturity only with post-Renaissance thought. This way of talking with and thinking about the world may be easily carried across the boundaries of cultures but, unlike in the case of music, the scientific way of reasoning and testing cannot be modified to match existing, established tastes and preferences. Instead, it is the locally preferred ways of thinking that must adapt to the methods of reasoning and testing of the sciences. This has, in fact, been happening. And, for very utilitarian reasons. Namely, because it is only through scientific conversations—with stones, palm trees, slugs, butterflies, bears, whales and people—that the well-being promised by industrial civilizations may be realized. But these conversations have their down sides. They bring with them disturbing spirits. They were let loose when Pandora's box was opened, with the understanding of the laws of nature in it.

I would like to nominate the figure of a 16th century fortune teller and magician to represent the temperament needed for negotiating with nature the scientific-industrial way. It is a legendary figure, named Faust. Through centuries of sculpting by great writers and poets, the figure of Faust changed from that of an itinerant confidence man to a humane, rational and powerful character, subject to the temptations of the flesh and to the demands and rewards of love. He has become the prototype of Western man who, in the words of Joseph Needham—to be cited in Essay No. 19—"suffered from the schizophrenia of the soul, oscillating forever unhappily between the heavenly host on the one side and the atoms and the void the other....." Faust could become the Wonder of the post-Renaissance world because, in the creation of a globalizing earth, he enlisted everyone who is—in the words of the Bard—"of imagination all compact."

The essay that follows concludes that Faust the Wonder, the magician, did build a new and immense world and that he is still on the road, trying to understand just what is it he achieved and how to live with it.

THE PROBLEMS OF EXPORTING FAUST[4]

Summary

For more than a millennium before the Renaissance, the Chinese were ahead of Europe in applying their knowledge of nature to useful purposes. Yet the most powerful methods of such applications, those of modern science, were not born in China. Reasons for this inversion in the rate and direction of development of scientific knowledge have been sought in different social conditions, cultural values, philosophical stances, languages, and attitudes toward history.

This paper approaches the question of why modern science was born in the West by focusing on the functional basis of natural science: its demands for the mathematization of hypotheses about nature and the validation of those hypotheses by experiment. It suggests that the demands for number and measure originated in certain closely allied teachings about time that were native to the West.

The modern concept of scientific law grew out of the metaphysical conviction that processes may be divided into those that are timeless (lawful, quantifiable, eternal, divine) and temporal (contingent, qualitative, passing, earthly). This time-timeless dichotomy, with roots in Pythagorean number mysticism, Platonic idealism, and Christian theology, has become a part of the unquestioned metaphysical assumptions of science.

An identically stark division may be found in Christian moral philosophy. It is between the timeless, divine rules set for human conduct and the temporal demands of human instincts. The conflict between the rules and the needs suffuses the character of the industrial West: tenseness, restlessness, a mechanical-analytic turn of mind, admiration of inorganic naturalism, and the love of number and measure.

What we know as the scientific method is a pragmatic synthesis of the intellectual heritage of the West with Christian morality, in the spirit of the Reformation and in the service of early mercantile capitalism.

The absence of indigenous natural science in China might therefore be partly attributed to the traditional Chinese preferences for organic naturalism and to the Chinese regard of history as the most exalted form of knowledge, in contrast to the West's regard for mathematics as the queen of the sciences and its high esteem for inorganic naturalism.

The figure of Faust is used as a symbol for that cultural, social, and philosophical ambience that gave rise to the favored identification of truth with the numerical and the experimental. This paper examines some of the problems caused by transferring Faust from his native habitat to the heterogeneous cultural world of our age. The difficulty is in the taming and subordinating of modern science and technology to the needs of people.

In the People's Republic of China, the task is that of integrating modern science and industrial productivity into a non-Faustian civilization. In the West, it is the integration of the powers of the scientific industrial state into Western humanistic tradition.

It is argued that the necessary integration cannot be based on a natural philosophy that divides the world into time and the timeless, because such a bifurcation does not allow for a continuity in nature between the atemporal, physical basis of the world at one end, and, at the other end, the demands of the human mind for free creativity, associated with noetic and social time.

As a replacement for the time-timeless dichotomy, this paper examines the hierarchical theory of time. Within the principles of natural philosophy appropriate to the hierarchical theory of time, a continuity among physical, organic, mental, and social processes may be traced in a manner resembling the ideas of evolution by punctuated equilibria.

The Platonic-Christian perception of time and the timeless corresponds to a conservative universe in static harmony. The hierarchical view of time corresponds to the contemporary perception of dynamic reality, organized along a hierarchy of increasingly complex structures and processes that remain, from the point of view of evolution, open-ended.

The ecumenical science of a contemporary Faust, it is concluded, must acknowledge at the basis of the universe the evolution of time itself.

Our epoch is characterized by a single, overwhelming economic fact: the needs of people for food, goods, services, and housing, judged in terms of immediate shortages and rising expectations, outrun what the productivity of nations can supply and the existing economic and political systems distribute. And if the productivity and means of distribution were available, it is not at all certain that the earth could keep on offering the necessary raw material and energy, in a sustainable manner. In spite of the phenomenal increase in the rates of production, the socioeconomic conditions at the end of the century demonstrate the continued validity of the principle of Thomas Malthus concerning population growth and the supply of goods.

The population explosion, together with increased expectations, has offset the increased rate of productivity. The ideas of Malthus, that nineteenth-century prophet of doom, are unpopular today because they do not conform to what has been called criminal optimism. The phrase stands for a slap-happy attitude popular with planners around the world, one that mistakes half-baked science fiction for social reality or else—or in addition—assigns magic powers to one or another ideology.[5]

The most frequently proposed remedy for catching up with the Malthusian lag and even for solving practically all other problems, from poverty and illness to war and pollution calls for more science and technology. The business concerns of the capitalist world build their promotion of more technology on the erroneous identification of the ascent of man with the history of science. As far as I can tell, the socialist countries do the same. It is only quite recently that these omnibus stances came to be challenged by calls for appropriate technology.

Assuming that increased scientific understanding of man and the world, without an accompanying humanistic-political control imposed by a major catastrophe, could in itself heal the ailments of our age, the question would still remain: What kind of science ought that healing knowledge be? How are we to select its principles and methods?

Joseph Needham has maintained a hope for the coming about of a single, ecumenical science as an accompaniment to the growth of organization and integration of human societies. He compared Chinese and Western mathematics, astronomy, physics, botany, chemistry, and medicine. He then constructed a graph, depicting the achievements of these sciences in a quasi-quantitative form, against a scale of centuries. The plot suggests the existence of "fusion points" in the development of the sciences into world-wide, ecumenical bodies of knowledge. He then remarks that "the more organic the subject-matter of a science, the higher the integrative level of the phenomena with which it deals, the longer will be the interval elapsing between the transcurrent point [of roughly equal sophistication] and the fusion point....[6] However, he goes on, given enough time, all knowledge will be subsumed "into the same oecumenical natural philosophy... until the coming of the world cooperative commonwealth which will include all people as the waters cover the sea."

What kind of natural philosophy, what kind of conceptual scaffolding could correspond to the ecumenical science of this beautiful vision? Could we look to science, as science is understood today, to provide the necessary multidisciplinary and intercultural system of thought?

I do not believe so, at least not until after epistemology ceases to be patterned upon those theories of knowledge that have their roots in the Platonic idealism of time and the timeless.

1. The Eternal and the Temporal

The separation of the changing from the permanent, the unpredictable from the predictable, distinguishing a perspective of becoming from a perspective of being in the human experience of time, has been a favorite game of philosophers ever since the pre-Socratics held their own time-society meetings along the shores of Magna Graecia and on the islands of the Aegean Sea.

Little did they anticipate that by the time of Newton a particular way of separating the lawful from the unpredictable would have become the formal basis of a new natural philosophy that, through its offspring, the Industrial Revolution, would change the patterns of Western life. Neither could they have guessed that the Pythagorean duality of sky-geometry on the one hand and earth-existence on the other would have become, through the labors of Plato, the paradigm of scientific thought concerning the nature of time.

The cutting of the pie of human time experience into no more and no less than two slices—the eternal and the temporal—proved to have been necessary for the formulation of the Galilean laws of motion and for all subsequent laws of physical science.

The success of the natural sciences, which all employ this dichotomy, appears to have demonstrated that the division of the world into time and the timeless was an undebatable fact of nature. The duality entered the metaphysical foundations of physics, and through it became a dogma of contemporary knowledge in general.

The thesis of this essay is twofold. First, I will argue that the sharp division of the world into what is temporal and what is timeless, sometimes identified with a division between what is judged physical and what is regarded as mental, has outlived its usefulness as a scientifically supportable natural philosophy of time. Second, I will maintain that only by replacing the antiquated dichotomy by a model of time appropriate to our understanding of the universe, to wit, by the hierarchical theory of time, can we hope to create an ecumenical system of knowledge.

Dividing the world into time and the timeless is an inadequate matrix for social history and elucidating social fact; it is too restricted a scheme to aid in understanding psychology; it is not a rich enough temporal background to evolutionary biology, and it cannot even accommodate the temporal behavior of inorganic matter, as revealed by contemporary physics.

Furthermore, the rigidity of defining time by its contrast to the timeless (to eternity) has prevented the development of a theory of knowledge that could subsume, in a systematic whole, the many ways in which we have learned to explore the relationships between man and the universe, among different forms of life, and among human beings.

Since the different sciences must ask their questions in different forms and set different criteria for the testing of truth, it is futile to search for a single, prescribed method of inquiry and a single kind of test for legitimating fact or theory. Because of these varying demands, the plurality of principles and modes of understanding, current in physics, biology, psychology, and social science, appear to be chaotic and without any evident, overarching pattern. The absence of an obvious, common perspective and mode of inquiry has come to be regarded as an unavoidable corollary of the scientific method and, therefore, a necessary hallmark of our culture. But a fragmented, inchoate view of the world is intellectually moribund because it fails to account for the striking unity that, in our experience, holds together all natural structures and processes.

What is needed is a new theory of knowledge that can identify a continuity among the many ways of understanding the world, one that could somehow join those phenomena that have been regarded as timeless with those that have been judged as temporal. Such an epistemology must be able to subsume in itself the hierarchy of causations, identities, beginnings and endings, and different regions of undeterminacy that the sciences of matter, life, man, and society have identified.

The evidence of these hierarchies suggests a new natural philosophy, one based on the unorthodox assumption that time itself is of a hierarchical character. As do causations and identities, so would time itself appear as a corollary of the complexities of processes and structures, revealed by contemporary knowledge along the different, stable integrative levels of nature. Since the dynamics of the universe is one of the evolutionary development of complexifying structures and processes, it would then also follow that time itself has evolved with the increasing complexification of matter. Should such an unusual postulate about time be made plausible by detailed argument, we would then have a conceptual scaffolding for an epistemology, appropriate for a truly ecumenical science.

It is easy to demonstrate that the concept of time has undergone many historical changes; but this is *not* my claim. My proposition is that time had its genesis in the early universe, has been evolving in a manner that resembles the dynamics of punctuated equilibria, and remains developmentally open-ended. I have dealt elsewhere with the conceptual difficulties and pragmatic consequences of such a theory.[7] Here we are concerned only with those aspects of the hierarchical theory of time that are significant for the theme of our conference: time, science, and society.

Together with his ideas of time and eternity, Plato's legacy to us includes the significance of quantitative thought. I would like to take advantage of this heritage and approach an understanding of the three-cornered relationship among time, science, and society through a critique of quantitative thought, because thinking in terms of numbers was necessary for the birth and growth of Western science and industrial civilization. Specifically, I will examine the origins, justification, and limitations of the peculiarly Western preoccupation with number.

In my reasoning I will appeal to the collective judgment and social ambience of the industrialized West because it is social values that guide—even when they repel and thereby guide—individuals in their search for knowledge. I agree with Ludwig Feuerbach that "the single man for himself possesses the essence of man neither in himself as a moral being nor in himself as a thinking being. The essence of man is contained only in the community and unity of man with man; it is a unity, however, which rests only on the distinction between I and thou."[8]

2. The Dominion of Number and Measure

Even a cursory examination of the history of Chinese science and technology will reveal the presence of great refinement in observation, ingenuity in classification, and a remarkable diligence in collecting systematized knowledge about nature. However, as Joseph Needham remarked some time ago, "what the Chinese did not develop indigenously was the method of testing by systematic experiment of mathematical hypotheses about natural phenomena."[9] Chinese science became eclipsed in the seventeenth century by the mathematization of physics in Europe. Nathan Sivin thus summed up that fateful change in the Western perception of nature: "The Scientific Revolution was a transformation of our knowledge of the external world.... It established for the first time the dominion of number and measure over every physical phenomena."[10]

The doctrine that all things and forms are number, that *harmonia*, a balance and order according to number, is the law of the universe, is associated with the name of Pythagoras and with the Pythagoreans. The chief task of the philosopher, in the Pythagorean spirit, was to discover the rules of number. Numbers performed a divine dance and exhibited through it the ultimate principles of the universe. Having found in the rules of number the key to absolute truth in a cosmic context, members of the Pythagorean brotherhood appealed to those rules for the salvation of their souls.

Benjamin Farrington, in his work on Greek science, has considered the Pythagorean use of number in its relation to knowledge and society. Mathematics, he wrote, not only provided a better explanation about the nature of things than was theretofore available, but it also kept

> the souls of the brethren pure from contact with the earthly, the material, and suited the changing temper of a world in which contempt for manual labour kept pace with the growth of slavery. In a society in which contact with the technical processes of production became ever more shameful, as being fit only for slaves, it was found extraordinarily fortunate that the secret constitution of things should be revealed not to those who manipulated them, not to those who worked with fire, but to those who drew patterns on the sand.[11]

The quasi-religious and mystical principles of the Pythagoreans came to maturity in Plato's theory of knowledge, a philosophy of great beauty and power. It is a way of looking at the world, and probably the most significant single contribution of Greek antiquity to the birth and development of natural science. But the Platonic and Aristotelian notions of time as a countable image of eternity could not have achieved their privileged status in the history of Western thought—they are unlikely to have been incorporated in the ways that the laws of nature came to be written—had these notions not been reinforced through certain beliefs and practices in the daily lives of the people of Christendom.

Reflecting upon the rise of Christianity, Needham wrote in his Herbert Spencer lecture that, "for the ancient Mediterranean thinkers, the world, which had neither beginning nor ending, was growing neither better nor worse. It has been powerfully argued ... that the major contribution of Christianity and one of the principal reasons why it vanquished its competitors among the religions of the Roman Empire was precisely that it introduced change and hope into the stagnating sameness of the ancient world."[12]

After the downfall of the Roman Empire, the message of revolution taught by Christ and his immediate followers metamorphosed into the theological edifice of medieval Christianity. The intellectual and spiritual labors of the Church Fathers and of most of the Schoolmen form a majestic mire that tends to alienate the modern reader. Yet, it was a working synthesis between whatever they saw as the eternal, unchanging laws of God and the temporal, unpredictable fate of man.

As part of that religious-philosophical synthesis, the respect for number and quantity makes its explicit appearance very early. Thus, St. Augustine, writing at the turn of the fourth century about the rules of number and the rules of wisdom maintained that "although it is not clear to us whether number is a part of or separate from wisdom, or whether wisdom is a part of or separate from number, or whether they are the same, it is clear that both are true, and immutably true."[13]

Johannes Kepler, in the preface to the reader in his *Mysterium Cosmographicum*, composed twelve centuries after the days of St. Augustine, asserted that "Nay, the idea of quantities have been in God from eternity. Quantities are identical with God; therefore they are present in all minds created in the image of God.... In this matter both the ancient philosophers and the Doctors of the Church agree." And again, in his *De Harmonice Mundi*, he maintained that "geometry is coeternal with the mind of God, it is identical with God himself; it served as a model for the Creation of the world and together with the image of God it was transplanted into man, and not simply received [by man] through the eyes."[14]

Yet another 350 years pass and we read in Einstein's autobiography that "in a man of my type ... he major interest disengages itself to a far-reaching degree from the momentary and merely personal, and turns toward the striving for the mental grasp of things."[15] For Einstein the mental grasp of things meant the Pythagorean geometrization of time into space-time. It is only a small step from here to C. W. Misner, K. P. Thorne, and J. A. Wheeler who go even further when they maintain that "the proper arena for the Einstein dynamics of geometry is not spacetime, but superspace."[16] A cut through superspace is "a leaf of history," which "describes the deterministic, dynamical development of space with time."[17]

By withdrawing into increasingly sophisticated mathematical abstractions, the contemporary scientific view of time—by which people invariably understand the views of time in physics—negates the significance of all events and processes that cannot be included in the numerical order. Parmenides and Plato would be proud of this brave denial of all that is generated, but we are stuck with a growing chaos of ideas concerning the nature of time. Each step of further abstraction offered as an elucidation of the temporal aspect of reality only moves the notion of time further away from experience and hence from the concerns and predicaments of life, man, and society.

Based on what and how St. Augustine, Kepler, and Einstein wrote on number and reality, there is little doubt that these men firmly believed that eternal, mathematical relationships constitute facts of nature entirely independent of man.

A metaphysical stance of such impressive continuity—from Plato to Einstein—could not have maintained itself if it had been detached from and independent of socially reinforced values implicit in the daily behavior of people. On the contrary, the identification of the eternal and timeless with number, and the simultaneous distrust of the free, the formless, and the unpredictable, maintained their hold upon the minds of those who created Western science, precisely because such opinions were organic parts of socially approved thought. In such matters, collective guidance works by embodying praiseworthy ideas among the teachings of praiseworthy conduct.

What among Christian practices and beliefs could have reinforced the high regard for number and geometry? What kind of behavior, expected of good Christians, has helped reinforce and perpetuate the dominion of number and measure in the Western understanding of nature?

Christianity in general and Protestantism in particular advocated that people control their most powerful drives, those associated with the preservation of the self and the perpetuation of the species, by means of sublimation rather than through satisfaction. With such teachings in the mind, the enterprise of life became a struggle between what people felt they really wanted to do and what they believed they ought to be doing.

In the ensuing call for self-control, the abstractness of geometry recommended itself for employment because geometrical forms and mathematical formulas are void of the temptations of "wine" and "women," though not totally of "song." The conspicuous absence of all human and animal figures from Islamic religious art and the ubiquity of geometrical forms in their stead is noteworthy. They demonstrate the intuitive realization that pure geometry does not tempt the flesh and seldom excites disturbing fantasies.[18]

In postulating laws of nature, deviation from geometrical perfection has been a taboo throughout the history of the exact sciences up to and including our own days.

The deep hold upon scientists by an aesthetic and even moralistic commitment to timeless forms has been described by G. J. Whitrow, following a term of Emile Meyerson's, as a trend for "the elimination of time." He saw it exemplified already in Archimedes' *On the Equilibrium of Planes*, a treatise that "attained the ideal, so earnestly sought in our days by Einstein and others, of reducing a branch of physics to a branch of geometry. . . ."[19]

Around the turn of the sixteenth century, Johannes Kepler challenged the Greek ideal that the orbits of the planets must be circular or at least curves generated by the relative rotation of circles. He asserted, instead, that the planetary orbits were elliptical, but he remained unhappy about his "ovals." Early in 1605, in a letter to the Danish astronomer Christian Longomontanus, he would only insist that they are less reprehensible than the ideas of the Ancients.

"You accused me," he wrote to Longomontanus, "of having sinned with my 'ovals,' yet you hold the ancients faultless for their 'spirals' [epicycles and helices]. If my ovals are but a cartful of dung, the spiral of the ancients are whole stables full of dung." He insisted that his is a very great improvement over the received views—but *unum carrum fimeti*, just the same.[20] The power of geometrical idealism, the intellectual ancestor of Francis Haber's "technological idealism" was and remained great, disturbing as this idealism has been against collectively held rules.

Early in the twentieth century the German sociologist Max Weber remarked that the origins of natural science ought to be sought in the decided propensity of Protestant asceticism for empiricism that was rationalized on a mathematical basis. He observed that

> the favorite science of all Puritan, Baptist and Pietist Christianity was . . . physics, and next to it all those natural sciences that used similar methods, especially mathematics. . . . The [mathematical] empiricism of the seventeenth century was the means for asceticism to seek God in nature. It seemed to lead to God, philosophical speculation away from him.[21]

The skillful use of number also brought the merchant increased prosperity, supporting his conviction that his belief in the immutable laws of nature and God were obviously correct. For the commercial and industrial capitalistic states, no less than for Plato, God was a mathematician.

Sir William Petty, the seventeenth-century founder of demographic statistics, wrote an essay called "Of Lands and Hands." In it he put forth the uses of number in the well-being of a nation.

> From the number of freeholders in each Country....
> From the number of other Electors....
> The numbers of people that are of every yeare old from one to 100....
> From the premises & the yearley interest of mony per cent....
> From the knowledge of the numbers of professors of all religionaryes who are distinguished by sensible marks....
> By knowing the number of working hands....
> By the knowledge of Naturall & Incurable Impotents....
> By the proportion between Maryd & Unmaryd teeming women....
> The number of persons living by seafishings....
> By the number of people & provisions of Cattle & Corne....
> By the number of ships yearly cast away & seamen drowned....
> By the number of Decrees, Verdicts & Judgements in all Courts...
> [may we judge the quality of our lives.][22]

The essay reads like a divine service, a litany, with the celebrant reciting the verses in praise of the efficacy of number, the faithful repeating the responsory. Indeed, in the history of the West, stemming from Greek antiquity and favored by Christianity, the absolute and ultimate principles of the universe had to be numerical and timeless. Time itself remained "but a walking shadow, a poor player/That struts and frets his hour upon the stage," and, after the teleological goal of Creation is reached in the Final Judgment, "is heard no more."

But the timeless cosmos, that Republic of Numbers, has not been without its troubles, which we will discuss, after a brief detour to the time of the first unification of the Middle Kingdom.

The Chinese have been very much interested in number but not in the quantitative. A notable exception was the ideology of the Legalists, a school of thought that rose to significance some time during the fourth century B.C. Theirs was a strict view of law and order, based on what may be called quantitative measures of justice.[23] They drew up a list of "Six Parasitic Functions" of people that sap the power of the authoritarian state. They included care for old age, beauty, love, and living without employment. They also had a list of Five Gnawing Worms, ruinous to the state, such as discussing benevolence and rightousness. The early meaning of their Chinese name signified "standard," or law fixed beforehand (that is, without reference to custom). The only emperor who attained power with the help of the Legalists, in 221 B.C., standardized not only orthography but also weights, lengths, the widths of roads, and the gauge of chariot wheels. Degrees of penalties were assessed by precise quantities, laid down in numerical detail.

It is tempting to speculate that this kind of a puritanical-mechanistic outlook, if maintained, might have established in China a preference for the quantitative in human experience. But in fact it did not, because the humanitarian resistance to tyranny and dislike of codification for behavioral norms made it impossible. In two decades, the rule of the Legalist emperor came to an end, giving way to the milder rule of the Han dynasty. Slowly but surely, in the words of Needham, Legalism was rejected by the Chinese people as the country returned to the traditional organismic, nonabstract, human-centered, qualitative rule of law.

3. Trouble in the Republic of Numbers

In the West, the power of quantitative thought, built into natural science, was pressed to serve a civilization that had been instructed by the God of the Old Testament to gain dominion over the earth. Those instructions were heeded, industrial technology was born and grew to be remarkably powerful. But just as genetic errors accumulate in a living body and lead to the crises of aging and death, even so certain peculiar errors of judgment have been accumulating in the body social of the industrial lands. From those lands they have been exported to all corners of the globe, carried, like pathogenic vectors, on the backs of promises for better lives for all.

The errors I have in mind come from a neglect—even a negation—of those biological and mental functions of people that cannot at all be related to the *harmonia* and numerical order of Pythagoras and Plato. Specifically, I am talking about the fact that beneath the immense creativity of our species lies the domain of most unharmonious, and essentially unresolvable conflicts. Unless these are carefully and shrewdly managed, they make humans as immensely destructive as they can make them constructive.

Watching the events of our own days and the attendant emotions bubble over, stripped of culturally imposed niceties, I cannot help observing that contrary to the way we have been classified, homo sapiens, we are only superficially reasoning creatures. Basically, we are members of a desiring, suffering, death-conscious, and hence time-conscious species. But this model of man has no place in geometrical idealism.

One consequence of having placed all our eggs of hope in the quantitative-rationalistic basket, as if people were really reasoning and wise creatures, is the increasing loss of relevance of scientific advances to the universal need for food, shelter, health, and a modicum of dignity.

To describe the way I perceive the situation, I wish to appeal to the metaphor of a computer that uses a certain language. People handling the computer must learn to speak, or at least to write that language. In the course of time, intimate and intense work with the computer will influence the language of its operators. Through language the computer will come to mold their ways of thought, their scale of values, and, in general, their judgments of the nature of reality.

What I just called a metaphor is not a metaphor at all. There is solid and extensive evidence that computers, with their peculiar demands and capacities, have already changed the value judgments of the computer states. Problems that call for qualitative judgments are increasingly assessed in terms of numerical calculations. Human behavior itself is being increasingly interpreted in terms of computer technology and concerns that should be framed in terms of historical man are increasingly framed in terms of the ahistorical present of machinery.

Joseph Needham's comments on the Legalists in the China of twenty-three centuries ago are uncannily valid for the scientific-computerized states of our own epoch.

> It is in connection with mathematics, geometry and metrology that we come upon the fundamental philosophical flaw in Legalist thinking. In their passion for uniformisation, in their reduction of complex human personal relations to formulae of geometrical simplicity, they made themselves the representatives of mechanistic materialism, and fatally failed to take account of the levels of organisation in the universe.[24]

Our own, contemporary changes in the collective temporal umwelt, reflecting the shift from the paradigm of historical man to the paradigm of ahistorical mechanical-electronic functions, is well represented by the introduction and success of digital clocks. Whereas dial clocks are models of the continuity of the revolving sky—Plato's moving image of eternity—digital time pieces identify as significant only the present. Literature reflects similar changes. The traditional novel with its many-tiered temporal structure is being replaced by journalistic reports about the present, marketed as books. In evolutionary terms, we are adapting to a man-made-machine environment, constructed according to the inorganic naturalism of Western science.

In a 1964 paper on science and China's influence on the world, Joseph Needham remarked that

> Europeans suffered from the schizophrenia of the soul, oscillating for ever unhappily between the heavenly host on the one side and the atoms and the

> void on the other; while the Chinese, wise before their time, worked out an organic theory of the universe which included nature and man, church and state, and all things past, present and to come. It may well be that here, at this point of tension, lies some of the secrets of the specific European creativeness when time was ripe.[25]

The Chinese are hardly alone in their search for the organizing principles of stars, beasts, man, and society, though they are probably unique in terms of the sophistication of their organic naturalism. In stark contrast, the Western vision of "nature and man, church and state, and all things past, present and to come" is one of mathematical order, that is, the very same system of thought that is responsible for the West's scientific and industrial progress.

Many of the fundamental works in the history of Western science display the same explosive but also painstakingly elaborate character as do the writings of certain schizophrenics. They suggest mental states that crossed a critical threshold of some kind. They witness inner tensions that can be lessened only now and then, and only through the ceaseless search for increasingly abstract and, presumably, increasingly timeless ideas.

These are symptoms of alienation from the flow of human time, untoward consequences of having identified the temporal umwelt of the human mind with the temporal umwelt of the physical world.[26] They demonstrate a retreat from human reality, in analogy to the retreat of schizophrenics from the responsibilities of a partly open future, into the narrow confines of an eternal present. In different words, it is an abnegation of human freedom.

The suggestion emerges that if such features of the Western temperament as "the schizophrenia of the soul," and the withdrawal into the present (a neglect of history) were necessary for the creation of natural science in its mathematized forms, then these sciences—and others patterned upon their methods and ways of thought—cannot be exported without the same genetic features in their makeup.

4. Faust in His Natural Habitat

The archetype of the man with the schizophrenic European soul whom Needham spoke about is the Dr. Faustus of Marlow's play and, in a more contemporary form, Goethe's Faust. It was the social environment of this latter Faust that bore and raised to adolescence the scientific-industrial civilization. Faust sought to comprehend the timeless aspects (laws) of nature, having preferred them to the uncertainties of feelings. It is not surprising that, instead of attempting to gain access to a woman directly, he had to conjure up an evil male character as a procurer, the same personage who used to bother Luther. The tragedy of Gretchen's infanticide follows inevitably from Faust's oscillations between "the heavenly host on the one side and the atoms and the void on the other."

By embracing Gretchen, Faust left behind the divine circles and epicycles of Ptolemy and Copernicus in favor of the irregular curves of human fate. Society's sentence upon his female half then drove him to his restless search for salvation. But no salvation on earth is possible for those who believe that violating taboos of a mathematical, divine order can leave nothing for the body and soul but a lesser cartful of dung than the Ancients had had. Faust's salvation came only in the uncanny beauty of Goethe's heaven, where the Chorus Mysticus informed him that time was no more and even males and females existed only in disembodied forms.

Perhaps that is the kind of heaven we should all try to reach. We cannot do so, however, unless we give up our societies, our minds, our lives, and even our dust, and return to the atemporal state of the photon. But such a journey is prohibited by the hierarchical organization of the universe, with its stable integrative level and metastable interfaces. Long before we could reach the atemporal state of the photon, we would have ceased to be organisms capable of the enlightened feelings of the mystical choir boys. In short, and there is nothing new in this, we are stuck between heaven and earth.

It is not easy to live on earth and imitate heaven; maybe we should not try so hard. In his essay on human law and the laws of nature, Joseph Needham tells us about a cock burnt alive in Basel in 1730 for the "heinous and unnatural crime" of having laid an egg. The cock was in good company, though. One of the accusations against Joan of Arc was that she often wore male clothing which, one would presume, was also unnatural by the same law of God. Regarding the cock in Basel, Needham goes on to ask whether the state of mind in which an egg-laying cock could be persecuted at law may not be just another facet of a collective view of man and the world which, in a different setting, would produce a Kepler. Needham concludes that "historically the question remains whether natural science could have reached its present state of development without passing through a 'theological' stage?"[27]

It is a matter of personal preference whether one sees science as having passed through a theological phase or Christian theology as having been secularized into science, preaching the absolute, numerical truth of God, with the scientist as the priest in possession of the sacraments of an invincible witchcraft. I would look at both as slightly different attempts to resolve some of the unresolvable, creative conflicts of humans. The element that is common to them—to Christian theology and Western science—is the separation of noetic time experience into the temporal and the timeless.

There exists a small but distinguished body of literature on the proposition that the different directions Chinese and Western science has taken is attributable to the different assessments of cosmic-historical time. The Chinese, we learn, tended to judge historical-cosmic time as cyclic; the Christian West, as linear.[28] Curiously, as far as I could ascertain, no one has as much as hinted that the crucial element of time in the birth of modern natural science was not the issue of historical linearity, but the metaphysics of timeless number and law versus temporal contingency.

There can be little doubt that the strong teleological (that is, linear) element of Christian salvation history did contribute to the rise of the industrial West. For the hope of a final redemption from toil gave divine authority to the Platonic vision of the Republic, naturalized first into St. Augustine's City of God, and later into Marx's, Engel's and Henry Ford's factory on earth. But I do not believe that this linear view of time has in itself been the major force beneath the creation of modern natural science. That came, as I have argued, through the placing of geometry into a timeless sky, by separating it from the temporal conditions prevailing on earth, and finally by perceiving in the natural order a manifestation of the heavenly, mathematized order.

In his paper "Time and Knowledge in China and the West," Needham asserted that "in Chinese culture, history was the 'queen of the sciences,' not theology or metaphysics of any kind, never physics or mathematics."[29] Perhaps the reason why modern natural science did not arise in China, independently of the West, was precisely because the traditional focus of Chinese intellectual interest has been life and man. Their sages and their pragmatic men, with exceptions such as the Legalists, have generally refused to perceive in geometrical order patterns universal enough to control the stars and heavenly hosts above and "all the creatures here below," or even to serve as paradigms of sorts for human behavior.

In the West, Plato's universe was emulated because it was so remarkably ordered and geometrical. But the serious Greeks have also allowed for a play of immortal gods and goddesses, behaving every bit like humans. However, the divinities of Olympus were later replaced by tensor equations, as was the star-throwing god of the Hebrew prophets. The starry sky, that image of eternity, is now a universe quite unrelated to the daily concerns of people.

"*The whole world, or any part of it, can now be treated as a unified system composed of nothing but dynamically curved space-time or 'super-space' or as nothing but gauge fields*, or Salam's 'elementary Particles.'"[30] This enthusiasm not withstanding, however, it is not possible to erect ethical and political systems on geometry, not even on curved space-time geometry. For, logical structures are eotemporal or, using an unexamined and misleading phrase, they do not respond to the direction of time, whereas our peculiarly human needs are sociotemporal, nootemporal, or biotemporal in their character. The distinction among the types of causations characteristic of these levels can be neglected only at the cost of generating unbearable tensions in the body social.

"The certainty and predictability of low-level phenomena cannot be found in the realms of 'free-will' at the higher levels," wrote Needham, paraphrasing an earlier critique of the Legalists.[31]

If the geometrical reduction of the quote from Gal-Or is unacceptable, as it must be, so are all propositions that would completely divorce the laws of lower integrative levels from those of life, mind, and society. For we ourselves share the hierarchical organization of the cosmos.

What an ecumenical science must seek, therefore, is an epistemology that can accommodate the many ways of human knowledge in a single system, yet allow for the different ways in which the various sciences and the different departments of the humanities must present their arguments and validate their evidence. A natural philosophy of time consistent with such an epistemology would allow us to trace a continuity from the primitive temporalities of the physical sciences, through the flow of biological time to the noetic and social temporalities created and experienced by members of our species.

A dynamic reconciliation of the multiplicity of the world in the unity of a hierarchical organization should ease the mind of Faust, and make it unnecessary for him to oscillate unhappily between the heavenly hosts of his Gretchen-experience and the atoms and the void of his laboratory work. Instead, he could just commute.

5. Legislating Order in the Republic of Numbers

What are the reasons of the "unreasonable effectiveness" of number in natural science, to borrow a phrase from Eugene Wigner's Nobel lecture?[32] I would like to sketch a possible answer, based on detailed arguments given in other writings of mine.[33]

In Book 6 of the *Republic*, Plato has portrayed all things of the intelligible and visible world as arranged along a vertical line, which we may imagine as connecting heaven to earth. On the top of the line are the eternal, timeless ideas and unchanging forms. In post-Platonic terms, these would include the circular planetary orbits and epicycles of Ptolemy and Copernicus, and the geometry of superspace. Underneath the timeless forms are the likenesses of them, such as geometrical figures or, in our age, the geodesics of space-time plotted by a computer. Further down on the Platonic divided line come animals, plants, and, I assume, the DNA molecule. This is also the region where objects made by man belong. Everything beneath the chalk circles are temporal.

The destiny of the soul, Socrates tells us, is to climb from the dark, the sensible, and the temporal toward the luminous, the intelligible, in short, the timeless.

I do not disagree at all with the Socratic assessment of human destiny: it is, or should be, a journey from darkness to light. But if so, then I must insist that the Platonic theory of knowledge is backward, the divided line of the *Republic* is upside down.

It is the atemporal—the perpetual motion of the photon, the ceaseless vibration of the electron—that represents the absence of choice, that is, the absence of freedom and, therefore (metaphorically speaking), the darkness of the mind. From the point of view of spiritual enlightenment it is the beam of light that stands for the biblical "darkness upon the face of the deep."

Only what is temporal is open to change and hence to improvement, having this potentiality in a degree proportional to the distance of its structures and functions along a scale of complexity, from the least complex forms of matter.

In Socratic language, the liberation of the soul consists of the journey from the timeless, along the stages of increasingly sophisticated temporalities, to the noetic and sociotemporal levels of nature. In terms that avoid reference to the soul, I have called this journey "time's rites of passage." The architecture of the world as we know it demands the turning of Plato's divided line right side up. The starting point is the atemporal, the goal is the increasingly freer, the more intensely temporal regions of structures and processes.

Living matter is certainly freer than those forms of matter whose actions are determined by the laws of physics only. Thinking life, by which I mean humans, is freer than those forms of life that follow biological and physical laws only. Finally, our societies, but especially the global fraternity, is frightfully free, as demonstrated by the crises of our days.

These remarks suggest the reason why quantitative methods are so "unreasonably effective" in understanding the behavior of matter in its least complex forms, and in the understanding of life and thought in their least characteristic functions. The lower integrative levels, whether in themselves or as parts of more advanced structures, are the least free or, in Socratic terms, the least enlightened regions of the universe.

The theory of knowledge that corresponds to the revised architecture of Plato's line may be represented by a hyperbole.

Plato directed our attention to the perfect, Darwin to the imperfect. But perfect, ideal, abstract forms cannot have history. All they need is an unchanging, eternal designer. Only the imperfect, actual forms of moving matter have the potential for change and can (and therefore do) define temporalities.

A year before he died, in the midst of the ravages of the Thirty Years War, Kepler wrote to his son-in-law, the mathematician Jakob Bartsch that, "when the storm rages and the state is threatened by shipwreck, we can do nothing more noble than to lower the anchor of our peaceful studies into the ground of eternity."[34] Kepler's desire for an absence of conflict, a wish we all share when tired from the labor of survival and individuation, may be used as an entry to an understanding of the new theory of knowledge, wherein temporality is the most elevated and timelessness is the most primitive aspect of the universe.

Eternity has been the paradigm of rest, continuity, permanent identity. Mathematicians assign eternity to the rules of number, the religious to the attributes of God. But the primordial world of radiation, that self-contradictory, atemporal universe, possesses nothing to which the ideas of rest, continuity, permanence, or identity could be applied. Neither time, nor space, nor connectivities (causations) of any kind exist in it—yet. All these emerge only through time's rites of passage from the atemporal to the prototemporal and beyond.

Continuity, rest, and identity are features of the eotemporal level of massive, aggregate matter and not of those integrative levels that have been and remain at the fundaments of the universe. One can construct a system of relative rest from systems in relative motion, but it is not possible to construct a moving system from elements of rest and no change. Motion and change are epistemologically and, therefore, ontologically prior to rest and no change.

Slicing sensory impressions and inner needs into the categories of permanence and change is a strategy for survival of living organisms. For reasons peculiar to the human mind, this dichotomy has expressed itself in the dividing of the human experience of time into permanence (the conservative, inertial trend) and change (that corresponds to the world of need). How these categories are conceptualized is a matter of social conditioning and linguistic creation. But this division of time into the eternal and the passing is only a first, very crude approach to an appreciation of time, even though it was useful up to about our own days, just as the Ptolemaic earth-centered system was useful as a first approach to the astronomical universe.

The Platonic dichotomy of time and the timeless is the simplest, and hence very appealing attempt to analyze the hierarchy of unresolvable, creative conflicts in nature, all of which are present in our experience of human time, because we ourselves are made of matter that moves, lives, and thinks collectively with its fellows. Nothing like this may be found in the Dialogues, of course, because to every epoch its own understanding of the world appears to be obvious and just about complete.

According to the theory of time as conflict, each integrative level is associated with a set of unresolvable conflicts peculiar to that level. These conflicts become more intricate in their character with the evolutionary complexification of matter. From atoms, to planets, to life, to man and society, the unresolvable, creative conflicts approach in their nature, and eventually become our own experiential conflicts. We may judge the physical world sometimes as peaceful, or the lives of birds as idyllic, only because between their conflicts and ours there is a qualitative difference that permits a projection of our hopes and dreams upon the immensity of the sky or the lives of doves.

The intensifying conflicts and increasing degrees of imperfection (as compared with idealized, abstract conditions) along the natural scale of increasing complexity are the very conditions that can give meaning to the Socratic call for enlightenment.

6. Ecumenical Science

With these ideas about a continuous line of knowledge from the atemporal to the sociotemporal, and a continuity in nature according to the principles of punctuated equilibria, we may now return to the question asked in the introduction to this paper: What kind of natural philosophy, what kind of conceptual scaffolding could correspond to Joseph Needham's idea of ecumenical science?

Giving the exact sciences the respect they deserve, one may point with pride to the fact that they are spread around the globe in almost identical forms. Mathematics is universal, the laws of physics are already ecumenical. Different understandings of atomic physics and astronomy are already fused into an international system. The same may be said about those branches of biology that may profitably mimic the methods of mathematical physics.

But medicine varies greatly around the globe and in certain of its branches there exists more than a single valid approach, with no particular practice being necessarily superior to others. Going from the life sciences to the sciences of the mind, we note that opinions concerning the nature of human behavior and the desirable methodology of psychology are legion. Finally, there is hardly anything that could be called ecumenical about views of history and social theory.

The reasons for this gradation may be found in the hierarchical organization of nature. Since laws (causations) are level-specific, so must be the formal principles of inference and methods of demonstrating truth. Light waves, particles, massive matter, life, mind, and society demand different logical precepts.

The laws of the eotemporal world have generally been taken as assuring precise predictability. This view is correct because in the eotemporal domain of pure (directionless) succession—in the world of the classical physicist's *t*—the direction of time can have no meaning. Nothing in that world can correspond to before-after relationships. Inductive generalization, which is the foretelling of the future, and deductive generalization, which is the summing up of the past, are indistinguishable. This privilege is limited, however, to the macroscopic behavior of inorganic matter.

The first breaks in the hegemony of mathematical predictions in physics came with the extension of physical science into the domain of particles where discrete mathematics had to replace calculus, and quantum logic had to replace classical logic.

Going upward from the eotemporal in the direction of organic processes, deductive and inductive reasoning separate. For living matter defines a present, a now, in terms of its necessary internal coordination and, with respect to that now, time acquires futurity and pastness. Rising from the biotemporal to the noetic and sociotemporal levels the ecumenical natural philosophy appropriate to Needham's vision must allow for an increasingly larger pluralism of scientific principles because the increasing complexity of structures and functions become increasingly unpredictable. By the time biologists, psychologists, and sociologists will be able to depend on mathematical models to the same degree as, for example, astronomers depend on them today, the queen of the science will include in her domain such methods that will allow for increasing undeterminacy of prediction and even for a range of ambiguity in logical relationships.

From these considerations on a natural philosophy for ecumenical science, we may now draw some speculative conclusions concerning the issue of exporting Western science to non-Western cultures. Again, we begin with the inorganic and work our way up toward the organic integrative levels.

The sciences of the inorganic were created by and seem to demand for their continued success the kind of collective personality that is usually associated with Protestant pragmatism: high esteem for order and fear of ambiguity. Physics, especially, demands strict uniformity and allows little or no variation in the precision of its principles. It is a thoroughly cross-cultural science because it deals with the least complex integrative levels of nature: elementary particles and gravitating masses behave the same way whether they function as parts of animals or people, and they certainly do not respond to differences in social environment.

The situation changes as one begins to deal with the sciences of the more complex stable integrative levels of nature. In the handling of their peculiar problems, the sciences of life, mind, and society must allow for an increasing plurality of methods and ambiguity in their statements of truth. Since this pluralism is not a sign of human ignorance but a symptom of levels of complexity, admitting them in the methodologies in the sciences of the organic is necessary for their healthy growth.

It follows that biology, psychology, and sociology, if freed from methodological commitments to physics, should be able to flourish in many and different ethnic, historical, and national forms as parts and parcel of Needham's ecumenical science.

7. Faust on the Road

We left the Faust of old in his laboratory, thinking about Gretchen and bargaining away the future of his soul. To make his character a useful metaphor, we must send him on two journeys. First he must go on a diachronic travel to the West of our epoch; then he has to take a synchronic trip to make him become a member of the worldwide community.

The life of Faust in its original setting was informed by a conflict between, on the one hand, his boundless search for knowledge, his love and fear of woman, and his desire for power, and, on the other hand, the ethical assets of his society which by today's standards were backward and icy.

At the end of his diachronic journey, approaching the turn of the twentieth century, he will find himself facing equally difficult though different problems. His new world is inundated with data, which is often equated with knowledge; his West possesses almost completely liquid ethical assets; and his new fellows have enough power to be classed with Plato's Demiurge as the creators of bodies and souls.

The search for knowledge changed from a qualitative to a quantitative proposition: more is better. This brought with it a new form of the Malthusian principle, mentioned earlier in its economic-biological significance. In an evolutionary developmental step, the principle crossed the noetic-sociotemporal interface and surfaced in a new context.[35] In complete analogy to the biological teachings of the theory, the rate at which information-as-knowledge is generated and the rate at which its transfer is demanded outrun by far the rate at which information can, in fact, be communicated and processed. Metaphorically speaking, more brain children are begotten and delivered than society can feed, house, educate, and integrate. The hefty tomes of Faust the alchemist are replaced by the information storage facilities of Faust the data banker. Withdrawal from those banks is performed by the read-out functions of computers. The processing of knowledge read out is also done by computing devices, now approaching their well-advertised "fifth generation." It is through the computer and communication world that the new form of the Malthusian principle enters.

In the United States, Japan, and western Europe one is continuously treated to the exhibit of faster, larger, and increasingly sophisticated machines which prove themselves inadequate to decrease the information backlog or break the traffic jam in data and instructions, because the amount of information the machines were supposed to have organized and processed has increased in quantity and scope while the devices were being made, installed, and tested.

Computer programs designed to facilitate the production and distribution of goods and services or the running of vast organizations often need hundreds of thousands or millions of instruction lines. These programs, unavoidably, come with genetic ailments which demand specialized computers for their repair. The scenario is analogous to the necessary complexification of organic evolution, with the introduction of unreducible, statistical uncertainties.

In the second part of Goethe's Faust the hero travels around the world. Alienated from his own age and from antiquity, he learns of the problems of his species (mostly through symbolic events) so that he can prepare his redemption and that of humanity. This enterprise is analogous to the synchronic journey that the contemporary Faust must take.

Getting on the road he will note, among other matters, that while some segments of the global society cry out for more knowledge in every field of endeavor, some others are choked by incoherent and therefore useless knowledge. He will also note that in spite of the tight control that some highly centralized governments have over the behavior of their citizens, the ethical norms of humankind are in a three-phase disequilibrium. Some are frozen, some liquified and hence fluid, some vaporized. From a distance he is likely to have a glimpse of his Gretchen, trying to find a dignified yet also practical leitmotiv for her and her child's life. And, if he examines the statistical surveys of the United Nations he will learn that Wagners,[36] witches kitchens, fighting soldiers, and slinking Mephistos are becoming democratically distributed over the surface of the earth.

It is in this world that the National Science Congress of 1978 in the People's Republic of China declared a new epoch: "Springtime for people, springtime for science." There followed a period of wild enthusiasm which subsided, however, as the immensity of the problems became evident. It was also realized that the importation of Western know how has brought with it certain ways of thinking and doing that had to be naturalized before they could become as effective in China as they have been in the West.

In an essay written four years ago, the China correspondent of the British *Nature*, Tong B. Tang, examined the influx of Western know-how into China and concluded that, "at the present there is a dynamic balance between the ideal of indigenous Chinese scientific culture and the influence of Western scientific ideas; only in the next five years, as these influences are absorbed, will the future pattern of the world's largest new scientific force become apparent."[37] The time scale of five years is obviously too short; twenty years may be a more reliable sampling period.

The "springtime of science" in the West was the eighteenth century. It was then that the philosophers, those early propagandists for the age of reason began to explore the social significance of the (then) "new natural philosophy." They identified it in the open horizons which science has offered for the control of man's own destiny. The change from wild enthusiasm to a realistic assessment of their problems that took the People's Republic of China only a few years was a fast-motion replay of the corresponding sobering up in the West that took over two centuries. It began with the Enlightenment and stretched to our own days when the realization that "Never has man held within his grasp so much technology beneficial to his welfare; never has he been so far from applying it to that end" became a broadly held opinion among responsible observers of the contemporary international scene.[38] Thus, while the Chinese are laboring to integrate the Faustian way of seeing and doing things into their non-Western civilization, the West itself, in a parallel task, must also integrate industrial productivity and new knowledge into its own humanistic tradition, if the forces of industry and science are to be constructive rather than destructive in their long-term effects. The many descendants of Faust must be educated and guided so that they can become useful members of a community that is more heterogeneous than that literary ancestor of modern Western man could have imagined.

The reasoning of this paper suggests, in sum, that in spite of its immense advances in technology and science our epoch remains uninformed because it has not yet established a way of addressing itself. In terms pertinent to our theme, there is a need for a coherent framework within which the significance of time can be traced from the atemporal roots of the universe to the peculiar freedoms of noetic and social time.

The contribution of this conference to the education of Faust's intellectual and cultural heirs is the exploration, through selected examples, of the relationships between science and society. This we propose to do by examining a few of the very many ways along which the reasoning and passionate faculties of twentieth-century people meet the challenge of noetic time. The foregoing arguments are intended to form a part of that examination.

19. BEING THE ONE AND ONLY

The Constitution of the United States, drafted in 1787, has the following preamble.

> We, the People of the United States, in Order to form a more perfect Union, establish Justice, insure domestic Tranquility, provide for the common defence, promote the general Welfare, and secure the Blessings of Liberty for ourselves and for our posterity, do ordain and establish this Constitution for the United States of America.

The call for a "more perfect Union" is one for changing an ill-defined collective identity to a well-defined one. "Domestic Tranquility" implies the existence of boundaries separating what is domestic from what is not. "Common defence" implies the existence of potential and/or actual enemies of a community located within boundaries that make it "domestic."

It would not be possible to compose a similar preamble to a constitution of a global humanity because, as far as we know, we are the one and only human species in the universe. Opinions about what is to be meant by being human vary immensely across cultural, religious, political and geographical boundaries, as do the moral tenets derived from those opinions. As an example, the assessment of the value of a human life and the consequent obligations of persons held by a Moslem fundamentalist, a liberal Englishman or a Buddhist monk differ substantially in many and mutually exclusive ways. The profound turmoil of our epoch pertains, partly, to the question of what we are to judge as the shared qualities—and hence shared needs—of "we the people" of the earth. A pragmatic but limited list would say that we share the desire for satisfying our physical and biological needs. But these leave unrecognized the depths of cultural needs. The Western view holds that each person is unique in the universe and hence, each may stand toe-to-toe with the Almighty, may bow, and bargain. But this view, shared only by a segment of our earth's inhabitants, carries the danger of leading to an earth without shared awe, hence shared goals, where there would be "nothing left remarkable/beneath the visiting moon."

TIME, GLOBALIZATION AND THE NASCENT IDENTITY OF MANKIND

Abstract

A person's identity is defined by the ways in which he or she is similar to and different from other persons in body, conduct, and thought. The identity of a tribe, nation or civilization is also defined by the ways in which it is similar to and is different from other social groups of its kind. A peculiar difficulty arises when the identity sought is that of a globalized mankind because there are no other mankinds with respect to which ours could establish its hallmarks of similarities and differences. A global society is a one-and-only system of its kind, as unique as is the object we call the universe or the idea of God, in monotheistic religions. To get around the difficulties posed by this uniqueness, the paper notes that all identities must include references to intentional conduct and hence to time. It then takes advantage of this necessary relationship and discusses some representative issues unique to a globalized mankind.

This article reflects on the apparently unique position of humans in the universe and uses our capacity of assessing reality in terms of time to help define the identity of a globalized mankind.

Personal and Communal Identities

Personal identity refers to the experience of an individual remaining himself or herself, different and distinct from others, despite changes in his or her conduct and psychobiological structure and functions. To help bypass, through simplification, a number of logical and philosophical difficulties associated with the concept, I propose a speculative journey back to the emergence of personal identity or self-awareness in our most distant human ancestors.

I imagine that as they roamed their countryside, they noted an object that always walked with them. It was an invisible companion that made birds fly away and tigers attack. Obviously, it shared the external world with the birds and tigers, that is why those animals could react by flight or fight. But it differed from all other objects of the outside world because, unlike those other objects, it was able to feel pleasure and pain. In its simplest form that invisible, permanent companion was the biological individual, to use Gerald Edelman's felicitous term (Edelman, 1989: 262–3). In its most sophisticated form, that permanent companion is the self of historical man.

Consider a person before a firing squad, shouting 'I don't want to die!' The identity of the person seems clear: there are the predators, and there is the prey. But if the same cry comes from a person dying of cancer, the self of the person cannot be his body because the body is hell-bent to die. The self is not a tangible continuity but a permanent companion to the biological individual, a symbolic entity that strives to secure its continued integrity. Yet, it can also order its own demise, such as in the case of a terrorist on a suicide mission: his body wants to remain alive while his self wants to die. Collective selves are also symbolic continuities. They are imaginary objects that share the world with other, similar objects external to themselves: other tribes, nations or civilizations. As does the individual self, the collective self also holds executive powers over collective conduct. It may strive to secure the continuity of the collective self or its destruction.

All identities must necessarily refer to time because their defining hallmark is intentionality, that is, goal-directed conduct. In the biological individual the purpose of goal-directed behavior is to assure the continuity of its life; in a person, its purpose is to assure the continuity of its selfhood. For both cases a present must be defined through inner coordination. It is with reference to them—to the organic and mental presents—that future and past may acquire meaning. (On the nested hierarchy of presents and their functional definitions, see Fraser, 1999: 34–5.)

A crucial step between the earliest humans and their immediate ancestors had to involve an expansion of temporal horizons. Human groups, so I imagine, began to be concerned with increasingly longer term futures and past, segments of time that included epochs beyond the deaths and before the births of individuals. This extension of temporal horizons made possible the formulation of intentionality directed toward symbolic goals, such as toward the benefits of post-mortem well-being. These were than added to the concrete goals of satisfying hunger for food and mate.

Keeping in mind the details of this sketch about the symbolic nature of personal and collective identities, I want to turn to some of the issues native to the emerging identity of a globalized mankind.

The Identity of a Globalized Mankind

Some time along the history of mankind, individual members of the species endeavored to escape some of the limitations of their biology through the use of their minds: having a mind made it possible for them to expand the boundaries of their temporal horizons for action. But, in their turn, individual minds, directing separate persons also had—also have—their limitations. Collectives of humans could—may—escape from some of those limitations by extending collective plans and memories beyond the deaths and before the births of persons.

How may the earth's nations and civilizations escape the limitations of their (now) separate identities? The answer, I believe, is this: by establishing a global identity.

The next question: what are the conditions under which a group, any group may acquire identity?

Consider the history of a married couple. Although each member has a history, as a married couple they can have a history only in respect to their communally created and maintained present. They can have no history stretching back to the years before they have met. Likewise, tribes, nations, and civilizations may have a history only from that epoch forward when they first establish their collective present.

The name given to the collective present of all nations of the earth is the 'global present'. It signifies a shared simultaneity among all nations, maintained through communication. It is a global time reference, from instant to instant to instant, with respect to which actions may be taken and, increasingly, more actions are being taken.

Only with a viable global present, collectively planned future, and collectively interpreted past, may we learn what it is that distinguishes mankind from a loosely connected group of tribes or nations. And only with that knowledge may the identity of mankind then be defined. That identity, if indeed established, will describe a symbolic continuity that will bear the same relation to the physical and biological functions and structures of globalized mankind as did the permanent companion to our hairy ancestor, the companion that prompted birds to fly away and lions to attack.

The Beginning of the History of Mankind

In 1992 much attention was paid to the idea of an 'end of history', proposed and defended by Francis Fukuyama. He maintained that insofar as there remained no viable alternative to some form of market-oriented democracy as a social system, the experimentations of history have effectively come to an end (Fukuyama, 1992).

I propose an opposite view. Namely, that it is only with the establishment of a global present that mankind as a single social entity may begin its history. And, it is not an easy beginning because, heterogeneous as mankind is, the time compact globe is entering the set of those systems of which there is only one each, such as the Universe, Nature, or in monotheistic religions, God. The difficulties of delimiting the boundaries of effectiveness and defining the identities of these one-and-onlies are notorious.

Group identities, whether tribes, nations, or civilizations, are established by noting similarities to and differences from other members of the same set of objects: other tribes, nations, or civilizations. But, there are no other humanities with which we could compare ourselves. There are no other mankinds which, while protecting their self-interests from our encroachments, could challenge us if we drift off into the unreality of shared fantasies. As a consequence any economic or military experimentation, if it involves a substantial part of mankind and if miscarried, may lead to socio-economic or cultural oscillations that can go out of control. Our age is a metastable transition between nation states and an uneasy global community.

The dynamics of globalization may be represented by one of the metaphysical poems of Boris Pasternak. In it, the poet is Hamlet himself as well as Hamlet the actor. This complex character reflects upon his predicament. 'I step forth or the boards' he says, and

> I strain to make the far-off echo yield
> A cue to the events that may come in my day.

On the time-compact globe the far-off echo is immediate and gives only cues about what may come during the next few seconds or hours. I propose, therefore, to listen only to three concerns of the present instant, serving *pars pro toto*.

1. *The Generalized Malthusian Principle*

The proposition put forth by Thomas Malthus 200 years ago was that the needs of people for food would always outrun what food production could deliver. It would follow that mankind was condemned to a chronic condition of hunger. The only way to lessen that misery, in his view, was through moral behavior, which he equated with sexual restraint. His ideas were instrumental in the formulation of the Darwinian principle of natural selection. Later, however, they were handled with a degree of contempt, for it was believed that technology and science would be able to take care of the growing needs of the world's population. But during the second half of our century, as some of the unintended consequences of technology and science have come into view, Malthus' ideas have gained new currency.

I want to take advantage of the power of the Malthusian principle and, in three steps, place it into a perspective beyond its original domain.

1. The needs of living organisms outrun what their inanimate and living environments are able to offer by way of satisfying those needs. This leads to the Darwinian 'preservation of favored races in the struggle for life'.
2. From the organic world in general, let us take a step up to humans. The needs of human life outrun what its living and inanimate environments can offer by way of satisfying it. This is the triage of mankind recognized by Malthus.
3. From man as separate organisms gathered in tribes and civilizations, let us step up to a global society.

The demands that the collectively functioning mental powers of mankind place upon each person around the globe outrun the learning capacity of any single individual. The hefty tomes of Faust the alchemist are being replaced by the information storage facilities of Faust the data banker. And here, the Malthusian principle surfaces in a new context.

Namely, the rate at which information is generated, and the rate at which its transfer into useful knowledge is demanded by a global society, outruns by far the rate at which that transfer may in fact be carried out. Cultures, religions, and national groups have difficulty responding to the challenges at the rate such responses are demanded. Metaphorically speaking, more brain children are begotten and delivered than society can feed, house, educate, and integrate. A consequence is a fragmentation of knowledge expressed in a cultural triage.

That triage is manifest in a revolt against all received teachings about human values, in cultural conflicts of substantial dimensions. These conflicts are nurtured by the development of science. Namely, the Copernican revolution removed the earth from the center of the universe; Kepler's recognition that the planetary orbits were elliptical broke the Aristotelian faith in the perfection of the heavens; Newton showed that it was possible to formulate mathematical laws valid everywhere and everywhen in the universe, laws that are open to human understanding; Darwin removed humans from the guaranteed pinnacle of all species; Freud removed men and women from the company of angels and devils and established them as nature's very unusual but not transcendental wonders; Einstein showed that the universe cannot have a center and bequeathed to mankind the finite but unbounded, inert, and inhospitable immensity of dumb matter.

I believe that the nested hierarchy of Malthusian stresses or triages, including those in the domain of information and knowledge, is forcing the development of a new scale of values pertaining to what mankind at large ought to judge as true, good, and beautiful. (On the revolution of human values, see Fraser, 1999.)

2. *The Reeducation of Eros*

By Eros is meant the aggregate of all impulses for which the power of sex is a paradigm. To gain a perspective upon that power, it is useful to begin with its evolutionary past.

All species control their fertility rates. Mammals do so by limiting the periods during which the animal is in heat. But humans, during most of their mature lives, are continuously in heat; in humans, the control of reproductive rates was transferred from a biological program to reasoned conduct. The effort necessary to maintain that control is a source of steady conflicts between instinct and reason. In turn, the need to manage these conflicts has been driving human creativity and destructiveness. What is novel to our epoch are the dimensions of the controls necessary: there are more people, more creativity, more destructiveness, more opportunities to make love and make hate. And there are other problems.

The power of being able to create new humans has been traditionally perceived as having its source in the universe. But conclusions drawn from that recognition, vary. Hinduism focuses on the organic of lingam and yoni; Chinese cosmology focuses on the principles of yin and yang; Christianity focuses on the sanctity of begetting children.

This sanctity found representation in Paul Claudel's *Le Soulier de Satin* (The Satin Slipper). In it the shadows of a man and a woman, merged into a single double shadow, are seen on a wall. But the man and woman whose double shadow it is have parted, leaving the shadow masterless in the land of shadows. 'Why,' asks the double shadow, 'having created me why have they cruelly severed me, me who am but one? Why have they carried off to the far ends of this world my two quivering halves?' (Claudel, 1931: 126).

This, the sacramental view of intercourse, has been fading rapidly in importance. As recreational and procreational sex are becoming easily separable in practice, Eros, like Claudel's double shadow, is seeking guidance on how to make recreational and procreational sex separable not only in practice but also in principle. Finding an acceptable formula is the first task in the reeducation of Eros.

In other, related matters, a few steps have already been taken. Examples include the welcoming of non-reproducing persons: gays, lesbians, dedicated loners. Also, the great increase in male impotence. Males, it seems, find it difficult to adapt to their changed position in the family, to the revised rules of their relations to women, that is, psychological causes, to which alcohol and drug abuse and the impact of environmental toxins may be added (Kimbrell, 1995: 8).

The reeducation of Eros includes changes in the position of children in society, in the conceptualization of childhood and in the consequences of those changes. In children's literature children are shown as less and less protected from the pressures and meanness and complexities of life; they are depicted as facing serious problems without substantial adult support.

A slogan of 1960s America, 'Make love, not war', was a call to support the biological over the ideological. It suggested the role of biologically fed outrage as an aid in bringing under control the mind run amok with ideas and ideologies.

In the emerging identity of mankind the biological is re-informing the mental. By that process it helps protect humanity from the dangers of the uncontrolled imaginative powers of the mind, while simultaneously it threatens civilizations by the raw power of untamed life.

3. *From the Comintern to the Internet*

Comintern was the name of an association of Communist parties called into congress by Lenin in 1919. Although it was officially dissolved in 1943, I am going to use the name to stand for the seven decades of the Soviet Union.

At the basis of Communist ideology there is a Christian Utopia, albeit without a divinity. It envisages a self-governing humanity of men and women who live well and in harmony with themselves and with their history. In practice, this Utopia turned out to be an all-intrusive system that wanted to know everything about everyone so as to be able to carry out its self-appointed task. But the system had difficulties entering the 20th century, let alone preparing for the 21st. After having inflicted immense suffering upon its followers and foes alike, it collapsed under the burden of its social backwardness and faulty understanding of human nature.

The society that served Marx and Engels as the model for the shortcomings of capitalism was 19th-century Victorian England. In the rigid class structure of its upstairs-downstairs architecture, stability was preferred to experimentation by rich and poor alike. By the middle of the 20th century, growing out of the temperament of the New World, the desire for stability was replaced around the world by radically increased expectations. And, by the end of the century, collective judgments almost everywhere came to prefer experimentation to constancy.

Globalized humanity is now informed by the belief that the increased expectations of everyone should, and could be satisfied through the free exercise of human potentialities. For that end, in the consumer societies of our age, information is now held not in secret government archives but in the behavior banks of large business concerns. They hold an immense amount of liquid data on all people within their economic domains, and these domains are not bounded by national borders. In the USA these data include information on health, credit, marital status present and past, education, employment history, legal involvements, real estate ownership and use, car ownership and use, travel history, the times and telephone numbers of every call made or received, the journals people subscribe to, all purchases if charged or discounted by discount cards, bills paid by check, cash or card, food shopping, and eating-out habits.

The network of behavior banks, and the homogenizing trend they promote, coexist with a fragmentation into what I have been calling Tribal Interest Cells or TICs. They are the commandos and raiding expeditions of the time-compact globe. In its old meaning and in its current anthropological meaning, 'tribe' refers to people held together by family and emotional ties. In the connotation to which I appeal, the tribes are held together by fanaticism and narrowness of purpose. They conduct local and transnational wars underground, on the ground and in the air, in support of ethnic, religious, economic, or political goals. They serve enterprising gurus no less than crime families, petty or grand. They often advocate historically ancient causes that have emerged from their lairs where centuries of civilizing efforts have tried to keep them.

The TICs, the globalized social institutions, globalized industry and commerce, military and political interests, diseases, knowledge, and ideologies, all focus on a narrow global present. Together, they are creating a new temporal context to daily life. The changes represent a transitional stage between the traditional aggregate of nation states and their globalized forms. Beliefs that used to be held as eternal truths now change rapidly and tend to break into shards.

One can hear the uncanny echo of Marx and Engels: 'All fixed, fast-frozen relations, with their train of ancient and venerable prejudices and opinions, are swept away, all new-formed ones become outdated before they can ossify.'

But the stage upon which this history is played out has changed into one which Marx and Engels did not and, probably, could not have envisaged.

The Spherical Stage of Instant Everywhere

Early Greek theaters could seat 15,000 people. The number of people who watch certain televised events is in the hundreds of millions. Those who can be reached by radio in the billions. Our amphitheater is that of the telecommunity of a spherical stage of instant everywhere. The Furies of our age are of the same kind as those of the Greeks but their number has increased by a division of labor. There is a much larger variety of anger because there exists a much larger variety of needs. Also, the guidelines by which good and evil are judged are more numerous than would have been acceptable to the philosophers of Athens or imagined by the Comintern.

Many of the great leaders of humanity have spoken about the 'brotherhood of man' and of a fulfillment of human destiny in that brotherhood. The reasons given why one should strive toward it have been powerful, but practical instructions about the path to be taken remained scarce and contradictory.

Does science offer truths upon which the people of the world may agree and upon which the identity of mankind may be based? I do not think so.

There is agreement around the globe about mathematics. In physical cosmology there are disagreements across cultural boundaries. Approaches to biology, especially to issues of organic evolution depend on ideological ambience. In psychology and sociology the differences are profound and seem irreconcilable among different cultural settings. Ethics, religions, philosophies, and ideologies are in open warfare.

Establishing a shared global identity is unlikely to come about by discovering correct answers to scientific truths about nature and man, because there are multiplicities and often contradictory, yet equally valid, answers to questions about what one ought to believe as true. If a worldwide agreement concerning the position of mankind in the universe does come about, it is more likely to have done so because one or another ideology will have succeeded in converting or else killing the adherents of other ideologies.

Writing by the Yalu River in 1895, Paul Valéry recorded an imaginary dialogue with a Buddhist sage. The monk reflected on the western ways of doing things. 'You have neither the patience that weaves long lines nor feeling for the irregular... You are in love with intelligence until it frightens you... [You] rage with desire for what is immediate and destroy your fathers and sons together' (Valéry, 1962: 372). The sage failed to foresee that a century later his criticism would apply to all cultures across the time-compact globe.

The human search for truths has been driven by the desire to identify things and ideas that are permanent and appeal to them as demonstrations of the unimportance of passage. Yet, the primary consequence of this search has been the generation and perpetuation of changes in what is believed to be true. The speedier those changes, the more intense the desire for continuity and permanence; the more intense the need and search for stability, the faster are the changes. The globalized world displays this feedback process at an ever increasing speed. The question of what people should believe as true—whether proposed by science, philosophy, religion, political ideology, the arts, or the letters—is being driven toward new crises of surely vast proportions but unknown resolution.

In his *Preparing for the Twenty-First Century*, the historian Paul Kennedy remarked that the forces of change are so 'far-reaching, complex, and interactive that they call for nothing less than the reeducation of mankind' (Kennedy, 1993: 339). As is the case with Eros, that reeducation is being carried out by many teachers, advocating many and different sets of values. Multinationals, ideologies, religions, cultures, alliances of all kinds are shifting around for their presumptive positions in a broader order. So are drug rings, mafias, terrorist movements, and smugglers of women, children, immigrants, weapons, and body parts. People everywhere are seeking some kind of order in this disorder. The ethos that will eventually conquer the minds of people will be the one that succeeds in creating an interpretation of history and of the significance of human life, upon which a believable and inspiring plan for the future of mankind may be constructed. The march toward that future is not to the tune of Dies Irae, nor to that of the Marseillaise or the Communist International. These turned out to have been no more than call signals of large tribes. The march is to something more elemental.

Let me appeal to the image of the solitary walker by remembering Jean-Jacques Rousseau's 'rêveries du promeneur solitaire'. At the turn of the 20th century a person may still reflect upon himself and the world as did Jean-Jacques. He may also sit at the side of lake and sing about time as did Lamartine:

Aimons donc, aimons donc! de l'heure fugitive,
Hâtons nous, jouissons!
L'homme n'a point de port, le temps n'a point de rive;
il coule, et nous passons!

But how a globalized humanity as a solitary walker in the universe will assess the nature of time, see its relation to future and past and define its identity, is yet to be seen.

References

Claudel, Paul (1931) *The Satin Slipper*. New Haven, CT: Yale University Press.

Edelman, Gerald (1989) *The Remembered Present: A Biological Theory of Consciousness*. New York: Basic Books.

Fraser, J. T. (1999) *Time, Conflict, and Human Values*. Urbana and Chicago: University of Illinois Press.

Fukuyama, Francis (1992) *The End of History and the Last Man*. New York: Free Press.

Kennedy, Paul (1993) *Preparing for the Twenty-First Century*. New York: Random House.

Kimbrell, Andrew (1995) *The Masculine Mystique*. New York: Ballantine Books.

Valéry, Paul (1962) *History and Politics*, trans. O. Foliot and J. Mathews. New York: Pantheon.

20. TURMOIL AT THE ANTHILL THRESHOLD

We became acquainted with the new post-Renaissance Wonder, a former magician named Faust, who is now roaming the earth. He even serves as the Pied Piper of our age. As we learned in the introduction to Essay 18. he is trying to educate himself so that he may continue his art: talking with stones, planets, ants, birds and people and addressing what he knows as humankind. To protect himself from the conditions that prevailed on the day of Hamlet's death—from "carnal, bloody and unnatural acts / Of accidental judgments, causal slaughters/Of deaths put on by cunning and forc'd cause..."—conditions that now fill the globe, he moved into a castle of a fellow-dreamer, the Prince of Denmark.[1] But today the setting and the dramatis personae have changed. The story now involves a virtual magician named Faust, a virtual castle named after Hamlet and is located in a virtual place, called cyberspace.

The scale of human values by which we judge the true and the good and appreciate the beautiful are now substantially different from what they were when Faust was an apprentice. Truth is now short-lived. The good is determined by whatever is of immediate advantage. The task of providing what folks judge beautiful was taken over by the high-tech primitive. The new conditions make it useful to turn to the Bard yet for further assistance. He is ready to send us Prospero, a wise and benevolent magician. In "The Tempest," as he gets close to having accomplished his goal of foiling evil intrigues, Prospero prepares to dismiss the world of his virtual reality, an ancestor of the virtual castle of Faust, located in cyberspace.

> The cloud-clapp'd towers, the gorgeous palaces
> The solemn temples, the great globe itself,
> Yeah all which it inherit, shall dissolve...
> And like this insubstantial pageant faded,
> Leave not a rack behind. We are such stuff
> As dreams are made of....[2]

Do we need Prospero to update the family of human values so that they may make our global intrigues survivable? The essay that follows says that we have Prospero whether or not we want him. After all, we are still "such stuff/ As dreams are made of" and hence we are at a boundary of the universe. We have the means of managing "the great globe" of the earth provided we understand the roles of seeking the true, doing what is right and admiring what is beautiful. The essay that follows reminds us that these values are revolutionary forces. They fire a cosmic arrogance and feed a slight shiver that characterizes our species. While Hamlet's old castle is crumbling, a new one is being built by the new set of human values. Welcome to cyberspace where Plato's former slaves are now so many Prosperos and Fausts. In the words of the essay that follows, the march today is "not to the tune of the Dies Irae, nor to the Marseillaise or to the Communist International....The march is to something more elemental."

HAMLET'S CASTLE IN CYBERSPACE*

Abstract

This chapter defines human values—truth, goodness, and beauty—in terms of their relation to the idea and experience of time. With the help of these definitions it then explores the effects of the increasing time-compactness of the globe upon those values.

It observes that in the global marketplace of ideas, all traditions about what constitutes and how to test for human values are being challenged.

The search for eternal verities is being replaced by truths of brief lives; conduct à la mode is preferred to stable ethical rules; attempts to create lasting beauty are replaced by constructions of the high-tech primitive.

Yet none of this suggests that human values have failed to carry on with their tasks. Rather, it demonstrates that they keep on doing whatever their cultural and evolutionary roles have been all along—namely, to serve as revolutionary forces that reflect the unresolvable conflicts that characterize humans and drive the remarkable creativity and frightening destructiveness of their formulators. Hamlet's castle has always been but a provisional one.

In this chapter I would like to reflect on some of the changes in human values that are being brought about by the increasing time-compactness of the globe. By human values I mean the family of collectively formulated guidelines concerning what is to be regarded as true, good, and beautiful.

My reasoning starts with comments on the essay's title.

The term "cyberspace" first appeared in a science fiction story in 1981. "Cyber" was borrowed, secondhand, from cybernetics, the science of control systems; "space" was borrowed from mathematics, where it means a container of any set of mathematical objects. In contemporary use "cyberspace" stands for the content of all electronic communication systems of the world at an instant. They carry music, voices, images, numbers, and such other signals as those of our garage door opener.

Cyberspace is a refined form of telephone space, which, when it was the most advanced technology for exchanging messages, carried voices. Telephone space was an offspring of telegraph space that carried Morse code. Telegraph space was preceded by the yodel space of the Alps, itself a descendant of the song spaces of birds, and the chemical spaces of bacteria. Obviously, cyberspace is not a space in the ordinary sense but a name for a new family of signals appropriate for communication in the technical and cultural setting of the age.

The figure of Hamlet may be traced to tenth-century Iceland, where he was Amlodi, a Norseman. But his character as a melancholy and intellectual young man and a revenger of evil may be identified in myths at places as far afield as Iran, India, and Polynesia. I will take Hamlet to stand for all people: for those who studied philosophy in Wittenberg, as did Shakespeare's Hamlet, as well as for the overwhelming majority who never heard of Wittenberg and never studied philosophy.

Hamlet's castle is said to be in Helsingör. Denmark. I prefer to think of it as a generalized castle, one that is both real and metaphorical. It includes all means—both structural and mental—that are necessary to protect men and women from the weather, from their enemies, and even from themselves. At the turn of the millennium these structures include the huts and hovels of the First, Second, and Third worlds, together with the techno-villas of new billionaires, built in cyber-nouveau. Among the mental castles I include all guidelines designed to help people select what they are to judge as true and false, as guidelines for conduct through judgments of right and wrong, and modes of organizing their emotions through judgments about what is beautiful and what is ugly.

My thesis is that all castles of tradition have become, if not useless, at least insufficient for protecting people from the dangers that emerge due to the time-compactness of the globe. They are all crumbling. What seems to be needed is a stronghold that resembles what the Psalmist had in mind twenty-six centuries ago: "A mighty fortress is our God/A bulwark never failing."

But what kind of fortress is it to be? What kind of God is it to be? And, what will be meant by "never" cannot be specified, because the fortress itself, the godhead, and ideas of time, are all in the process of reconstruction through a revolution in human values.

Background

To set the background for an appreciation of that revolution, we must first deal with the contents of cyberspace. Specifically, we must look at the emergence of the global present, the narrowing of temporal horizons, the dynamics of a nested hierarchy of Malthusian selection principles that has become evident, and the question of mankind's identity.

The contents of *cyberspace* is a high-tech version of the "Library of Babel" in the story of the Argentinian poet Borges bearing that title. The Library of Babel is "a sphere whose exact center is any one of its hexagons and whose circumference is inaccessible." It contains everything about the universe and man: future, past, and present, including many "books of apology and prophecy which vindicated for all time the acts of every man...." People rush around madly in search of the volumes that contain their vindications. They proffer dark curses and strangle each other, and some of them go mad.[3]

There are enough similarities between that library and cyberspace to give one pause. But there is also a fundamental difference.

The readers of that library see themselves immersed in the continuity of history; the users of cyberspace do not. Their temporal horizons have been contracting to the limited perspectives of today, tomorrow, and yesterday.

This narrowing is a corollary of two closely related but distinct causes. One is the establishment of the global present; the other is the immense acceleration in the speed of daily life and of social and cultural changes. With the help of a natural philosophy of time, known by the name, the hierarchical theory of time, I will first consider *the global present*, then *the speed of life*.

Presentness or nowness in the physical integrative levels has no meaning. In the organic, mental, and social integrative levels, presents are created through the coordination of level-specific processes in the service of the continued identity of a living organism, a functioning mind, or a viable society. The phenomenal manifestations of these coordinations in service of different identities constitute the nested hierarchy of the necessarily simultaneous organic, mental, and social presents.

A new member of the nested hierarchy of presents is now being created through the coordination of social processes across cultural, national, and ethnic boundaries. It may be called the *global present*. Its construction is driven by the needs of economic and cultural survival and made possible by advances in technology.

What is meant by the global present?

It is a communal reference instant with respect to which actions across the surface of the earth may be and, increasingly more often, are being taken. In people's daily lives, the global present intrudes as images of instant everywheres or, by a visual metaphor, as split-screen presents. This metaphor was suggested in 1991 by the split-screen television images of the Iraqi preparations for the bombing of Jerusalem on the left of the screen, and preparations of air raid shelters in Jerusalem on the right. Since then, split-screen techniques have become part and parcel of television reporting and have been offering to hundreds of millions of people a God's-eye view of events around town and around the world.

It is possible to imagine a time-compact globe whose people are as sensitive to their positions in history as were the people of the Christian centuries. But this is not the way our globe actually is, because the technology that makes the creation and maintenance of the global present possible is a part of an immense acceleration in the rate of social, political, and cultural change. There seems to be a relation between the speed of life and rate of forgetting. It is not the global present, now being defined, but the speed of life that narrows the temporal horizons of cyberspaced humanity to that of the experiential instant.

I would like to represent the constellation of the global present, the narrowing of temporal horizons, and the acceleration of life by a stanza from a poem by Robert Louis Stevenson. Well over a century ago he expressed in charming cadences the fascination of the dangers of the night, as seen from the safety of a Scottish home.

> Whenever the moon and stars are set,
> Whenever the wind is high,
> All night long in the dark and wet,
> A man goes riding by.
> Late in the night when the fires are out,
> Why does he gallop and gallop about?

The answer is now at hand. The riding man is the Paul Revere of the nonstop society: he carries news of the needs, joys, horrors, and promises of the nonlocal half of humanity.

Let me next turn to a nested hierarchy of *Malthusian selection processes*. They form a family of temporal stresses unique to our epoch.

Malthus's proposition, formulated two hundred years ago, was that the needs of people for food will always outrun what food production can deliver, and hence, mankind is condemned to a chronic condition of hunger and misery. The only way to lessen that misery, in his view, was through moral behavior, which he equated to sexual restraints. His ideas were instrumental in the formulation of the Darwinian principle of natural selection, although later they were handled with a degree of contempt. It was then believed that technology and science would be able to take care of the growing needs of the populace. But during the last few decades of the twentieth century, Malthus's ideas, at least as they apply to economic needs, gained new currency.

In a number of steps I would like to generalize the Malthusian principle.

1. The needs of all life forms outrun what their living and inanimate environments are able to offer to satisfy those needs. The result is organic evolution through "the preservation of favored races in the struggle for life," in the mid-nineteenth-century words of Darwin.
2. The needs of human life outrun what its living and inanimate environments can offer to satisfy them. This is a ceaseless triage, the condition noted by Malthus.

3. The demands that collectively functioning mental powers place upon the individual members of any group outrun the learning capacity of any single individual. This is a cultural triage that leads to a division of mental labor and the development of civilizations.
4. Finally, the contemporary interaction of social, cultural, and economic systems around the globe demand changes in traditional human values. And, it demands them at rates to which the historical, more or less isolated systems have difficulty responding. The result is a malaise around the globe, concerning human values.

The functioning of this many-tiered hierarchy of Malthusian selection processes, with each tier proceeding at different rates, is being recognized under the umbrella concept of time ecology.[4]

Let me now turn to the *identity of mankind.*

It has been the task of human values to tame and guide hunger for food, for mate, and even for knowledge. But it is difficult to formulate such guidelines because a referent—a single, integrated humanity that could adopt such guidelines—does not yet exist. Together with its gods and fortresses, the identity of humankind is only now being constructed. What do I mean by this?

Consider a married couple. Each member has a history, but as a married couple they can have a history only in respect to their collectively maintained present; they can have no history stretching back to the years before they met. Likewise, a tribe, a nation, or a civilization can have a history only from an epoch forward when it first established its social present. Only then can it acquire its collective identity.

What is generally understood as the history of mankind is a set of unrelated or loosely related stories of groups, believed to have had common origins. It is only with the establishment of a global present that we can begin to think about collective planning for the future. Once such planning has begun—if it does—there will surely follow a shared reevaluation of the past. With a viable global present, collectively planned future, and collectively reinterpreted past, an identity for mankind may then become defined.

Individual and collective identities are established through a delineation of boundaries and by noting similarities to and differences from other members of the same class. But there are no other mankinds with respect to which the identity of this mankind could be defined. There are no cohorts that could wake us from our collective fantasies and help us define the boundaries of our identity. Heterogeneous as mankind is, the time-compact globe is entering the set of those systems of which there is only one each, such as the universe, nature, or in monotheistic religions, God. The difficulties of delimiting the boundaries of effectiveness and defining the identities of such one-and-only entities are notorious.

With these various thoughts as background, let me sample the shifts in human values.[5]

A Sampling of Changes

When Shakespeare's Hamlet returns from Wittenberg, he remarks that "Denmark is a prison." A courtier responds, "We think not so." To which the prince replies, "Why, then 'tis none to you; for there is nothing either good or bad but thinking makes it so; to me it is a prison."

All the world's Hamlets, their castles, their yodel—and cyberspaces, and their libraries evolved together in a virtuoso performance, characterized by cosmic arrogance and a slight shiver. Value judgments—which are beneath both the arrogance and the shiver—have their roots in man's constant bargaining with time or, more precisely, with the inevitability of death. When the collectively created temporal horizons open up, such as during the Renaissance, or when those temporal horizons narrow down, such as in our own days, modifications of value judgments follow.

Human values are not eternal ideas, but mutable rules. And fortunately so, because the fortifications needed to protect ourselves from the unintended consequences of time-compactness must be appropriate for global conditions that never existed before.

The Brief Lives of Truths

Let truth as a human value be defined as the recognition of permanence in reality.

This simple definition raises a simple question to which there is no simple answer. Namely, what is to be meant by "permanence"?

The dictionary meaning of permanence is a state or object that is everlasting, that endures without change. The problem is that there is nothing in nature to which this meaning can correspond.

What we do find, instead, may be called permanence substitutes. They are different forms or manners of connections among events, different forms of causations. Each form is appropriate to the temporality of the integrative level where it is found.

They are, from bottom up: absolute chaos, probabilistic causation, deterministic causation, organic intentionality, mental intentionality, and collective intentionalities. By the definition of truth as a recognition of permanence, these different forms of causation stand in for the idealized notion of permanence and determine the character of what I have called the canonical forms of truths. What are these forms?

We recognize truths that are unpredictable. This in itself is nothing new: traditionally they are described as examples of becoming. The most primitive levels of being are probabilistic and deterministic truths; as we enter the integrative levels of life, of the mind, and of society, we find truths that relate to different types of intentionalities.

These qualitatively different types of truths are hierarchically nested. For instance, truths of the noetic world necessarily and always have elements of mental intentions, as well as elements of organic intentionality, elements of determinism, probability, and chaos. Truths are not monolithic. They possess time-related structures. The quality of any specific claim of truth depends on the nested hierarchy of the canonical forms of truths of which it is constituted.

For instance, the apparently simple statement that "today is Monday" includes collective intentions, mental intentions, organic goal-directed components, deterministic components from the physical world, and probabilistic components from the physical, organic, personal, and social worlds, as well as components of chaos from all integrative levels.

Christian theology favors deterministic truths for the world at large, with allowance for human choice. Here is a hymn from around 1796.

> Praise the Lord for He hath spoken,
>   World his mighty force obeyed,
> Laws which never shall be broken
>   For their guidance he hath made.

It was this belief in determinism that helped give rise to the notion of scientific truth. And it was not until the twentieth century that the domain of scientific truth came to be enlarged to include probabilistic and, later, chaotic forms of truths.

Those interested in man and society but not in the epistemology of truth in general may well ask, Why is it necessary to recognize the time-related structuring of truth, if our interest is an understanding of the position of truth on a time-compact globe? The answer is, because different social and cultural conditions, different public moods, different epochs favor different types of truths. Here are two examples that illustrate the narrowing of the horizons of truths.

Early in the morning of April 15, 1865, Abraham Lincoln died on a bed that was too short for him. His secretary of state, Edwin Stanton, then uttered a phrase that has come to be identified with tragic greatness: "Now he belongs to the ages." One hundred and twenty-four years later, *The New York Times* carried a full-page advertisement. "Graffiti-covered subway trains are now history. We cleaned them up yesterday." The ages, which used to be thought of as stretching from Creation to the Final Judgment, deteriorated into yesterday, today, and tomorrow.

The means of disseminating truth has also changed in ways that narrow the temporal horizons. It shifted from printed sacred texts of great age to the displays of cyberspaced media. The character of that media does not permit the past to be comprehended as a connected set of events responsible for the present, because the present is ceaselessly updated in such a peremptory manner as to make what has gone by appear not only irrelevant, but also somewhat uncouth.

Could books help secure stability for some ideas?

Writing arose from earlier skills of drawing and painting. It was a step useful in the communication of articulate thoughts, inarticulate feelings, and prosaic demands. Pictures on cave, castle, and church walls were means of reaching those who could not read, and complemented texts of discourse for those who could. During the second half of the twentieth century, pictures came into ascendancy over the written word as the favored method of mass communication because, as we learn from "The Rape of Lucrece," "To see sad sights move more than hear them told/For then the eye interprets the ear."

Images in cyberspace abound: from photos taken from spy satellites to baby pictures, mug shots, scientific graphs, murdered bodies, and nude people very much alive. Instead of appearing on cave, church, and castle walls, they are now along the walls of boardrooms, bedrooms and classrooms, in places of work and entertainment, in houses of healing and slaughter. And instead of being stationary, they continuously wiggle. Dancing images on the screen have become domesticated animals, but they are not yet housebroken. They are colored shadows of reality. They report about a world that does not stand still long enough to suggest stability as an important aspect of human life. The pictures favor speed and stress the primacy of the current instant.

These images of frenzy keep company to a world of restless intellectual search.

In mathematics, the Pythagorean/Platonic view of numbers, and relations among numbers as eternal verities, has been broadly challenged. Interpretations based on human perceptive and cognitive faculties are in ascendancy.

As to religious truths: process theology now considers God him- or herself as an incompletable being. The resemblances between that idea, and the incompletable quality of mathematics revealed in proof theory, are too patent to be dismissed.

What kind of truths do the political practices of mass cultures favor?

First, those of chaos or unpredictability, the sensational and the ephemeral, pertaining to the instant. Next, organic intentions—that is, truths that acknowledge future and past but only with limited horizons, such as those of hunger or calamity.

It is appropriate now to recall the definition of truth—as the recognition of permanence—then reflect on what has been said about the increasingly briefer lives of truths.

As people search for permanence and for the (presumably) timeless, their search generates changes in what is to be believed to be true. The speedier those changes, the more intense is the need for continuity, permanence, stability. The more intense the need for stability, the faster the changes. This feedback process is displayed in our age at an ever-increasing speed. Our cyberspaced world now seems to seek permanence with such vehemence that it creates a whirlwind of change. The question of what people should believe as true—whether proposed by science, philosophy, religion, or political ideology—is being driven toward new crises of vast proportions but yet unknown resolution.

Conduct à la Mode

Let good as a human value be defined as an assertion that certain conduct, intent, or a certain character trait will promote stable balance and harmony in the mind of a person and in the affairs of society.

The question immediately arises: How long must balance and harmony exist before it may be declared stable and hence the conduct, intent, or character trait that supported it, as right?

Traditional time scales in terms of which stabilities were judged have been those of the cosmos or its earthly kin, history. But the relationship between humans on the one hand, and the cosmos and history on the other hand, has undergone substantial changes.

The Copernican revolution removed the earth from the center of the world; Kepler's recognition that the planetary orbits were elliptical broke the Aristotelian faith in the perfection of the heavens; Newton showed that it was possible to formulate mathematical laws that were universally valid yet subject to human criticism; Darwin removed humans from the guaranteed pinnacle of all species; Freud removed men and women from the company of angels and devils and established them as nature's very unusual but not transcendental wonders. Einstein showed that the universe cannot have a center and bequeathed to mankind a finite but unbounded and inhospitable immensity of dumb matter.

In early understandings, the universe was seen governed by divinities whose value judgments were unmistakably human. The nature of the cosmos that houses the cyberspaced humanity is qualitatively different from those traditional models. It is a mathematical relation of the kind that Friedrich Gauss, with remarkable prescience, described as one of "astral geometry." To that universe the cosmologist must bring the idea of time's passage, as well as his value judgments, because neither the experiential passage of time, nor guidance in matters of good and evil can be obtained from the type of knowledge that is now used to describe the cosmos. The consequence is a cosmic lostness, which encourages conduct à la mode.

In the words of the historian Paul Kennedy, what we are witnessing is not a new world order but a troubled and fractured planet. A society

> which desires to be better prepared for the twenty-first century... will have to retool its national skills and infrastructure, challenge vested interests, alter many old habits, and perhaps amend its governmental structure. But this assumes long-term vision at a time when most politicians... can hardly deal with even short-term problems.[6]

He added that the forces of change are so "far-reaching, complex, and interactive that they call for nothing less than the reeducation of humankind."

I would like to mention a few examples of that reeducation, where time-compactness appears to be instrumental.

Social institutions, such as *religion* and *patriotism*, are being reconstructed to fit the needs of societies in transition to a cyberspaced world. Five billion Hamlets are rearranging the forms and contents of their social institutions. They are changing them from being appropriate for an earth upon which human groups interacted only slowly, to an earth upon which social institutions interact instantaneously and continuously.

These interactions promote the sharing of the material and spiritual aspirations of all people, but they also reveal in image, word, and quality of life, the unequal means and skills available for the support of those aspirations.

During this period of reeducation—to use Kennedy's words—violence and spectacle combine in uncountable ways with the enforcement of justice, which itself has become spectacle and violence. This frightening show is driven by alliances among guns, connected, as everything else, through cyberspace. English Romantic writers of the eighteenth century compared God with clockwork, while allowing for clockwork devils. Considering the popular penchant for salvation through science and technology and damnation through techno-science Godzillas, we may add computer and cyber gods and devils to the problems of good and evil upon the time-compact globe.

Eros is also getting reeducated but it did not graduate yet. In a world that holds immediate need satisfaction as a feature of civilized life, and where recreational and procreational sex is easily separable in practice, Eros is seeking guidance on how to make the two separable not only in practice but also in principle.

A related condition bears a nasty name: *surplus people.*

Bringing forth vastly more offspring than the environment can support and societies of insects, fish, birds, or people can feed, educate, and integrate has been the policy of life throughout its 3.5-billion-year history. The term "surplus people" does not refer to a fixed number but to ever-changing conditions that are jointly determined by societies, their living and inanimate environments, and by cultural judgments of right and wrong. It seems to me that morality, pertaining to life and death, is being reconfigured as death control is joining birth control.

Part of this effort involves *artificial selection in humans.* Since the techniques of this selection are the products of the human intellect, such as genetic engineering, artificial selection in humans became indistinguishable from natural selection.

The reconstruction of religion and patriotism, the reeducation of Eros, and the handling of surplus people are all taking place against the backdrop of *environmental problems.* The details are broadly known. What does need stressing is that remedies are difficult because the causes of environmental problems have ambiguous social positions. Atomic waste comes from plants that generate electricity; the deadly organisms in vials, threatening public health, were used in medical research; the garbage of the world is made of things left over from activities that helped people survive. The satellites that crash like so many falling Lucifers made cyberspace operable. Many of the current enemies of the environment were formerly its friends.

It is reassuring that the periodic regression of humans from phenotype to genotype is still with us in the life cycle. As we read in *The Space Child's Mother Goose,*

Foundations shake,
Computers break
And science goes be-pop,
But baby's joy
Is still the toy
With foolish ears that flop.[7]

At least where children do have such toys. In an unforgettable 1996 photograph, three small, bare Macuxi Indian children of Brazil are seen playing with the head of a butchered cow.

Meanwhile, multinationals, ideologies, religions, and alliances of all kinds are shifting around to find their niches in the cyberspaced world. So are drug rings, mafias, terrorist movements, and smugglers of women and children, immigrants, weapons, and body parts.

It is my guess that ideas of right and wrong on the time-compact globe will be based on an interpretation of history that will permit the construction upon it of a believable and inspiring plan for the future.

In the 1860s the American poet James Russell Lowell sensed the historical significance of the incipient Civil War and wrote a long poem called "The Present Crisis." I want to borrow a line from that poem, as a summation of this section. "New occasions teach new duties," he wrote, "Time makes ancient good uncouth."

The High-Tech Primitive

Since Plato, the ordinary interpretation of beauty has tended to involve this question: What is the light that shines from certain objects and events, illuminating the lives, loves, and sufferings of men and women? I would like to exchange the source and the sink and ask this: What is the nature of the light that informs certain feelings and thoughts of men and women, illuminating the impersonal and uncaring universe by making some of its features, sometimes, appear beautiful?

In other words, I propose that aesthetic judgments derive from the quality of feelings mobilized by external or internal, real or imaginary events, conditions, or processes. If the quality of feelings makes one desire its perpetuation, then whatever is believed to be responsible for it is said to be beautiful. If the quality of the feelings makes one desire its absence, than whatever is believed to be responsible for it is said to be ugly.

What kinds of feelings would I want to perpetuate? Those that suggest timelessness in time. The foremost crafts for inducing such feelings are the arts and letters, but conditions that bring forth such feelings are not limited to them, nor are the arts and letters limited by the demand that they induce such feelings. It is this freedom that permits the arts and letters of our age to mutate, mate, and in many of their forms, die at a dizzying speed.

Contemporary art crosses all borders among different traditional arts, both in form and content. We now have sculpture that sings, poetry that dances, and architecture that is movable, as artistic skills are transferred from the artist to the judgments of the people who make and sell the marketable product. The panorama at the turn of the century is that of a teeming and boisterous chaos. Except for motion pictures, masterpieces are rare if present at all because the very idea of masterpiece has become unacceptable. Although there are many examples of ingenuity, skill, and spiritual travail, the overall impression I have is that of seeing or hearing the high-tech primitive.

Let me recall the definition of aesthetic judgment I proposed.

If the quality of feelings makes one desire its perpetuation, then whatever is believed to be responsible for it is said to be beautiful. If the quality of the feelings makes one desire its absence, then whatever is believed to be responsible for it is said to be ugly.

But it then follows that in an age when people are uncertain as to what feelings to perpetuate and what to reject, the arts and letters are going to shift continuously in search of stable form and content. Aesthetic judgments, then, make for appropriate company to short-lived truths and experimental morality.

The contemporary arts and letters serve—as they have served all along—as collective Rorschach tests to which artists and audiences alike are trying to attach meaning. But this is no reason for alarm, because it is through this process of signification that they carry on with the traditional task of aesthetic judgments. That task is the blazing of paths into worlds that did not exist before those paths were blazed. Here I am repeating an idea by Fred Turner. The paradoxical nature of this claim is the paradox of creativity itself; it describes becoming as the emergence of the unexpectedly new from the chaotic substratum of the universe.

New forms of beauty and ugliness do indeed bubble up around the world, visit for a while, then become moribund as other, newer forms arise. Today, tastes divide and are celebrated by age, sexual orientation, and ethnic background. Radio stations follow formats appropriate to the ethnicity, sex, religion, income, and daily schedules of their listeners. The media, serving as beat and shape banks, are being pushed and shoved by vast business interests that guide and are guided by the pushes and shoves of a restless public taste.

The United States now has a National Poetry Month, which in 1997 was also Soy Products Month. It brought forth a defense of poetry from Robert

Pinsky, that year's poet laureate. He cited certain parallels between poetry and technology and spoke of poetry as a digital computer data device. Had he been interested in the visual arts, he might have noted that anyone wanting to see a contemporary version of a cubist papier collé may do so by working with Windows 2000.

All past examples of great art shared the desire to record for posterity such feelings and thoughts as would guide people's value judgments for ages to come. Although examples of excellent writings continuously appear, it is difficult to imagine a literary work today that could remain significant long enough to serve as a catalyst for a stable, collective view of man and the world.

I would like to propose a unifying perspective to this fertile and emotional chaos, so rich in potentialities and so poor in lasting accomplishments. My perspective derives from an understanding of tragedy as the most general statement that can be made about the human sense of time and human freedom.

Tragedy weighs obligations, memories, hopes, and fears in terms of prevailing values. It then leads to decisions, made with the help of steady reflections upon future and past. Because of the decisions made and actions taken, the number of remaining alternatives narrow until there is no choice left but the denouement. On the organic level, death by aging makes it possible for new and newer generations to differ from their ancestors in a fashion that is advantageous for the continuity of life at large. Likewise, the tragic in human life makes it possible for new and newer generations to differ in their values in a manner that is advantageous for the continuity of civilization. What death by aging is on the organic level, tragedy is on the noetic and social levels: both are unavoidable and necessary.

What for the tragic hero are the challenges to his ideas are, in the life of each and every person, the consequences that follow from his or her awareness of time's passage. This knowledge demands deliberate choices. It is a selection pressure that works from within. It is more implacable than any external selection pressure and more tyrannical than any social convention or political tyranny. The knowledge of time thus generates a steady tension between the dream of permanence and the biological facts of passing, lending to human life its irreducible aspects of the tragic.

I submit that before aesthetic judgments can mature to be appropriate for a new high culture, they will have to admit the irreducibly tragic dimension of human life as an expression of the hierarchy of unresolvable creative conflicts in the evolutionary process of nature at large.

The problem is that tragedy demands open temporal horizons. But such horizons do not sell well in mass-market technocracies that see human existence not as a continued historical struggle but as a series of problems to be solved here, now, and rapidly.

Human Values

In the turmoil of Hamlet's crumbling castles, have human values failed us?

Not at all. What has happened is that the role of human values in the household of civilizations has been grossly misunderstood.

As forms of judgments, they have been traditionally seen as conservative influences that promote permanence, continuity, and balance in the affairs of the mind, heart, and society. This may sometimes be the case for the short term. But for the long, historical term, the role of human values has been the opposite. They have served as revolutionary forces that promote change by giving rise to and maintaining certain unresolvable conflicts.

Thus, the search for truth, in its many forms, has as its primary consequence the creation and perpetuation of changes in what is believed to be true.

Moral judgments create and maintain conflicts concerning conduct and, through them, keep alive a steady revolt against whatever principles happen to be guiding people's behavior.

The aesthetic faculties generate conflicts between the world as we find it to be and that other world of imagination to which art gives transient forms and changing names.

These conflicts combine in innumerable ways. Together, they help perpetuate the insecurity of the species that comes from our awareness of the finiteness of life, held up against the dream of eternity. Attempts to lessen that insecurity by emphasizing permanence, such as by following human values, have been driving the remarkable creativity and frightening destructiveness of humankind.

An Integrated Understanding of Time

In *The Library of Babel*—which I have already cited—Borges remarks that at one time it was "hoped that a clarification of humanity's basic mysteries—the origin of the Library and of time—might be found."[8]

The international society for the study of time has been busy with both—with the study of time and with the infinite capacity of the library, that is, the apparent infinite capacity of the human mind. In pursuit of these inquiries, we have had the temerity to inquire into many different fields of learning.

In 1966, at the concluding session of a conference at the New York Academy of Sciences, I outlined the need for a systematic study of time, listed its advantages, and pointed to its methodological problems.[9] I reasoned that

> our knowledge of the nature of time must remain partly obscured until we learn to benefit from insights that stem from all of our rational, introspective, and experiential knowledge [of time] as expressed in the sciences and the humanities....

Also, that such an approach

> amounts to a call for an epistemology acceptable at least for our era and adequately encompassing the idea and experience of time.

Such an epistemology is now a part of the hierarchical theory of time. In the same 1966 paper I remarked that

> the important element which must be sought is a freshness of vision and rebelliousness of mood.... The way to grant [the study of time] its charter resides in...supporting an intellectual climate where creativity common to all forms of knowledge is permitted to bring forth their synthesis by interacting through the common idea of time.

In attempting to achieve an integrated understanding of time we are doing more than studying the nature of time: we are pioneering the practice of "reciprocal literacy" on the professional level, to be employed for many and different purposes.

The cyberspaced world, with its ruins of Hamlet's castles, with its need for reciprocal literacy, and with its search for new fortresses to protect its inhabitants from the metaphysical winds that blow through the universe, is very much on the march.

That march is not to the tune of Dies Irae, nor to the Marseillaise or to the Communist International. These turned out to have been no more than call signals of large tribes. The march is to something more elemental. We are in a metamorphosis where the moral creature, searching for truth and beauty, is being driven by the unresolvable conflicts of the present social systems toward a new level of complexity, with its own unresolvable creative conflicts.

This vast historical change happens to coincide with the passing of the responsibilities for running this Society from an older to a younger generation. This rite of passage displays its own changes and continuities.

The changes may be represented by the transition from letter writing to sending e-mail.

As to the continuities, I would like to represent them with the words of three men.

Since our host is the Evangelische Akademie, it is appropriate to start by thinking of Martin Luther. This is from one of his letters to Melanchton. "Pecca fortiter, sed credo fortius." Sin bravely, but have even more faith.

Four centuries later similar advice came from Oliver Wendell Holmes, justice of the United States Supreme Court. "If I were dying," he wrote, "my last words would be, 'Have faith and pursue the unknown end.'"

The third remark comes from Bertrand Russell. I like it because he identified better than I could have done the forces that have been driving my own work in founding and directing this Society. This is what he wrote:

> Three passions, simple but overwhelmingly strong governed my life: the longing for love, the search for knowledge, and unbearable pity for the suffering of mankind.

21. WHOSE PAST IS OUR PROLOGUE?

The king of Naples was returning from the wedding of his daughter in Tunis, when his ship was caught in a tempest and broke in two. All on board became jetsam and flotsam. The storm—so we learn—was raised by a magician who lived on an island along the course of the ship and who wanted to set old wrongs right. Confusion, good will, evil follow. They are all liberally intertwined with innocence and young love. All of which, told with consummate skill, make for a garden of divine folly and human joy as *The Tempest* blows to its denouement.

> Earth's increase, foison plenty,
> Barns and garners never empty:
> Vines with cust'ring bunches growing;
> Plants with goodly burden bowing[1]

Early in the story, just after the shipwreck, the King and the brother of the usurping Duke of Milan have a discourse about time and history. The Duke's brother makes his point.

> We all were sea-swallow'd, though some cast again,
> And by that destiny to perform an act
> Whereof what's past is prologue, what to come
> In yours and my discharge[2]

They knew their past, as well as the past may ever be known. The character of that knowledge may be represented by the report of a child. "It all started, teacher, when Jimmy hit me back."

Our earth, in the process of globalization, is deciding with guns and words, who started the bloodshed by hitting back. The essay that follows concludes that the current havoc of humanity is a struggle to decide on whose interpretation of the past, on whose formulation of the prologue to globalization, are the plans for humanity's future to be based.

REFLECTIONS UPON AN EVOLVING MIRROR[3]

This paper suggests that the violent turmoil of our age is a symptom of an identity crisis of humankind at large, precipitated by globalization. For an understanding of that identity crisis, the evolutionary origins and uses of intent, memory and identity are sought and interpreted. This interpretation is then applied to our global laboratory in which many, incompatible needs demand fulfillment. In that perspective, the identity crisis may be seen as a struggle to decide upon whose understanding of the past, upon whose collective memories are the plans for the future of mankind to be based.

1. About Final Conflicts

The reasons that led to the founding of this Society had nothing to do with anyone's interest in the nature of time. They had to do with the puzzlement in the mind of a man of twenty-one who, in the autumn of 1944, found himself on a mountainside between two vast armadas. Behind him was the armed might of Nazi Germany, in front of him the immense masses of the Soviet Union. He knew that he was watching a struggle between two ideologies, each of which was convinced that it, and it alone, was destined to fight and win the final conflict of history. The Soviets had their creed summed up in their revolutionary anthem: "This is the final conflict / Let each stand in his place / The international party / Will be the human race...."[4] The official Nazi march said the very same thing, in different words. "This is the final bugle call to arms / Soon Hitler's flag will wave o'er every single street. / Enslavement ends / When soon we set things right."

Having been aware of both dogmas, I came to wonder whether there does exist a final conflict in history. Perhaps the buzz bombs the Nazis kept on sending over London were not the ultimate weapons they were claimed to be. But, being hungry, cold and miserable, I did not pursue the puzzlement. All I did was to promise myself that if I ever got out of that hell alive, I would enroll in Plato's Academy and report to it about the wisdom of Robin Goodfellow, "Oh what fools these mortals be!".[5]

Nine months after I witnessed the clash of those final conflicts, I stood in an almost empty St. Peter's in Rome, in front of Michelangelo's early Pietà, a piece of Renaissance marble transfigured by human feelings. I saw two sculptures in it: a heavenly and an earthly one, joined by the two natures of the female figure: the *mater dolorosa* and the *amante dolorosa*, the grieving mother and the grieving lover.

The heavenly sculpture showed the Virgin holding the dead body of her son. In it Michelangelo asserted that the suffering of the Redeemer freed man from his earthly conflicts and opened up the way to a fulfilled, everlasting life.

I could not help but observe that the Virgin's figure was that of a woman much younger than the man whose dead body she held. I did not then know that her youth, compared to that of the man, had a veritable literature and that Michelangelo himself was asked about it.[6]

In the earthly sculpture, the youth of the female figure was no problem. She was Michelangelo's Italian model, real or imagined. She was also Dante's Beatrice, murmuring "L'Amor che muove il Sole e altre stelle" I gave the age difference an interpretation that made its way into my writings. This is from *Time, Conflict, and Human Values*.

> Just out of the havoc of World War II, I was ready to jettison all received teachings. I failed to see the Virgin holding the body of Christ. What I did see was a young woman of exquisite beauty holding the body of her man, murdered by the powers of law and order. Her face is one of infinite sadness as the irreversibility of his death permeates her unbelieving mind. Her beauty suggested to me that she was with child for I believed that women were most beautiful when they were pregnant. In a melodrama the woman of the statue would faint. In the Roman Pietà she bears up because she carries the child of the man whose body she holds.[7]

During my visit, Michelangelo reassured me about the affinity between Eros and Agape. He also told me that if one is pregnant with life or with an idea, one cannot afford to faint.

A year later I was on board an American troop ship en route to the United States. On a foggy September morning I sailed by the Statue of Liberty in New York harbor.

> Give me your tired, your poor,
> Your huddled masses yearning to breath free,
> The wretched refuse of your teaming shore.
> Send these, the homeless, the tempest-tossed to me
> I lift my lamp beside the golden door.

Homeless and tempest-tossed. That was me. The words "Give me your tired, your poor" joined in my mind a much earlier invitation: "Come unto me, all ye that labor and are burdened, and I will refresh you."[8] My only concern was that the door was made of gold. First, because that earlier invitation did not say, "Come unto me all you that labor, and I will give you lots of gold." Second, because for me gold was only a dead metal, atomic weight 197 and, having survived the dictatorships of the true believers in final conflicts, I did not want to become subject to the censorship of the true believers in gold as the final arbiter of all things human.[9] I arrived in the land of my dreams: of Buffalo Bill, Thomas Alva Edison and Thomas Jefferson. I was where I wanted to live, love, die and be buried. And what a privilege it was to be among people who did not worry about final conflicts but were committed, instead, to a permanent revolution.

2. Why Study the Nature of Time

Soon after my arrival, the promise I made to myself called me to task. How was I going to tell people about that awesome stage upon which I was an insignificant walk-on? War stories were coming out in great profusion and I thought of contributing to the flood. One day, while browsing in a bookstore in New York's Greenwich Village, I came upon a comic book called, "The Nazis and the Invisible Man." I did not then know the poetry of T. S. Eliot. If I had, I would have thought of "Go, go, go, said the bird: human kind / Cannot bear very much reality."[10] Even without Eliot, I gave up the idea of writing or even speaking about my war experiences.

Instead, I began to search for a vehicle that could carry "Oh what fools these mortals be" as well as Beatrice's planetary theory of love. It had to be a subject of universal interest, yet one that demanded clarity of thought and exposition, so as to protect it from the merchants of bogus scholarship and wooly science.[11]

My memory obliged. I remembered that one morning in gymnasium we learned about the mathematical pendulum and how it may be used to measure time. That evening I saw a movie in which people danced around a fire. The subtitle said that they danced to help them forget the passage of time. The next day, all across town, there were crowds, ecstatic with hatred and love. They marched in ways that looked to me like dancing around the fire.

Obviously, the pendulum was used to measure something people wanted to forget. If I could trace a connection between the swings of a pendulum and the desire to forget whatever it measured, then I could bracket both the foolishness and the greatness of the species. My theme, then, could serve as did the images on the shield of Achilles: an illustrated encyclopedia in which people could see themselves both as heavenly and earthly.

Four years after I sailed by the Statue of Liberty, while finishing my work for my first degree, I wrote a paper called "A short essay on time." It won a national humanities award for science students. This encouraged me to search the literature of time—which led me to the writings of S. G. F. Brandon, then professor of comparative religion at the University of Manchester.

He maintained that the human knowledge of time is a powerful tool in the struggle for life because, with the help of memory, it makes preparations for future contingencies possible. But, it is also the source of "an abiding sense of personal insecurity" which inspires people to seek such forms of refuge as represent their ideals of safety from all they fear and help conserve all they desire.[12]

Brandon's lines met in my mind my memories of the war and came to be expressed in the Introduction to *The Voices of Time* (1966).

> Watching the clash of cultures and the attendant release of primeval emotions stripped of their usual niceties, I could not help observing that man is only superficially a reasoning animal. Basically he is a desiring, suffering, death-conscious and hence, a time-conscious creature.[13]

I realized with a pleasant shock that the question I posed many years earlier, namely, whether there can exist a final conflict in history, was too crude to be fundamental. Namely, it is possible to imagine a world without mass murders but it is not possible to imagine humans who will not declare, in innumerably many ways, "Death, be not proud...."[14] because the conflict that gives rise to such a rhetorical command—the conflict between the knowledge of an end of the self and the desire to negate that knowledge—is at the very foundation of being human. This conflict is unresolvable because if it ceases, personhood collapses. A man or woman may well remain alive but only with impaired or absent mental identity. For this reason I came to regard that conflict as constitutive of personhood and came to see all other, overt conflicts as derivative from the fundamental one. Also, with my interest in the natural sciences, I began to wonder how such a merely human conflict fit the dynamics of nature at large?

In agreement with Brandon, I came to believe that the efforts to be able to live with that unresolvable conflict drive both the immense creativity and the frightening destructiveness of the species.

I wrote to Professor Brandon. He replied kindly and suggested that I write to Joseph Needham in Cambridge. By and by I was guided to an impressive group of British scientists and scholars and through them to their colleagues in Germany, Switzerland and France. Through these people, whose writings and letters awed and inspired me, I found my way back to my fellow countrymen, starting with David Park, a physicist at Williams College and George Kubler, an art historian at Yale.

3. ISST

Correspondence with an increasingly larger group of scientists and scholars convinced me that the intellectual climate was ready for a collective enterprise directed to an interdisciplinary, integrated study of time. To prepare for such an undertaking, it appeared useful to publish a volume about the significance of time in the major academic disciplines. *The Voices of Time* (1966) is that survey.[15] The title had two sources. One was St Paul, "There are . . . so many kinds of voices in the world, and none of them is without significance."[16] The other was Simon and Garfunkel's 1964 lyric, "The Sounds of Silence."

I was ready to report to Plato's Academy about the study of time as an ideal means to explore the greatness and the foolishness of our species, but I could not find the Academy's address. I decided, therefore, to help create a Grounds, Buildings and Services Department to the Academy.

In January 1966 the New York Academy of Sciences held a conference on the theme "Interdisciplinary Perspectives of Time." Participants included Gerald Whitrow and Brian Goodwin of England, Olivier Costa de Beauregard of France and Georg Schaltenbrand of Germany. In my talk I remarked that the way to an interdisciplinary study of time must include support for "an intellectual climate where creativity common to all forms of knowledge is permitted to flourish, and aspects of reality previously separately understood are permitted to produce their synthesis by interacting through the idea and experience of time."[17]

After the conference Gerald Whitrow, then professor of the history and application of mathematics at the University of London, Satoshi Watanabe, who taught quantum theory at Yale, and I sequestered ourselves to a quiet corner and, following my proposal to form a professional group for the study of the nature of time, we declared the International Society for the Study of Time as having been founded.[18]

There was a strong personal reason for my proposal. I needed a lookout tower. Let Sam Walter Foss, a New Hampshire poet, supply the words. The title of the poem is "The House by the Side of the Road."

> Let me live in a house by the side of the road
>   Where the race of men go by—
> The men who are good and the men who are bad,
>   As good and as bad as I.[19]

I had high hopes for such a Society, provided it did not collapse into mediocrity and provincialism. The challenge of finding people who could articulate the similarities and differences among a crowbar, a candy bar and a kilobar, was still ahead.

4. Time, the Evolving Mirror

My inquiries began to bring books and articles by the drove. The themes people judged essential for a study of time extended from the iconography of Renaissance art to information conveyed by the bees' dance, from medieval poetry to the entropic measure of human migrations. With the flow of ideas the problems of any interdisciplinary dialog became evident. Namely: different disciplines employed different jargons, had different criteria for testing for truth and maintained different, unstated assumptions about reality. Also, opinions about which field of knowledge was the most appropriate one for studying the nature of time, though widely divergent, were always accompanied by deadly parochialism.

I had the privilege of discussing the problems of interdisciplinary exchanges with Joseph Needham. He responded by giving me a copy of his Herbert Spencer lecture "Integrative levels: a revaluation of the idea of progress."[20] "See, Fraser, whether this will help," he said. It helped immensely.

The idea of integrative or organizational levels extends from Plato and Aristotle to the Christian Platonists and Aquinas, to Hegel, Marx and Bertrand Russell. It occurred to me that recognizing in nature a nested hierarchy of stable integrative levels, distinct in their complexities[21] and languages,[22] could accommodate the different epistemologies necessary for dealing with the worlds of radiation, particle-waves, solid matter, life, the human mind and human society. And, for that reason, it could serve as a framework for an interdisciplinary, integrated study of time.

I did not realize until many years later that the reasons of the remarkable appropriateness of the nested hierarchical model of nature for integrating the epistemologies that an interdisciplinary study of time must accommodate, may be found in the logical structure of Gödel's incompleteness theorem.[23]

Indeed, the model could accommodate Brandon's recognition of the conflict between (1) the human knowledge of time as a weapon and (2) the "abiding personal insecurity" of our species, together with the dynamics of the life- and the physical sciences. It gave rise to the theory of time as a nested hierarchy of unresolvable, creative conflicts. I will now appeal to that theory to help us learn about the evolutionary origins and roles of intent, memory and identity.

The theory employs an operational definition of reality. Specifically, it extends the biologically based definition of reality, formulated by Jakob von Uexküll a century ago, to all forms of human knowledge: experiential, experimental and abstract. What emerges is an understanding of reality as a relationship between the knower and the known. Applying this understanding of reality to the diverse material that must enter an interdisciplinary study of time leads to the conclusion that what, in ordinary use is called "time," has a structure, that it comprises a nested hierarchy of qualitatively distinct temporalities.[24]

Let me introduce the two major dramatis personae of the theory. They are evolving causations and evolving temporalities.

First, let me attend to causation and name its evolutionary stages. They are: chaos, probability, determinism, organic intentionality, noetic intentionality and collective intentionality. Next, let me visit each separately and identify the steps in the evolution of temporalities.

The primeval chaos is without any connections among events. It supports no causation, its world is without any features that may be associated with time. Absolute chaos or pure becoming is atemporal.

The organizational level of nature above chaos, known through quantum theory, is that of particle-waves. In that world distinct instants do not yet exist, only probabilistically distributed likelihoods of instants do. Time is not yet continuous. That world is prototemporal.

The next step in cosmic evolution was the coming about of the galaxies that form the astronomical universe of solid matter. Instants in that universe are well-defined. They are connected through deterministic relations, as embodied in both Newtonian and Einsteinian physics. That level-specific time is eotemporality. It is one of pure succession, without preferred direction. The reason why we cannot find purely deterministic processes is that, because of the nested hierarchical organization of nature, there can be no deterministic processes without probabilistic and chaotic components.

Michael Heller, physicist and Catholic theologian, has shown that the physical world is time orientable, that it allows for two directions of time, but it need not be so oriented, that it is complete and intelligible without directed time.[25] Heller's conclusions are consistent with P. C. W. Davies' assertion that "The four dimensional space-time of physics makes no provision whatever for either a 'present moment' or a 'movement' of time"[26] and that, "It is a remarkable fundamental fact of nature that all known laws of physics are invariant under time reversal."[27]

By the Principle of Parsimony, the hierarchical theory of time maintains that what physics reveals about time in the physical world is the total truth. The reason why the laws of physics do not make provisions for a present and for the flow of time is not because physical science is incomplete in that respect but because the temporalities of the physical world are incomplete when compared with what we, as living and thinking beings experience and think of as time.

For Einstein, according to his friend and biographer Paul A. Schilpp, "there was something essential about the Now which is just outside the realm of science."[28]

The parochialism of equating "science" entirely with physics, dooms an understanding of the "now" in terms of natural philosophy. But, if biology is included among the sciences as it must be, then, a scientific, operational basis of the "now" may be identified. To explain how life gives rise to the "now," to intent, to memory and to identity and how the flow of time acquires meaning through the life process, we have to think ourselves back to the primeval chaos of pure becoming.

David Layzer gave good reasons in support of the claim that chemical order was not present at the initial universe but was created by cosmic expansion.[29] It is necessary to add that ordering, created by the expansion of the universe, has been opposed all along by disordering, governed by the Second Law of Thermodynamics. Also, that ordering and disordering define each other.[30]

In the physical world ordering and disordering are random, uncoordinated. In sharp contrast, within the boundaries of an organism, ordering and disordering—growth and decay—go on simultaneously and are coordinated from instant to instant. The life process is identically equivalent to securing that events which must happen simultaneously do so happen, events that ought not happen simultaneously, do not. When coordination between growth and decay fails the conflict ceases, the organism dies. For this reason, the coordinated processes of growth and decay are the constitutive conflicts of life. They are necessary and sufficient to define the life process. The instant to instant coordination that maintains the life process, introduces nowness into the nowless universe of nonliving matter. With nowness it defines conditions of non-presence. Nowness is, a local phenomenon. This is the reason why, in the words of Whitrow, the Special Theory of Relativity

> denies the universal simultaneity of spatially separated events. Consequently, the simultaneity of events throughout the universe becomes an indeterminate concept until a frame of reference (or observer) has been specified.[31]

In Shakespeare's "As you Like It" we learn that "from hour to hour we ripe and ripe / and then from hour to hour we rot and rot."[32] The Bard had youth and age in mind. But the life process, as I suggested, is identically equivalent to simultaneous ripening and rotting. Or, one may speak of simultaneous and coordinated entropy decreasing and entropy increasing processes. Their conflicts, as I mentioned, are unresolvable in the sense that if they cease, the organism dies. Ripening and rotting are, as I mentioned, the constitutive conflicts of life.

That nowness has no meaning in the physical world does not mean that our experience of the present is a figment of human imagination as are leprechauns. No more so than the death of a man or woman is a figment of his or her imagination just because galaxies or crystals do not die.

Let me turn to biogenesis as the evolutionary origin of intent and memory.

Living systems are thermodynamically open. They demand matter and information from the world external to them to be able to maintain their constitutive conflicts. With that demand, need is born. Need directs behavior toward need satisfaction, known as intentionality. Memory, I suggest, was selected for because of its usefulness for guiding intentionality. The subject of our conference could and should have been, "Time, Intentionality and Memory."

Into the directable but not directed temporality of the physical world intentionality and memory introduced distinctions between two non-present conditions. Imagined non-present conditions that relate to intent, driven by desire for need satisfaction, are said to be in the future. Imagined non-present conditions that suggest usefulness in the pursuit of need satisfaction are subject to classification, through the complex process of reality testing as memory or as fantasy.[33]

With future and past referred to a present, a flow of time acquired meaning. When post-mortem life became imaginable, so did a future of unending time. And with it, I would think, open-ended prepartum time as well.

It took eons for the human brain to reach the degree of coordinated complexity that could display minding and, supported by memory and fantasy, project intentions to increasingly distant futures. During those eons noetic time became established as a part of human reality. Biotemporality was already a part of that reality because thinking humans are also alive. So were the physical temporalities, because we are made of matter.

This natural history clarifies the Kantian view of time as a form of pure intuition which is also empirically real. As a pre-Darwinian thinker, Kant had difficulty reconciling the two.[34] The hierarchical theory of time accommodates the empirical reality of time as having become, through evolution, a form of intuition.

In summary: to the probabilistic and deterministic causations of the physical world, life added organic intentionality. It is driven by the needs of organisms, guided by whatever forms of memory the organism possesses.

Let us step up from the biological to the noetic individual. The constitutive conflicts of personhood, as I proposed, are those between, on the one hand, a person's awareness of the end of her or his self in death and, on the other hand, her or his ceaseless efforts to escape from that ending through biological, intellectual and social offspring.

As human beings we function with the nested hierarchy of all forms of causations and in all the different temporalities. Also, our dynamics subsumes all the constitutive conflicts of matter, life and the mind. For this reason, our ideas about our experience of time, to use a poetic turn of phrase, amount to reflections upon an evolving mirror.

We may sum it up with Tennyson.

> I am a part of all that I have met;
> Yet all experience is an arch wherethrough
> Gleams that untravelled world, whose margin fades
> For ever and for ever as I move.[35]

5. The Global Laboratory

Having thus sketched the evolutionary origins of intentionality and memory and with them, the emergence of the notion and experience of time's passage, we are ready to examine the thesis of this paper. Namely, that the violent turmoil of our age is a symptom of an identity crisis of humankind and that the crisis involves a struggle to decide upon whose interpretation of the past, upon whose collective memory are the plans for the future of mankind to be based.

Let us begin by recognizing the steps in the evolution of identity.

The simplest identity, that of the biological individual, resides in the peculiar manner in which it applies its memory to serve its intentions. And, does this so as to maintain its constitutive conflicts and their phenomenal manifestations, which is the living body. The instant by instant coordination of ripening and rotting, as I proposed, defines the organic present.

The noetic identity of a thinking human resides in the peculiar manner in which it applies his or her memory to serve its intentions and apply it so as to maintain his or her constitutive conflicts and their phenomenal manifestations, which are his or her mental processes. The instant by instant coordination, as I proposed, defines the mental present.

I submit that the constitutive conflicts of a human society are those between its collective needs and the needs of its members. By a practical shorthand, these opposing trends may be described as those of gathering and scattering that is, between favoring the needs of the collective or favoring the needs of individuals.

The identity of a society, then, resides in the peculiar manner it applies its collective memories to serve its collective future so as to maintain its constitutive conflicts. The instant by instant coordination between gathering and scattering defines the social present.

How does this model of identity apply to a globalized humankind?

Globalization itself is not new. It was already evident in the 4000 mile Silk Road that connected Rome with China in the 2nd century B.C. The process of globalization acquired a theological foundation in the belief that in God's eyes all humans are equal. These foundations gave rise to ideas which assumed globalization as a natural and desirable condition of humankind. This assumption turned into political convictions. I am thinking of the mercantile capitalist desire for cheap labor and rich consumers, as well as the Communist dream for the workers of the world to unite.

What is new to 21st century globalization is that the electronic global present did not come about through evolutionary selection favoring the continuity of a global community, as did the lower order organic and social presents favoring life and nations. Instead, it came about through technological developments sui generis. Humans are now selected for by their own, self-created environment. We have to learn to live with, what Daniel Boorstin called the "global instant everywhere." Where do we stand in that selection-and-learning process?

In 1881 Walt Whitman saw America as a land of "contraltos…carpenters…duck shooters…deacons…spinning girls…freaks…patriarchs [and] opium eaters".[36] If, in this kaleidoscope, we replace the pieces with merchants of Einstein bombs, with happy rappers, with merchants of slaves, with dealers in sperms and in wombs-to-let, smugglers of drugs and of stolen art, and with an increasing separation between the haves and have-nots, a globalized humanity comes into view. We even see Faust on the road, peddling his discoveries to the highest bidder.

Whitman's America succeeded better than many other regions on earth in reconciling its contraltos, carpenters and duck-shooters. It did so under the motto, "E pluribus unum" or "Out of many, one." What could possibly serve as the "unum" for a worldwide bazaar of wheelers and dealers in life and death and things and stuff?

In 1933, the philosopher-mathematician Whitehead explored humanitarian ideals which, in his view, could be shared by all people. He found them in religions. He did not say which religion it ought to be but he did maintain that, in the long run, "that religion will conquer which can render clear to popular understanding some eternal greatness incarnate in the passage of temporal fact."[37]

On the globalized earth, according to S. P. Huntington, there are three great ideologies that struggle for the mind of man or, to employ Whitehead's words, address "the passage of temporal fact." Each offers its dogmatic interpretations of life, death, history and the right rules of conduct. They are the Judeo-Christian West, Confucian-Socialist China, and the Islamic world.[38] Each of these ideologies offers a different view of man's position in the universe and, consistently, a different story about the past of humankind and of life.[39]

I believe that an amalgamation of these assessments of the past is already in the works. To find out what it is, I propose to attend to the witness of the dramatic arts.[40] because, as we learn from Claudia Clausius, drama "enacts the cultural history of a people at the same time as it defines its own contemporary self-consciousness."[41]

Specifically, I want to point to some important changes that have taken place in the ways tragedies depict the uses of the past in the service of the future. My reasons for selecting tragic drama are these. In tragedies obligations, memories, hopes and fears are weighed with steady reflection upon future and past.[42] Also, tragedy on the social level has the same role as death-by-aging has on the organic level. The usefulness of both is to enable new generations to differ from their ancestors in a manner that is advantageous not for persons but for the community. The tragic, no less than death-by-aging, pays for social change with human suffering.

In terms of my definition of collective identity as the mode of applying collective memory to collective intent, tragic dramas articulate the identities of the communities in which they are set. I would expect that a tragic drama, appropriate for globalized humankind, would reflect humankind's identity-in-the-making.

I propose to examine three great tragedies by three great dramatists, associated with three different epochs. Then, compare the ways their protagonists employ the past in service of their future.

The first one was written in 1602. It is the story of a man who is informed about the past by a supernatural agent. This makes him confront his destiny.

"The time is out of joint;" he says, "O cursed spite / That ever I was born to set it right!"[43] Yet, that is exactly what he begins to do. The vibrant sensitivity of this Shakespearean character is recognized in Pasternak's poem, "Hamlet." In it, the Prince of Denmark, an actor acting himself, talks to himself. "I stand alone. All else is swamped in Pharisaism. To live life to the end is not a childish task."[44]

With that realization, he begins to force the future. He engages a company of actors to recreate that past and takes action to repair that past through sacrifice. Hate and love converge to a denouement of "Good night, sweet prince" echoing in an otherwise empty universe. After a heartbroken farewell and military salute, the hero is laid to rest.

Hamlet is set in the court of a king. The protagonists are a small group of the privileged. The audience—during its early existence—were those who fit in the Globe or other small theaters. The conflicts of the plot are between the finity of human power on the one hand and the infinity of human ideals on the other hand. I class Hamlet, together with Goethe's Faust, as tragedies in the Greco/Western mode of the drama of redemption, leading to a denouement of mission accomplished.

The second tragedy I have in mind is a painful register of hope lost. It was written in 1939. Its protagonists wanted to change the world the way they thought would make it better, but they failed. Now they are trying to recover their personal identities in the hope of reconciling their enthusiastic past with an uncaring present. As these efforts also falter, they drift into the slow death of derelicts. When one of the them jumps to his death from a fire escape, only a single voice says, "God rest his soul in peace." All the other voices celebrate a birthday by singing and shouting in wild cacophony.[45] The central character stares in front of himself, "oblivious to the racket."[46]

"The Iceman Cometh" is set in a bar. In it we are watching a collective identity crisis in a petri dish. The protagonists are men and women, down and out. The tragic tension is between the memories of the characters and their assessments of their present. That tension dies in the empty, futureless life of a flophouse. I class "The Iceman" together with some of Beckett's plays as tragedies of impotence. They are dramas of worlds where there is nothing left either to live or die for. Their moods remind me of those religious views which, in Brandon's words, "reject…the consciousness of the self…as an illusion of dangerous consequences."[47]

During the sixty years after "The Iceman" the world has changed immensely. The new epoch has no patience for character development. They must be immediately legible as are the characters of the Audio-Animatronics figures of Disneyland. The plot must also be simple because the vast, worldwide audience to which the mass media caters, shares only the most primitive of human concerns, which are the spilling of blood and of semen.

In a keen and sensitive recognition of the profundity of the human drama, Paul Harris wrote of "a sense of sweeping change in the nature of being human, a feeling that we are reaching the end of an epoch in the history of our species."[48] In the same paper, he also wrote that, as a professor of literature, confronting a hypertext universe, he finds himself "oscillating between a kind of naïve technophilia and the frustrated rage of a luddite."

I believe that globalized humanity is confronting a hypertext world that goes much beyond the boundaries of wired communication. I have been calling it the anthill threshold. It is an incoherent community whose members have incompatible scales of values. They could coexist as long as there were distinct cultural boundaries. But now they live in a cohabitation enforced by a tight communication network and are all subject to the information rampage. They have all been thrown off balance by the interpenetration of financial and military empires where each community, in itself, is a powerless subject of an uncritical amalgamation of human values.

The consequent tensions are expressed in the conflicts between a technophilia of unrealistic hopes on the one hand and, on the other hand, a frustrated rage due to unfulfilled expectations. The crisis gives rise to a form of tragic drama that is neither in the Greco/Western redemptive category, nor in the category of impotent fading away. Instead, it is a tragic drama of globalization-in-process.

A dramatist who wrote such a tragic drama, is a woman. She is known to carry a trumpet to announce fame, a book in which she records events and a clepsydra to tell time. She is the muse of communal memory that is, of shared beliefs about the past. She is the muse of history. Her name is Clio.

Today her trumpet is the worldwide media. Her book is the Internet with its 420 billion pages of writings. The premiere of her new creation was performed on the spherical stage of a global theater. Her viewers numbered a few hundred million people, all of whom were both protagonists and audiences. At that first performance her clepsydra, calibrated in Gregorian chronology, showed 9–11–2001.[49]

In a fine essay, Anne Lévy wrote about "an ongoing American tendency to create temporal bubbles in which only the stimulating but solvable incidents rise pell-mell to the status of great meaning, while true predicaments and tragedies are whisked out of sight."[50] Clio's latest plot, in which she made four airplanes stand for humankind, cannot be whisked out of sight. There are no temporal bubbles left in which anyone can hide. The denouement, which we have not seen yet, pertains to a decision about the identity and future of humankind.

The conduct of humans and gods in Greek tragedies set the tone of Greek cultural identity. The tragedy of Christ, reenacted in the Mass, helped form the cultural identity of the West. The tragic spectacle of murder in service of a savage King Ludd, using low-tech to annihilate high-tech, is appropriate in its form for a globalized humankind. For that reason, it is likely to shape the identity of the emerging community and likely to remain, at least for a while, a deed to be "acted o'er / In states unborn and accents yet unknown."[51]

The events of 9–11 involve conflicts between the Dionysian and Apollonian trends innate in history which, Nietzsche maintained, give rise to tragedy. In a different perspective, they are also conflicts between the biotemporal and nootemporal assessments of reality. The "global instant everywhere," in cahoots with social advances, have lifted the lid off the inner turmoil of people everywhere, allowing the reptilian brain to act out its desires.

I asked earlier what, for the case of globalized mankind, may serve the same role as the "unum" did in "e pluribus unum?" When, in 211 B.C. the armies of Carthage reached Rome, the cry went up in the City: "Hannibal before the gates!" The cry appropriate for humankind in globalization "Clio before the gates."

History is now at all gates, everywhere and all the time.

If, it is indeed the case that memory evolved as an aid to intent and, if one agrees with the idea that mankind's identity resides in the manner it employs its collective memory to serve its future, then, before any realistic plans for the future may be drafted, it is the past, it is an agreement on history that will have to be negotiated.

I do not mean agreement on the dates of one or another king. I mean an agreement about the origin and evolution of man, the origin and evolution of life, and the origin and evolution of the universe. But views on these issues differ, depending on whether they are based on critical scientific reasoning, on revealed religion or on mythology. Consistently, they lead to different recommendations for future actions. Also, they recognize different and mutually incompatible needs. The chances of reaching an agreement about those needs are very slim. Yet, in my view, until we do agree on the origin and evolution of man, life and the universe, we shall have to live in Blake's "London" of 1794.

> In every cry of every Man,
> In every Infant's cry of fear,
> In every voice, in every ban,
> The mind-forg'd manacles I hear.[52]

6. The Magic Show

The evolving mirror I spoke about is the capacity of the human brain to assess reality in categories of open-ended futures and pasts, referred to a present. That present is defined locally by life, by the mental processes and by society.

The sources of this ability of the human brain are unclear and intriguing enough to have given rise to a question by Walter Russell Brain, later Lord Brain. He was President of the British College of Physicians and one of the scientist-scholars who helped place the interdisciplinary study of time and ISST on the map.[53] If memory corresponds to a brain state, he wrote, and sense impressions to another, and since these coexist, how are they told apart?[54] The formulation of this question surely changed since 1963 but, to my knowledge, it has not been satisfactorily answered.

To memory and sense impressions listed by Lord Brain I would add a third family of brain states, namely that which corresponds to intentions, expectation and hope. Then, I would ask: how are these three categories of time told apart?

I hope that this question will be considered at this conference, with the benefit of some useful constraints. For instance, the answer cannot appeal to an unambiguous physical future, past and present in which biological, psychological and social processes unfold and to which the individual's mental faculties, somehow, adapt.

My talk began with reference to a young man standing between two immense armadas, sixty some years ago. It is not inappropriate to conclude by narrowing the cosmic vistas we visited to the limited boundaries of a single life.

Here is a poem by the American poet Robert Hillier; "The Wind is from the North."

> And now at sunset, ripples flecked with gold
> Leap lightly over the profounder blue;
> The wind is from the north, and days are few
> That still divide us from the winter cold.
> O, it was easy when the dawn was new
> To make the vow that never should be old,
> But now at dusk, the words are not so bold,—
> Thus have I learned. How fares the hour with you?
> A heron rises from the trembling sedge,
> His vigil at an end. Mine too is done.
> A late sail twinkles on the watery edge,
> And up the shore lights sparkle one by one.
> Seasons will change before tomorrow's sun,
> So speaks the dune grass on the windy ledge.[55]

22. EXPANDING THE UNIVERSE

The Latin word *universum*, from *unus*, "one" and *verto, -ere, versum* "to turn" literally means "to be turned into unity," to be turned into a single whole. The Greek *kosmos* means "order" as well as "good behavior, decency." It also means "ornament, decoration," especially of women. *Kosmetike* is the art of dress ornaments. Both words, "universe" and "cosmos" imply harmony and beauty. So does Biblical language when it reports that at the time when the foundations of the earth were laid, "the morning stars sang together, and all the sons of God shouted for joy."[1] Hidden in all these words is the shared belief that there is orderliness in the big everything.

Contemporary cosmology recognizes structural and functional examples of orderliness, conditions that were not only unsuspected earlier but were outright unimaginable. We dealt with two examples. One was the expansion of the universe. Another was the existence of well-defined edges to the cosmos: boundaries to length, to periods of time, to velocity, temperature and complexity. We gave reasons why the complexity boundary is that of the complexity of the human brain. It may, indeed, be said that human minding, human creative capacities actively partake in the enlargement of the cosmos. They do not do so by increasing the volume or temperature of the universe or the range of possible velocities but by bringing about conditions that could not have been brought about, were it not for the human agency. This makes us partners in the enterprise of expanding the scope of the universe. It is a unique partnership, not without its problems. Here is one of them, already mentioned but without elaboration: a qualitative change in the texture of human life.

Upon the globalizing earth, the march of humans is no longer to the tune of any of the great, historical calls. It is not to the tune and text of *Dies Irae*, that powerful medieval hymn about the Last Judgment: "Death is struck and nature quaking / all creation is awaking / to its judge an answer making." Nor is it to the music and words of the Marseillaise: "Allons enfants de la patrie / Le jour de gloire est arrivé." Nor to the words of the Communist International, preaching its secular religion: "This is the final conflict / Let each stand in his place." These are passé. The march is now to something much more elemental than any of these calls-to-arms. To something that compares in its dimensions with such earlier examples of emergence as that of life from inanimate matter and that of the human minding abilities from the matrix of living species.

What we are witnessing is the emergence, from the nation states, of a single, global, densely wired community whose nervous system is now being assembled. That assembling process has been described, enthusiastically and with abandon, by those who work on it. Here is an example. "We Are the Web.... At its heart [is] a new cultural force based on mass collaboration... open source, peer to peer: behold the power of the people." It sounds like one of the 20th century's ideas of final conflicts, discussed in the preceding essay. Surely, a warning. But one that cannot overlook the power of these links. The total number of Web pages exceed 500 billion. "AI [artificial intelligence] will emerge not in a stand-alone supercomputer... but in the vast digital tangle of the global Machine.... This gargantuan Machine... will evolve into an integral extension not only of our senses and bodies but our minds."[2]

But, whereas the two earlier examples of vast events in evolution—that of the coming about of life and of the human mind—have taken eons, the current emergence of a globalized humankind with its global sociotemporality must happen rapidly because of the metastability of evolutionary interfaces, a condition I discussed elsewhere.[3]

But the wiring—in its many aspects—is only a part of a process in which we are actors and of which we are the dancing, copulating and bleeding audience. It is a global Grand Guignol, a sensational, horrifying, violent play. Thousands of magazines and newspapers report about the stage that is being wired. They speak about many wars. Some of these are between nations, others among ethnic groups and/or religions. Others are among mercenaries fighting for TICs, for Tribal Interest Cells, the commandos and freebooters of the time-compact globe. TICs conduct local and transnational wars underground, on the ground and in the air in support of ethnic, religious, economic and political goals. They serve enterprising gurus no less than crime families, petty or grand.[4] While technology, economic needs and ecological pressures are pushing the globe toward homogenization, the armed conflicts generate and maintain fragmentation in a turbulent act of creation.

There have been world wars—so called—before. But the direct clash of cultures of our epoch is, historically, the first truly worldwide conflict. Religions, where they are in cahoots with the states, reanimated murderous drives that, obviously, have been dormant. Yet, where civilized religions are not permitted to serve as the opium of the people, opium tends to become the religion of the people. The great cultural continuities of humankind—the arts and the letters—tend toward the high-tech primitive and are being broadly prostituted through the aliteracy of the presumably educated. Scientific knowledge is marketed for its usefulness, seldom, if ever for its beauty. In this wheeling and dealing without borders, the destructiveness of our species is becoming increasingly efficient. It keeps up, as different but equal, to remarkable examples of readiness for love and sacrifice.

That the world is now "wired" by trillions of miles of wire and wireless links is, indeed, a new challenge. But it does not help in the working out the goals and rules of conduct that could be shared by humankind. On the contrary. In the past, distinct cultural regions, though often at war, maintained their different families of values. By the early 21st century each of those regions is in continuous and intimate contact with all others, as cited above. A consequence is a tempest of culturally based and fed fears and hatreds. Upon the tightly wired, time-compact planet, nightmare-like fantasies are played out and, by that act, become acceptable forms of conduct in individual, national and international affairs. In the global supermarket water, food, women, men, children, software, drugs and body parts flow in all directions. Women rent their wombs to provide the daily needs of survival for their families. During the heydays of slave traffic from Africa to the New World, the traffickers transported eleven million people. It is estimated that at the Anthill Threshold, two-hundred-million people are in the hands of criminal traffickers.[5] It is conservatively estimated that almost thirty-million of them are actually in slavery.[6] While the 600 billion web pages happily exchange their messages through the use of advanced technologies, eighty-one percent of women in Bangladesh marry before the age of 18.[7] Every nation makes and transports arms with no regard to their origins, destinations or intended use. In this global wheeling and dealing the once strong and creative middle classes shrink, the differences between the haves and have-nots increase.

Good reasons may be given in support of the idea that if a significant portion of the earth's population reaches a certain level of social complexity, the global socialization and evaluation of time will subsume the office of the individual as the primary measure and measurer of time. Identifying the temporal organization of such a world is difficult because we do not have a platform from which it could be beheld and described. We are like the fish, contemplating the Beyond in Rupert Brooke's poem, "Heaven,"

> Fish say, they have Stream and Pond;
> But is there anything Beyond?
> ...somewhere beyond Space and Time,
> Is wetter water, slimier slime.[8]

These concluding reflections of *Time and Time Again* are written at Manitou Chapel, New York. The Hudson River is barely visible to the northwest;

behind its silvery lines the Catskill Mountains wait, surrounded by lazy clouds. Except for that view, we see only the forest which contains us in its lonely eminence in a continuous now, in an epilogue and prologue to a present. Downhill, the Appalachian Trail passes on its way from Georgia to Maine. This is that part of the late afternoon when the warmth of the sun spreads tranquility on the rich, brown leaves of autumn, and gaiety on the still green shrubs. Slowly but quite perceptibly the sun is journeying westward; elsewhere a spring day is about to begin. Once in a while the long, deep voice of a riverboat drifts up here, followed sometimes by an echo from the mountains. Otherwise the only noise on this clearing comes from the rustling of bushes and of high trees, and from the motion of small unseen creatures beneath the leaves. In such a setting, the making of analytical conclusions appear not so much meaningless as insufficient. It favors, instead, an attitude expressible only through that involvement of man in life which gives rise to the problems of time. They are issues of human life and death. Composing pithy summaries of such extensive, heavy and intricate matters is the task of poets. Here are two such summaries. The first one is an expression of hope.

> Foundations shake,
> Computers break
> And Science goes Be-bop,
> But Baby's joy
> Is still the toy
> With foolish ears that flop.[9]

The other summary has the form of accomplished poesy. It is about the struggle between the knowledge of passing and the feeling of beauty. They are the lines of an unidentified bluegrass poet I heard, and wrote down, many years ago on Mount Desert Island in James Bay, Ontario, Canada.

> Scarlet of sunset fades slowly from the rippling water
> And the twilight falls.
> The lap of wave on hard sand
> Grows muffled and faint.
> If I could know that there will be
> Beyond life's end a sure rebirth....
> But how can I know that time will not bring
> Only dusk and galling night,
> Only a soft blotting out of all I love
> Like twilight on Desert.

Hickory Glen, Connecticut
September 10, 2006

NOTES

1. The Change Ringing Cosmos

[1] The mass of the universe is estimated to be 10^{54} grams. (C. W. Allen, *Astrophysical Quantities*, London: Athlone Press, 1955, p. 245.) The electron's mass is 10^{-27} grams. (*The European Physical Journal*, section of *Particles and Fields*, 15 (2000), p. 73). These quantities yield $10^{54} / 10^{-27} = 10^{81}$ as a speculative but useful figure for the number of particles in the universe.

[2] For a discussion of what is meant by the expansion of the universe, see Edward R. Harrison's superb *Cosmology: The Science of the Universe*. Cambridge: Cambridge University Press, 1981, pp. 201–230.

[3] For relevant material visit their web site, "The Neurosciences Institute" and click on "Publications."

[4] Gerald M. Edelman, *Bright Air, Brilliant Fire—On the Matter of the Mind*. New York: Basic Books, 1992, p. 148. By the same author, *Wider than the Sky: The Phenomenal Gift of Consciousness*. New Haven: Yale University Press, 2004, p. 14, "The human brain is the most complicated material object in the universe."

[5] J. W. S. Pringle, "On the Parallels between Learning and Evolution," *Behavior* 3:174–215.

[6] H. A. Simon, "The Architecture of Complexity," *Proceedings of the American Philosophical Society* 106:467–82.

[7] C. H. Waddington, *Tools for Thought* (London: Jonathan Cape, 1977).

[8] John von Neumann, *Theory of Self-Reproducing Automata*, ed. and completed by A. W. Burns (Urbana: University of Illinois Press, 1969), 79–80.

[9] P. T. Saunders and M. W. Ho, "On the Increase in Complexity in Evolution," *Journal of Theoretical Biology* 63 (1976): 375–84; Saunders and Ho, "On the Increase of Complexity in Evolution II," *Journal of Theoretical Biology* 90 (1981): 515–30.

[10] Issues related to complexity in biology first appeared in J. T. Bonner's book *On Development* (Cambridge, Mass.: Harvard University Press, 1974), 65–126. Fourteen years later they became the central theme of his *Evolution of Complexity by Means of Natural Selection* (Princeton, N.J.: Princeton University Press, 1988). With regard to different trends in evolution, Bonner sees complexification as the result of selection for an increase in the diversity of the parts of an organism (a division of labor) simultaneous with a collateral increase in the sophistication of the control systems necessary to ensure coherence. (The reader may recognize in the concept of coherence the demands for the definition of the organic present.) The two trends together lead to a refinement of organic functions manifest in the differentiation of cell types. The diversity among the cells may then serve as an index of complexity of an organism.

[11] John Maynard Smith and Eörs Szathmary, "The Major Evolutionary Transitions," *Nature* 374 (1995): 227–32.

[12] M. Hazenwinkel, ed., *Encyclopedia of Mathematics* (Boston: Kluwer, 1995) s.v. "Complex systems."

[13] Ibid., s.v. "Hierarchical theory."

[14] G. J. Chaitin, "Algorithmic Information Theory," in *Encyclopedia of Statistical Sciences*, ed. Samuel Kotz and N. L. Johnson, 8 vols. (New York: Wiley, 1982), 1:38–41. See also his "Randomness and Mathematical Proof," *Scientific American* 232, no. 5:47–52.

[15] This formula takes it for granted that the person or computer following the instructions knows how to handle temporal relationships of the before/after type.

[16] This is the number of symbols in the program for calculating by the Ramanujan formula written in Octave, a mathematically oriented language. I am indebted to Mr. David Felsenthal for writing the program. Those interested may note with well-deserved amazement that in three iterations the series converges to twenty-three significant digits.

[17] For an entry into discussions pertaining to the equations of science as examples of compressibility in nature, see Michael Heller's "Chaos, Probability, and the Compressibility of the World," in *Chaos and Complexity*, ed. R. J. Russell, Nancey Murphy, and A. R. Peacocke (Berkeley, Calif.: Center for Theology and the Natural Sciences, 1995), 107–21.

[18] Quarks are structural units from which other particles are formed. They are kin to the five geometrical forms—the five regular solids—Plato used to represent the four elements and their relationships.

[19] A few decades ago the number of known elementary objects (particle-waves) could be easily listed and unambiguously counted. At the end of the century the listing of such objects and the identification of their properties takes up 708 pages of a special issue of the *Physical Review D* ("Review of Particle Properties," *Physical Review D: Particles and Fields*, 3d ser., 54, part 1 [1 July 1996]). The numeral in the text—500—is a ballpark figure of what has been called the "particle zoo."

[20] David R. Lide, ed., *CRC Handbook of Chemistry and Physics*, 78th ed. (Boca Raton, Fla.: CRC, 1997).

[21] John A. Dean, ed., *Lange's Handbook of Chemistry*, 13th ed. (New York: McGraw-Hill, 1988).

[22] The two figures are from V. H. Heywood, ed., *Global Biodiversity Assessment* (Cambridge: Cambridge University Press, 1995), 118 and 202.

[23] E. O. Wilson, "The Current State of Biological Diversity," in *Biodiversity*, ed. E. O. Wilson (Washington, D.C.: National Academy Press, 1988), 5. See also Michael L. Rozenzweig, *Species Diversity in Space and Time* (Cambridge: Cambridge University Press, 1995), 3.

[24] Wilson, "The Current State," 7.

[25] Laurent Keller and Michel Genoud, "Extraordinary Lifespan in Ants: A Test of Evolutionary Theories of Ageing," *Nature* 389 (1997): 958.

[26] I assumed that an average ant is 10 mm^3 and an average microorganism (bacteria and viruses) is 10^{-3}–10^{-9} mm^3. Microorganism size is taken from a table in Bonner's *Evolution of Complexity*.

[27] Prof. David Pimentel, private communication.

[28] C. W. Allen, *Astrophysical Quantities* (London: Athlone, 1955), 106.

[29] Prof. David Pimentel, private communication.

[30] Allen, *Astrophysical Quantities*, 106.

[31] Prof. David Pimentel, private communication.

[32] Allen, *Astrophysical Quantities*, 244. The number of particles in the universe has been estimated as 10^{81}.

[33] Gerald M. Edelman, W. Eimar Gall, and W. Maxwell Cowan, eds., *Synaptic Functions* (New York: Wiley, 1987), 1. Blinkov and Glezer give 10^{10}–10^{18} neurons (Samuil M. Blinkov and Ilya Glezer, *The Human Brain in Figures and Tables: A Quantitative Handbook*, trans. Basil Haigh [New York: Plenum, 1968], 201–2).

[34] Paul Churchland refines the type of calculation I gave here. Instead of calculating conditions with neurons on or off, he allows for ten different strengths of connections and obtains a figure of $10^{10^{14}}$; see his book *The Engine of Reason, the Seat of the Soul: A Philosophical Journey into the Brain* (Cambridge, Mass.: MIT Press, 1995), 4–5.

[35] Edelman, Gall, and Cowan, *Synaptic Functions*, 1.

[36] Under the sponsorship of the National Center for Ecological Analysis and Synthesis in Santa Barbara, California, thirteen ecologists and economists joined forces to estimate the annual value of the services that the earth's ecosystem provides. "We have estimated the current economic value of 17 ecosystem services for 16 biomes, based on published studies and a few original calculations. For the entire biosphere, the value (most of which is outside

the market) is estimated to be in the range of US$16–54 trillion ($10^{12}$) per year, with an average of US$33 trillion per year. Because of the nature of the uncertainties, this must be considered a minimum estimate. Global gross national product is around US$18 trillion per year" (Robert Costanza et al., "The Value of the World's Ecosystem Services and Natural Capital," *Nature* 387 [15 May 1997]: 253–60).

The estimates and figures I offer belong in the same category of calculations, except that they are rooted in natural science.

2. From Timelessness to Time

[1] Benedict de Spinoza, *Improvement of Understanding*. Washington, D.C.: M. Walter Dunn, 1901, p. 3.

3. Reality as Examined Appearances

[1] See "Universes of Perception" in J. T. Fraser, *Of Time, Passion, and Knowledge*, Princeton: Princeton University Press, 1975, pp. 75–6.

[2] *The Problems of Philosophy*. Oxford: Oxford University Press, 1959, p. 9.

[3] Newton had this to say about propositions:

> In experimental philosophy we are to look upon propositions inferred by general induction from phenomena as accurately or very nearly true, notwithstanding any contrary hypotheses that may be imagined, till such time as other phenomena occur, by which they may either be made more accurate, or liable to exception. (Newton, in Cajori 1973: Book 3, 'Rules of reasoning in philosophy', rule 4).

[4] The theory, first suggested in 1970, has been worked out and critically examined in some fifty papers and in four books (Fraser 1978, 1982, 1987, 1990). A summary of, and introduction to this family of publications may be found in Fraser (1999: 21–43) and its references (1999: 255–260).

[5] This observation formalizes and carries to its end-of-the-century status Uexküll's reflections on the different Umwelten of the astronomer, deep-sea researcher, chemist, etc. (Uexküll 1957: 76–80).

[6] See Appendix B, 'Complexity and its measure', in Fraser (1999: 235–242) and the related discussion in the text.

[7] See note 32 in Fraser (1999: 258).

[8] See Appendix C, 'Entropy: Its uses and abuses', in Fraser (1999: 243–245).

[9] See note 34 in Fraser (1999: 259).

4. What Kind of a Universe to Expect?

[1] For a summary of what we know of him, see "Pythagoras of Samos" in G. S. Kirk and J. E. Raven, *The Presocratic Philosophers*, Cambridge: At the University Press, 1957, pp. 217–231.

[2] Thomas Browne, "The Garden of Cyrus or the Quincunciall Lozenge" (1658) in *Religio Medici and Other Works*, ed. by L. C. Martin. Oxford: Clarendon Press, 1964, p. 174.

[3] Reprinted in Eugene Wigner, *Symmetries and Reflections*, Bloomington: Indiana University Press, 1967. Chapter 17.

[4] Baruch Spinoza, *Improvement of Understanding* (Washington, D.C.: M. Walter Dinn, 1901), 3.

[5] Immanuel Kant, *Kant's Inaugural Dissertation and Early Writings on Space*, tr. John Handyside (La Salle, Ill.: Open Court, 1929), sec. 3, para, 14, item 5.

[6] Albert Einstein, *Geometry and Experience*, in Samuel Sambursky, ed. *Physical Thought from the Presocratics to the Quantum Physicists. An anthology.* (New York: Pica Press, 1975), 492.

[7] This is discussed in my paper, "The Problems of Exporting Faust," in J. T. Fraser, N. Lawrence and F. C. Haber, eds., *Time, Science, and Society in China and the West.* (Amherst: University of Massachusetts Pres, 1986) 1–20.

[8] Johannes Kepler, *Gesammelte Werke*, Walter von Dyck and Max Caspar, eds. (München: C. H. Beck'sche Verlagsbuchhandlung, 1938–75), 8:30n8 & 6:223.

[9] Albert Einstein, "Autobiographical notes," in *Albert Einstein, Philosopher-Scientist*, P. A. Schilpp, ed. (New York: Tudor, 1949), 7.

[10] This may be found in the correspondence between Ferdinand Karl Schweikert, a professor of law and Gauss, in Carl Friedrich Gauss, *Werke.* (Leipzig: B. G. Teubner, 1900) 8:177–82 and 10, pt. 2 31–35.

[11] John von Neumann, *Theory of Self-Reproducing Automata*, A. W. Burns, ed. (Urbana: University of Illinois Press, 1969), 47–48 and 51.

[12] In Ernest Jones, *The Life and Work of Sigmund Freud*, (New York: Basic Books, 1953) 2: 419.

[13] The best way to enter this difficult but fascinating work is via the introduction by R. B. Braithwaite to Kurt Gödel's *On Formally Undecidable Propositions of "Principia Mathematica" and Related Systems*, tr. B. Meltzer (Edinburgh: Oliver and Boyd, 1962).

[14] For an introduction to the hierarchical theory of time and a guide to its thirty year development see "Perspective on Time and Conflict," in J. T. Fraser, *Time, Conflict and Human Values*, (Urbana and Chicago: University of Illinois Press, 1999). 21–43.

[15] Wallace Stevens, *Poems by Wallace Stevens*, (New York: Random House, 1959), 73.

[16] David Park, *Classical Dynamics and its Quantum Analogues*, (New York: Springer Verlag, 1979).

[17] Albert Einstein, *Out of my Later Years*, (New York: Philosophical Library, 1950), 96.

[18] For a discussion for the process and state description of complex systems, such as the human brain, see J. T. Fraser, *Time as Conflict—a Scientific and Humanistic Study*, (Basel: Birkhäuser Verlag, 1978). 117f.

[19] *Time, Order, Chaos—The Study of Time IX.* J. T. Fraser, Marlene P. Soulsby and Alexander J. Argyros, eds. (Madison, CT: International Universities Press, 1998). 9.

[20] John Fowles, *Mantissa*, (Boston: Little, Brown and Company, 1982). 149.

5. The Beginning or Origin of Time

[1] See the superb chapter, "The Dawn of Creativity," in: S. G. F. Brandon, *Creation Legends of the Ancient Near East.* London: Hodder and Staughton, 1963, 1–13.

[2] *The City of God*, tr. Marcus Dodd, NY: Modern Library, 1950, 350.

[3] An entry may be had through the chapter "Perspectives on Time and Conflict", in J. T. Fraser, *Time, Conflict, and Human Values.* Urbana: University of Illinois Press, 1999, 21–43.

[4] English and English, *A Comprehensive Dictionary of Psychological and Psychoanalytical Terms.* New York: David McKay, 1964, s.v. "Umwelt".

[5] J. T. Fraser, "The Extended Umwelt Principle—Uexküll and the Nature of Time." In: *Semiotica*, v. 134–1/4 (2001), 263–73. Also, chapter on "The Principle of Temporal Levels,"

in: J. T. Fraser, *The Genesis and Evolution of Time: a Critique of Interpretation in Physics*. Amherst: University of Massachusetts Press, 1982, 18–36.

[6] On the absence of self-contradiction in the idea of expanding space, see the section "Expanding rubber sheet universe," in the chapter "Expansion," 207–30 in Edward R. Harrison's excellent *Cosmology—the Science of the Universe*, Cambridge: Cambridge University Press, 1986.

[7] On this division of cosmogenesis see K. A. McGrath (ed.), *World of Physics*. Farmington Hills, MI.: Gale Group, 2001, s.v. "Big Bang."

[8] Einstein believed that there is something essential and very important about the "now" but whatever that is, it lies outside the realm of science. By "science" he meant physics. In: P. A. Schilpp (ed.), *The Philosophy of Rudolf Carnap*. LaSalle, Ill.: Open Court, 1963, 37. But if biology is included among the sciences as it should be, the "now" ceases to be outside the realm of science. See, "The Organic Present," in J. T. Fraser, *Time, the Familiar Stranger*, op. cit., 128–30. Also 'Time and the Origin of Life," in: *Time, Conflict and Human Values*, op. cit., 253–4.

[9] "Entropy: its Uses and Abuses," in: *Time, Conflict, and Human Values*, op. cit., 243–245.

[10] I am using a table, "Inferences about hominid behavior and ecology," in Steven Jones [et al.] (eds.). *The Cambridge Encyclopedia of Human Evolution*, Cambridge: Cambridge University Press, 1992, 333.

[11] The idea of the mental present in terms of neuronal coordination was first formulated in J. T. Fraser, *Of Time, Passion and Knowledge*, 2nd ed. Princeton: Princeton University Press, 1990, 80–91. Later in *Time, Conflict, and Human Values*, op. cit., 33–39.

[12] "Timelessness in time," in: *Time, Conflict, and Human Values*, op. cit., 128–32. Also, J. T. Fraser, "Human Temporality in a Nowless Universe," in: *Time & Society*, 1(2), 1992, 159–73.

[13] Roger Penrose, *The Emperor's New Mind: Concerning Computers, Minds, and the Laws of Physics*. New York: Oxford University Press, 1989, 302. Also, *Time, Conflict, and Human Values*, op. cit., 32–33 and its references.

[14] A spacetime manifold "is a suitable arena for physical process if and only if any of its points distinguishes between two time directions. However… "there is no criterion to determine which of the two directions is past and which is future. For the arrow of time one must look elsewhere." M. Heller, "The Origin of Time," in: J. T. Fraser [et al.] (eds.), *The Study of Time IV*, New York: Springer Verlag, 1981, 90–93. The quote is from 91.

[15] For a rich and authoritative work, see R. J. Russell, Nancy Murphy and C. J. Isham (eds.), *Quantum Cosmology and the Laws of Nature*. Vatican City State: Vatican Observatory and Berkeley, California: The Center for Theology and the Natural Sciences, 1993.

[16] Edward R. Harrison, *Cosmology—the Science of the Universe*. Cambridge: Cambridge University Press, 1986, 358. Details: Planck length is the distance within which all conventional physical laws break down. Planck time is the time taken by a photon to travel that distance: it is, therefore, the shortest meaningful period. Paul Davies (ed.), *The New Physics*. Cambridge: Cambridge University Press, 1989, 500.

[17] J. T. Fraser, "From Chaos to Conflict," in: *The Study of Time IX*. Madison, CT: International Universities Press, 1998. J. T. Fraser, M. P. Soulsby and A. J. Argyros (eds.), 3–17.

[18] "The canonical forms of time," in: J. T. Fraser, *The Genesis and Evolution of Time*. Amherst: The University of Massachusetts Press, 1982, 29–31, also 33–39, *Time, Conflict, and Human Values*, op. cit.

[19] Argyros, Alexander in S. L. Macey (ed.), *Encyclopedia of Time*. New York: Garland, 1994, s.v. "Fraser, J.T."

[20] D. Corish, "The Beginning of the Beginning in Western Thought," in: J. T. Fraser, N. Lawrence and D. Park (eds.), *The Study of Time IV*. New York: Springer Verlag, 1981, 34–45.

[21] S. L. Goldman, "On the Beginnings and Endings of Time in Medieval Judaism and Islam," in: J. T. Fraser [et al.] (eds.), *The Study of Time IV*, op. cit., 59–72.

[22] J. T. Fraser, "Temporal Levels and Reality Testing," in: *Int. J. Psycho-Anal.* 62, (1981), 3–26, esp. 5–7. Also, J. T. Fraser, "Timelessness in Time," in: *Time, Conflict, and Human Values*, op. cit., 128–131.

[23] "In principio creavit Deus Coelum & Terram... (justa nostra Chronologiam)...quae XXIII diem Octobris... in anno periodi Julianae 710... Anno ante eram Christianam 4004." James Ussher, *Annales Veteris... Prima Mundi Origine Deducti Bremae*, MDCLXXXVI, 1.

[24] G. J. Whitrow, *The Natural Philosophy of Time*. Oxford: Clarendon Press, 1980. 2nd ed., 375.

[25] Plato, *Timaeus*, 37c–e.

[26] Physical phenomena are said to be symmetrical if the equations describing them remain invariant under all changes of the system that are of interest to the physicist. The appearance of any form of orderability breaks that symmetry.

[27] T. S. Eliot, *Selected Poems*. New York: Harcourt, Brace & World, 1936, 80.

[28] A. Cornelius Benjamin, "Ideas of Time in the History of Philosophy," in: J. T. Fraser (ed.), *The Voices of Time*, 2nd ed. Amherst: University of Massachusetts Press, 1981, 8.

[29] G. J. Whitrow, *The Natural Philosophy of Time*. Oxford: Clarendon Press, 1980. 2nd ed., 1.

[30] Vide, Ilya Prigogine's *From Being to Becoming*. San Francisco: Freeman, 1980.

[31] The usual interpretation of Creation as a step from being to becoming parallels the usual, inverted and hence wrong formulation of the problem of Zeno's paradox of the flying arrow. Namely, rest (no change) is universally assumed to be ontologically more fundamental than motion and change. But the world is the other way round. Motion is ontologically prior to rest. The question about the flying arrow is not how its motion may be constructed from elements of rest, but how an imagined state of rest for an arrow—which is never at rest—may be constructed from elements of motion.

[32] The Origins of Being and Becoming," in: *Of Time, Passion, and Knowledge*, op. cit., 311–13.

[33] Allen, C. W. (ed.), *Astrophysical Quantities*. London: The Athlone Press, 1955, 245.

[34] "Quantum Cosmology," in: Kimberley A. McGrath (ed.), *World of Physics*. Farmington Hills, MI.: The Gale Group, 2001, 592.

[35] Paul Davies, *The Cosmic Blueprint*. New York: Simon & Schuster, 1989, 125.

[36] In his "Marriage of Heaven and Hell," in: *The Portable Blake*. New York: The Viking Press, 1969, 256.

6. Constraining Chaos

* The Founder's Lecture, July 3, 1995.

[1] This is from Hesiod's *Theogony*, but is believed to be of much earlier origin (Kirk and Raven, 1957, p. 24).

[2] A derivation of this equation may be found in David Park's paper, "The fortunes and misfortunes of Laplacian determinism" in J. T. Fraser, M. P. Soulsby and A. J. Aryros. eds. Time, Order, Chaos—The Study of Time IX, Madison, Connecticut: International Universities Press, 1998, p. 150.

[3] To trace that history, enter through Kellert (1993, p. 122).

[4] Examples of objects with noninteger dimensions include a curve called the Triad Cantor Set; it has a dimension of 0.6309. Its space is richer than that of a point but poorer than that of a line. It is not a curve in the ordinary sense; it is more like dust particles distributed along a line (Hilborn, 1994, p. 397). Another such object, one that looks like a snowflake, called the Koch curve, has a dimension of 1.26. It is more than a line but less than a surface (Hilborn, 1994, p. 398). Yet another object, the Menger Sponge has a dimension of 2.7268: it is too rich for a plane, too poor for a volume (Stewart, 1989, p. 220).

[5] *Of Time, Passion, and Knowledge* (Fraser, 1975) remains the main reference. The dialectical nature of the theory was elaborated in *Time as Conflict* (Fraser, 1978). The theory's implications for physics were the subject of *The Genesis and Evolution of Time* (Fraser, 1982). A good first entry is *Time, the Familiar Stranger* (Fraser, 1987).

[6] The integrative levels, when expressed in terms of ideas appropriate for nonlinear systems, remind one of stable attractors with their basins separated by chaotic transitions.

[7] The numerical measures so calculated display distinct ranges, offering support for the importance of qualitative differences which are easily recognized but are difficult to measure. See Fraser (1982, pp. 154–156, 167).

[8] Prigogine and Stengers (1984, p. 285) maintain that there are systems in nature that show reversibility "[b]ut most systems of interest to us, including all chemical systems and therefore biological systems, are time-oriented on the macroscopic level.... Irreversibility is either true on *all* levels or on none." The hierarchical theory of time does not identify irreversibility with time and reversibility with the timeless. Rather, it maintains that all processes above chaos are temporal, but the quality of temporality changes with the complexity of the integrative level considered.

[9] Originally, umwelt meant the circumscribed portion of the environment meaningful and effective for an animal species. In this definition "environment" stands for our human umwelt. In the hierarchical theory of time, umwelt signifies the level specific realities of nature as revealed through scientific experiment, theory, and critical humanistic reflection.

8. How to Use a Clock

[1] See, for instance, William J. H. Andrews' rich *The Quest for Longitude*, Cambridge, Massachusetts: Harvard Department of Collection of Historical Scientific Instruments. 1996.

[2] G. J. Whitrow, *The Natural Philosophy of Time*, 2nd ed. Oxford: Clarendon Press, 1980, pp. 224–245.

[3] This often cited but seldom referenced reflection is from the conclusions of Kant's epochal theory of ethics, *The Critique of Practical Reason*, London: Longmans, 1909, p. 260.

[4] For a sampling of the problems, see "Toward an Integrated Understanding of Time," in J. T. Fraser, ed. *The Voices of Time*, Amherst: University of Massachusetts Press, 1981, 2nd ed. pp. xxv–xlix.

[5] Eric Kincanon, "Misuses of Physical Models in Understanding Time," *KronoScope* 4/1 (2004) pp. 69–73.

[6] This oft-quoted line is from *1 Henry IV*, V–iv–82.

[7] An entry to the literature about space and time may be had through Milic Capek, ed., *The Concepts of Space and Time*, Boston Studies in the Philosophy of Science, v. xxii. Dordrecht and Boston: Reidel, 1976 and Charles Sherover, *The Human Experience of Time*, New York: New York University Press, 1975.

[8] A. Einstein, "On the Electrodynamics of Moving Bodies," in A. Einstein et al., *The Principle of Relativity*, New York: Dover, 1923, pp. 37–71. Quote from p. 37.

[9] A. Einstein, ibid., p. 41.

[10] Gerald Holton, *Thematic Origins of Scientific Thought*. Cambridge: Harvard University Press, 1973, p. 362.

[11] In *The Collected Papers of Albert Einstein*, John Stachel ed. Princeton: Princeton University Press, 1989, v. 2, p. 254.

[12] Albert Einstein, "The Problem of Space, Ether, and the Field in Physics." (1954) in *Space from Zeno to Einstein*, Nick Huggett, ed. Cambridge: The MIT Press, 1999, p. 258. The Lorentz transformation is a set of equations that transform the coordinates of an event

measured in one inertial franme to the coordinates of the same event measured in another inertial frame.

[13] On this perspective and its implications, see "Reaching for the Permanent," and "Number Born from the Search for Invariance," in *Time, Conflict and Human Values.* Chicago: University of Illinois Press, 1999, pp. 18–20 and pp. 50–2.

[14] *Sir Isaac Newton's Mathematical Principles of Natural Philosophy and His System of the World,* trans. E. Motte, ed. F. Cajori, Berkeley: University of California Press, 1966, p. xiii.

[15] See Section 2.2, "A Nested Hierarchy of Causations, Languages, Temporalities and Conflicts," in *TC&HV.*

[16] Distinguishing between real and imaginary numbers has good logical and historical reasons. But the names should not be taken any more literally than the names of some of the quantum numbers of quarks, called "color," "strangeness" and "flavor."

[17] "Space and Time," in A. Einstein et al., *The Principle of Relativity,* New York: Dover, 1923, p. 88.

[18] For a keen and powerful inquiry into this relationship, see Eugene Wigner's Nobel lecture, "The Unreasonable effectiveness of Mathematics in the Natural Sciences," in his *Symmetries and Reflections,* Bloomington: University of Indiana Press, 1967. Also, J. T. Fraser, "Mathematics and Time," *KronoScope* 3/2 (2003), pp. 153–167.

[19] For the intricacies involved in the nature of space-time intervals see the section "Intervals between Events" in W. G. V. Rosser, *An Introduction to the Theory of Relativity,* London: Butterworths, 1964, pp. 125–130. Also the magistral chapter on "Relativistic time' in G. J. Whitrow's *The Natural Philosophy of Time.* 2nd ed. Oxford: Clarendon Press, 1980 and the chapter on "Space-time and Four-Vectors" in Wolfgang Rindler's, *Essential Relativity,* New York: Springer Verlag, 1977, pp. 61–74.

[20] For a conceptual framework, see *TC&HV* pp. 30–32. Also, J. T. Fraser, "The Secular Mystery of the First Day," in Walter Schweidler, ed. *Zeit: Anfang und Ende,* Sankt Augustin: Academia Verlag, 2003, pp. 431–445.

[21] Lawrence Sklar, *Space, Time, and Spacetime,* Berkeley: University of California Press, 1974. p. 356.

[22] *TC&HV,* pp. 26–44.

[23] "Space and Time," in A. Einstein et al, *The Principle of Relativity,* New York: Dover, n.d. 1975. pp. 75–91.

[24] *The Natural Philosophy of Time,* 2nd ed. Oxford: Clarendon Press, 1980, p. 275.

[25] "The Astral Geometry of Gauss," and "The Astral Geometry of Einstein" in J. T. Fraser, *The Genesis and Evolution of Time: A Critique of Interpretation in Physics,* Amherst: The University of Massachusetts Press, 1982, pp. 116–126.

[26] For an introduction to the intricacies of non-Euclidean spaces and their employment in physics, see "Space and Time" in G. J. Whitrow's *The Structure and Evolution of the Universe.* New York: Harper & Brothers, 1959. For mathematical details visit http://mathworld.wolframcom/

[27] "Physics constitutes a logical system of thought which is in a state of evolution, and whose basis cannot be obtained through distillation by any inductive method from the experiences lived through, but which can only be attained by free invention." Albert Einstein, *Out of my Later Years,* New York: Philosophical Library, 1950, p. 96.

[28] G. J. Whitrow, *The Structure and Evolution of the Universe,* New York: Harper & Brothers, 1959, p. 101.

[29] Arthur Eddington, *Space, Time, and Gravitation,* New York: Harper, 1959. p. 48.

[30] Herbert Dingle, "Time in Relativity Theory: Measurement or Coordinate?" in J. T. Fraser, ed. *The Voices of Time,* 2nd ed. Amherst: The University of Massachusetts Press, 1981, pp. 455–472. Quote from p. 471.

[31] P. C. W. Davies, *The Physics of Time Asymmetry,* U. Calif. Press, 1976, pp. 21 and 26.

[32] P. A. Schilpp, ed. *The Philosophy of Rudolf Carnap* (La Salle: Open Court, 1963) pp. 37–38.

[33] P. C. Davies, op. cit. p. 2.

[34] David Park, "The Myth of the Passage of Time." J. T. Fraser, F. C. Haber and G. H.

Müller, eds. *The Study of Time*, Berlin and New York: Springer Verlag, 1972, pp. 110–121. Quote from p. 112.

[35] G. J. Whitrow, *The Natural Philosophy of Time*, 2nd ed. Oxford: Clarendon Press, 1980, p. 254.

[36] E. J. Zimmerman, "Time and Quantum Theory," in J. T. Fraser, ed. *The Voices of Time*, 2nd ed. Amherst: University of Massachusetts Press, 1981, p. 492.

[37] Ibid. p. 253. See also "Conventionality of Simultaneity," in *Stanford Encyclopedia of Philosophy*, Also: http://plato.stanford.edu/entrieds/spacetime-comvensimul/

[38] A. Einstein, *Relativity: The Special and the General Theory*. London: Routledge, 2001. pp. 28–29.

[39] See "From Simultaneities of Chance to Simultaneities of Need" in the chapter on "Biogenesis and Organic Evolution," in *The Genesis and Evolution of Time: A Critique of Interpretation in Physics*, Amherst: The University of Massachusetts Press, 1982. pp. 144–154. also "Biotemporality: the Cyclic Order of Life," in J. T. Fraser, *Time, the Familiar Stranger*, Amherst: University of Massachusetts Press, 1987, pp. 112–130.

[40] For details, see "Time and the Origin of Life," in J. T. Fraser and M. P. Soulsby, *Dimensions of Time and Life, The Study of Time VIII*, (Madison, CT.: International Universities Press, 1996, pp. 3–17.

[41] This view emerges from the integrated study of time. See Subsection 7 of the Section on "A Nested Hierarchy of Causations, Languages, Temporalities and Conflicts," pp. 33–39 in *TC&HV*, See also its index, s.v. "Presents." And J. T. Fraser, "Reflections upon an Evolving Mirror." *KronoScope*, 4–2 (2004), pp. 210–211.

[42] For a definition of reality as the term is used here, see J. T. Fraser, "The extended umwelt principle: Uexküll and the nature of time." *Semiotica*. Volume 134–1/4 (2002), pp. 263–273. Also: "Temporal Levels and Reality Testing," *Int. J. Psycho-Anal.* 62, (1981), pp. 3–26.

[43] For detailed reasoning see "Perspectives on Time and Conflict" in *TC&HV*, pp. 21–43.

[44] Michael Heller, "The Origins of Time" in J. T. Fraser, N. Lawrence and D. Park, eds. *The Study of Time IV*, New York: Springer Verlag, 1981, pp. 90–93.

[45] The lyric cadence representing noetic time is from Sec. IV of Henry Wadsworth Longfellow's "The Golden Legend." The line about fish is from K. M. Gowell, ed. *Rupert Brooke: the Complete Poems,* London: Sidgwick & Jackson, 1977, p. 132.

[46] Barbara Adam, *Time & Social Theory*, Philadelphia: Temple University Press, 1990. p. 66.

[47] Nigel Thrift, "Time and theory in human geography." Part II. *Progress in Human Geography*, 1/3 (1977) pp. 413–457. Quote from p. 448.

[48] Nigel Thrift, *An Introduction to Time-Geography*, No. 13, (1977), p. 4.

[49] Tommy Carlstein, "Innovation, Time Allocation and Time-Space Packing." In Tommy Carlstein, Don Parkes and Nigel Thrift, eds. *Human Activity and Time Geography*, (*Timing Space and Spacing Time*, v. 2.) London: Arnold, 1978, pp. 146–161. Quote from p. 161.

[50] Nigel Thrift, "For a New Regional Geography 3" in *Progress in Human Geography* 17/1. (1993) pp. 92–100. Quote from p. 93.

[51] Aharon Kellerman, *Time, Space, and Society—Geographical Societal Perspectives*. Boston: Kluwer, 1989, p. 106.

[52] A. J. DeLong, "Phenomenological space-time: Toward an Experiential relativity." *Science*, 213 (1981) pp. 681–683. Quote from p. 682.

[53] A. Pellionisz and R Llinás, "Space-Time Representation in the Brain. The Cerebellum as a Predictive Space-Time Metric Tensor." *Neuroscience*. 7 (1982), pp. 2949–2970. Quote from p. 2969.

[54] Patrick Baker, "Space, time space-time and society" *Sociological Inquiry*, 63/4 (1993), pp. 406–424. Quotes from pp. 406 and 420.

[55] Wyndham Lewis, *Time and Western Man*, Boston: Beacon Press, 1957. Quote is from p. 419.

[56] For an assessment of this ferment, see J. T. Fraser, "Time, Globalization, and the

Nascent Identity of Mankind." In *Time & Society*, v. 9, (2000), pp. 293–302 . The same theme appears, in relation to human values, in *TC&HV* "The Global Laboratory," pp. 165–228. Also in "Reflections upon an Evolving Mirror." *KronoScope*, 4–2 (2004), pp. 210–211.

[57] This is the theme of *TC&HV*.

[58] "Toward an Integrated Understanding of Time," 2nd ed. J. T. Fraser, ed. *The Voices of Time*, Amherst: University of Massachusetts Press, 1981, pp. xxv–xlix.

[59] Even natural languages may be said to have personalities. "Escapement" is a toothed wheel used in clocks. It controls the motion of the hands of the clock. The word carries a connotation of gaining freedom. The German equivalent in *Hemmung*. It means "restraint" or "inhibition." It carries a connotation of losing freedom. In describing a mechanical and presumably emotionally neutral device, two languages perceive in its function two diametrically opposing states of human condition.

[60] Alexander J. Argyros, "The Minimal Epistemological and Ontological Conditions for a Theory of Systemic Interdisciplinarity." *Philosophica*, 48 (1991, 2), pp. 57–74 Quote from p. 57.

[61] Ibid. p. 72.

[62] Ibid. p. 59.

[63] Ibid. p. 73.

[64] An entry may be had through *TC&HV* or through the Argyros paper cited.

[65] Translated by Joseph Needham in his *History of Embryology*, Cambridge: Cambrige University Press, 1959, p. 93.

[66] *TC&HV*, pp. 26–43.

9. Coordinated Clock Shops

[1] *Genesis*: 1–11.

[2] The Founder's Lecture, July 4, 1992.

[3] The hierarchical theory of time, also known as the theory of time as conflict, has been elaborated in the author's books and published papers. A convenient entry may be had through Fraser (1987, 1990).

[4] Though often spoken of as new life, the birth of a child does not amount to the creation of life; it only illustrates a particular method of reproduction. The same holds for test-tube babies: they attest to technical virtuosity in stealing human eggs and sperms. One would be closer to demonstrating biogenesis if one could construct a phoenix from chemicals on the shelf. But the genetic alchemist would still have to convince others that his bird can reproduce by some known means and not by self-immolation and resurrection, and that it is a member of a species that could have been, but was not, created by organic evolution.

[5] Pasteur did not say anything about the origins of life. What he asserted, with reference to experiment and general scientific knowledge, was that there were no known circumstances in which it could be affirmed that microscopic beings came into the world without germs, without parents similar to themselves (Pasteur, 1864).

[6] "Democritus... says that of all the shapes the spherical is the most mobile, and that this is the shape of the particles of both fire and mind," Aristotle, *De Anima*, 405a–11. He regarded soul as something distributed throughout the body, resembling what today is loosely called human life.

[7] For guidance to recent work on long-range correlations that guarantee coherence, see the editorial "Long-range correlations within DNA" (Maddox, 1992, p. 103). The DNA molecule is structurally stable in spite of its wild dance because potentially reactive subgroups of the nucleotide bases are tucked inside and immobilized by its geometrical tightness.

[8] Here are some samples. "Biological clock: an innate mechanism by which living organisms are able to perceive the lapse of time" (Wallace, King, and Sanders, 1986, p. 1138).

"The internal mechanism of an organism that regulates circadian rhythms and various other periodic cycles" (Toothill, 1981, s.v. "Biological Clock"). Biological rhythms are "periodic biological fluctuations in an organism that correspond to, and are in response to periodic environmental change.... The internal mechanism by which such a rhythmic phenomenon occurs and is maintained even in the absence of the apparent environmental stimlus is termed a biological clock" (*Encyclopedia Britannica Micropaedia*, 1983a), s.v. "Biological rhythm"). "Living things are extraordinarily well adapted to their rhythmic environment and have become periodic in diverse aspects of their physiology" (*Encyclopedia Britannica Macropaedia*, 1983b, s.v. "Periodicity, Biological").

[9] The fastest ticking clocks are the molecules of the body, such as those of the skin. They respond to ultraviolet light at 10^{16} Hertz. Retinal cells respond to light between 10^{15} and 10^{14} Hertz. Photosynthetic processes that capture light energy and change it into forms of energy useful in the synthesis of organic compounds involve cyclic reactions with periods of 10^{-12} seconds. Periods of insect wing beats range between 10^{-4} and 10^{-3} seconds. Human vocalization occupies a frequency range between about 20,000 and 100 Hertz; that is, they have periods between 10^{-5} and tenths of a second. Periods of neural signals are between 3 and 10 seconds. The fastest growing bacteria reproduce every 600 seconds, cells divide at rates from 10^3 to 10^5 seconds. All living organisms, probably down to the genes, display circadian periods just below 10^5 seconds, hundreds of thousands of species show lunar periods of about 10^6 seconds, circannual rhythms of 10^7 seconds, and there are plants which flower every 13 or 17 years; that is, with a period around 10^8 seconds; for a graphical summary see Fraser (1987, p. 127).

[10] Einstein avoided dealing with the "now" by maintaining that although there is something essential about it, identifying what that essential feature is, is not within the tasks of science (Schilpp, 1963, p. 37). He took time to be absolute as far as its flow is concerned, though relative in its measure. Building on the intuitive obviousness of time's passage and excusing himself from dealing with the "now," he constructed Special Relativity Theory by defining what one is to mean by now at a distance, provided one already knows what to mean by the "now" here. This was a way of smuggling biotemporality (and the higher temporalities) into the description of physical change.

[11] Joseph Needham, writing about mesoforms that occur between successive stable levels of organization, noted that between the living and nonliving realms the crystalline represented the highest degree of organization (1944, p. 255). See also his *Order and Life* (1968, p. 158 and passim). Bernal called for a generalized crystallography as the key to the biology involved in the origins of life (1951, p. 34; 1967, p. 192). A. G. Cairns-Smith carried these arguments further by maintaining that the ancestors of life were, in fact, crystalline structures that stood in for the later DNA-RNA-protein system of modern biochemistry (Maddox, 1985, p. 197).

[12] This evolutionary change has surviving examples. For instance, a tree needs a few years to notify its roots that its head has been cut off whereas, in advanced species, the notification proceeds rapidly. The tree's organic present is much broader than that of a man.

[13] This is analogous to the cost-of-reproduction argument of J. T. Bonner (1974) introduced on p. 68 and recurring throughout the book.

[14] Sexual union keeps reshuffling the genes of all successful individuals so that a virtually infinite number of combinations is produced, making the species prepared for a large variety of contingencies. Asexual reproduction remains useful in stable environments. A reliable introduction to the issues involved may be found in the works of John Maynard Smith.

[15] The essence of sexual reproduction is the exchange of genetic material. The development of two different forms of germ cells, a large, stationary one with food material and a small, mobile sperm is a later development. With the coming about of sexual reproduction the cyclic order ceased to be the only form of life; to it was added the linear or aging order of life.

[16] Specific formulations of entropy principles exist for such diverse uses as steam tables, transfer of messages along radio links, learning behavior of rats, population pressure, and the economics of commodity production. There is no single unit or even physical dimension of entropy that would be common to these different uses.

[17] Information theory saves the idea of using entropy as a measure of directed change by using a sleight of hand. It is a particular representation of the future. It postulates an ensemble of possible future events and determines current entropy in terms of the uncertainties associated with them. This approach has proved itself useful in handling messages in languages whose statistical properties are known, but has little usefulness for calculating the entropy of a living sheep, for the statistical properties of sheep yet to evolve are not known.

[18] The actual changes inside the closed container are governed by the laws of the physical, biological, mental, and social worlds existing within, including an increase in negentropy. Since the Second Law is a statistical statement, it allows for transient trends of decreasing entropy for the processes taking place in the container.

[19] Why is it that the literature of time in physics, with very few exceptions, insists that our sense of time must be attached to increasing disorder? One can think of at least three reasons: (1) the provincialism and metaphysics of physical science that holds that anything as important as time must necessarily derive from physical phenomena; (2) a cultural setting which is much more impressed by the inorganic than by the organic; and (3) the association of time with the basically tragic view of life implicit in Christianity.

[20] We learn from Shakespeare's *As You Like It* (1600 ca.) that

> And so from hour to hour, we ripe and ripe,
> And then from hour to hour, we rot, and rot,
> And thereby hangs a tale.
>
> (Act II, scene 7, line 26)

Although youth is taken to be the age of ripening and old age that of rotting, in the life process ripening and the rotting take place simultaneously, kept coherent by the synchronization of its clockshops.

10. From Puppy Love to Faithful Love

[1] "The Tempest," iv–i–188. For this remarkable archeology, including the role of Francis Galton in the history of nurture vs. nature debate, I am indebted to Matt Ridley's *Nature via Nurture* (New York: Harper, 2003). See its chapter, "A Convenient Jingle."

[2] Recognition of the importance of selfhood is not to be confused with the political concept of individualism.

[3] Compare this with the random distribution of menstrual periods in women, entrained to circa-lunadian rhythm.

11. Logos at the Edge of the Cosmos

[1] Relativity theory is silent on the origins of time. It only gives instructions on how to measure time and remain consistent with motional and positional variations of clocks, whose nature also remains unanalysed. Newtonian (absolute) and Einsteinian (relativistic) time are identical in that they are both independent of the quality and complexity of the observer. This fact is never noted in physics texts because of the temperament of physical science, a projection of the unresolved conflicts of the physicist (Fraser, 1975). If the interpretation of time in physics, as given by physicists, is uncritically accepted and the vocabulary of relativity theory is carried over to psychoanalytic theory (e.g. Schneider, 1948), confusion results. Even such a meticulous scholar as Abraham (1976) mistakes the relativistic metaphor of the variability of the time metric for a qualitative, rather than quantitative, statement about time.

[2] The German term conjures up a dynamic image best rendered into English as 'fellow traveller' rather than the usual 'double' which corresponds to a static image. Here I wish to emphasize the open and dynamic rather than the static and closed nature of this eccentric projection.

[3] 'Time travels in diverse paces with diverse persons. I'll tell you who Time ambles withal, who Time trots withal, who Time gallops withal, and who he stands withal.' (*As You Like It*).

[4] J. J. R. Tolkien in his delightful epic *The Lord of the Rings* gave a literary representation of the difference between ideational and biological tempos. The trees of the story spoke 'old Entish' which, as one tree put it 'is a lovely language, but it takes a very long time to say anything in it, because we do not say anything in it, unless it is worth taking a long time to say and listen to it'.

[5] Ungebändigt immer vorwärts dringt.' Goethe's *Faust I*, Scene 4.

12. Unbounding Society

* In its first form, this paper was read at the meeting of the American Society for Eighteenth-Century Studies, organized and conducted by Samuel L. Macey at the University of Victoria in May, 1977. At that time, when teaching the history of science, I noted the absence of source material on what the great figures of Enlightenment science had actually said about time and the age and size of the world, and wondered how their views related to the literary currents of the epoch. Through the courtesy of the late Dr. Bern Dibner, founder of the Burndy Library in the History of Science, then in Norwalk, Connecticut, I was given the opportunity to read the works of Kepler, Huygens, Euler and other great figures of seventeenth- and eighteenth-century science in their original languages. The resulting survey is offered in honor of Professor Macey, who inspired the adventure.

[1] In *De Stella Nova in Pede Serpentarii*. See Johannes Kepler, *Gesammelte Werke*, ed. M. Caspar (München: Beck'sche Verlagsbuchhandlung, 1938), Chapter xxi, 251–57.

[2] Christian Huygens, *Cosmostheoros*. See Book II of the English translation: *Celestial Worlds Discover'd* (London: Childe, 1698).

[3] On this, see Alexander Koyrè, *From Closed World to the Infinite Universe* (Baltimore: The Johns Hopkins Press, 1957). Also, G. J. Whitrow, "Kant and the Extragalactic Nebulae," *Quarterly Journal of the Royal Astronomical Society*, 8 (1967): 48–56.

[4] In *Philosophical Transactions*, 29 (1714–16): 390–92. Cf. his two papers in 1720, "Of the Infinity of the Sphere of Fix'd Stars," *Philosophical Transactions*, 31, pp. 22–26. In these he considered the infinity of space as an unquestionable truth and assumed, though held debatable, that matter does fill space totally. The papers are reprinted in S. L. Jaki, *The Paradox of Olber's Paradox* (New York: Herder and Herder, 1969) 249–52.

[5] See his *Essai de Cosmologie*, Part III, "Spéctacle de l'Univers," also *Discours sur les Différentes Figures des Astres*. Both are in *Oeuvres* (Lyon, 1768).

[6] *The Principia* (1734), trans. J. R. Rendell and Isaiah Tansely, (London: The Swedenborg Society, 1912). See volume 2, chapters II and IV.

[7] Thomas Wright, *An Original Theory or New Hypothesis of the Universe Founded upon the Laws of Nature*... (London: Chapelle, 1750). Appropriately, Wright begins by quoting Pope from *An Essay on Man* (I.24–28):

See worlds on worlds compose one universe,
Observe how system into system runs,
What other planets circle other suns,
What vary'd being peoples ev'ry star,
May tell why Heav'n has made us as we are.

[8] H. G. Alexander, ed. *The Leibniz-Clarke Correspondence* (Manchester: Manchester University Press, 1956).

[9] The issues involved are numerous. See, for example, "Time, Space, and Free Will: the Leibniz Clarke Correspondence," by Denis Corish, in *The Study of Time III*, J. T. Fraser, D. Park and N. Lawrence, eds. (New York and Heidelberg: Springer Verlag, 1978). For the reflection of the clockwork universe in Enlightenment literature see Samuel L. Macey, "Poets and the Clock Metaphor," in his *Clocks and the Cosmos: Time in Western Life and Thought* (Hamden, CT.: Archon Books, 1980) 123–66. Also, "The Watchmaker God's Clockwork: Black or White" in his *Patriarchs of Time: Dualism in Saturn-Cronus, Father Time, The Watchmaker God, and Father Christmas* (Athens and London: University of Georgia Press, 1987) 77–110.

[10] See Euler's essay "Reflections on Space and Time" (1748), reprinted in Arnold Koslow, ed. *The Changeless Order* (New York: Braziller, 1967) 115–25. Also, C. Truesdell, "Leonhard Euler, Supreme Geometer" in H. E. Pagliaror, ed. *Irrationalism in the Eighteenth Century* (Cleveland: Case Western University Press, 1972) 51ff.

[11] *An Essay Concerning Human Understanding*, ed. A. C. Fraser (1894; rpt. New York: Dover, 1959) Book II, Chapter 14.

[12] *On the Heavens*, Book I, Chapters 5–8. *Physica*, Book III, Chapter 4, 203–b–15 through Chapter 8, 208–a–26. Cf. Plato, *Philebus* 15 b, on the unlimited.

[13] *Ars Conjectandi* (1713) at the close of the section "Tractatus de Seriebus Infinitis"; reprinted in *A Source Book in Mathematics* (New York: McGraw Hill, 1929), ed. D. E. Smith 271 (trans. J.T.F.).

[14] See D. J. Struik, "Kepler as a Mathematician" in *Johann Kepler 1571–1630* (Baltimore: Williams and Wilkins, 1931) 39–57.

[15] In the *Brouillon Project* (1639), part of *Oeuvres des Desargue*, ed. M. Pudra (Paris: Leiber, 1864) I: 99ff.

[16] John Wallis, *Arithmetica Infinitorium* (1655) in *Opera Mathematica* (Oxford, 1695), I: 410.

[17] *Paradoxien des Unendlichen* (1850). In English, *Paradoxes of the Infinite*, trans. D. A. Steele (London: Routledge and Kegan Paul, 1950). The quotation is from par. 25 of the original, p. 101 of the translation.

[18] See the chapter, "Chronology: the Years of the World," in Samuel L. Macey, *The Dynamics of Progress: Time, Methods, and Measure* (Athens: University of Georgia Press, 1989) 41–61. Also my entry, s.v. "Calendar: Western" in *Encyclopedia of Time*, ed. Samuel L. Macey (New York: Garland, 1994).

[19] For the remarkable observations and deductions of Xenophanes about fossils (and for details of his theory of life, sea, and earth) see G. S. Kirk and J. E. Raven, *The Presocratic Philosphers* (Cambridge: Cambridge University Press, 1975) 175–79.

[20] On the "fossil dilemma" see F. C. Haber, "Fossils and Early Cosmology" and "Fossils and the Idea of Process of Time in Natural History," in Bentley Glass et al., eds., *Forerunners of Darwin, 1745–1859* (Baltimore: Johns Hopkins Press, 1968). Also, The Darwinian Revolution in the Concept of Time," in J. T. Fraser, G. H. Müller and F. C. Haber, eds. *The Study of Time I* (New York and Heidelberg: Springer Verlag, 1972).

[21] *Telliamed, or Discourse between an Indian Philosopher and a French Missionary on the Diminution of the Sea, the Formation of the Earth, the Origin of Man and Animals* (London: Osborne, 1750), 280.

[22] Ibid., 175.

[23] See F. C. Haber, *The Age of the World* (Baltimore: The Johns Hopkins Press, 1959), 115–36 or, directly, Comte de Buffon, *Epochs of Nature*, in his *Natural History, General and Particular*, trans. William Wood (London, 1812) II: 207–39.

[24] James Hutton, 'Theory of the Earth; or an Investigation of the Laws Observable in the Composition, Dissolution, and Restoration of the Land upon the Globe," *Transactions of the Royal Society of Edinburgh* 1 (1788) 304.

[25] From a paper read to the Académie des Sciences de Paris in 1744. Quoted, without

further identification of source, in Wolfgang Yourgrau and Stanley Madelstam, *Variational Principles in Dynamics and Quantum Theory* (Philadelphia: Saunders, 1968) 19.

[26] Bentley Glass, "Maupertuis, Pioneer of Genetics and Evolution," in Glass et al., *Forerunners of Darwin*, 58.

[27] See the two superb papers by F. C. Haber: "The Darwinian Revolution in the Concept of Time," in Glass et al., *Forerunners*; and "The Cathedral Clock and the Cosmological Clock Metaphor," in J. T. Fraser and N. Lawrence, eds. *The Study of Time II* (New York and Heidelberg: Springer Verlag, 1975) 393–413. See also the chapter on "Philosophers and the Clock Metaphor" in Samuel L. Macey, *Clocks and the Cosmos* (Hamden, CT.: Archon Books, 1980) 65–101.

[28] Cf. "Special Session on Timekeepers and Time," in *The Study of Time II.*

[29] On the intellectual history of timekeepers, see J. T. Fraser, *Of Time, Passion, and Knowledge*, 2nd ed. (Princeton: Princeton University Press, 1990) 47–71.

[30] My source is *The Borghesi Astronomical Clock (in the Museum of History and Technology)* by S. A. Bedini (Washington, D. C.: Smithsonian Institution, 1964), Paper 35. The long subtitle of the book is so telling that I quote it here *in toto*:

> The Most Recent Perpetual, Astronomical Calendar Clock: Theoretical—Practical: by means of which besides the hours, the minutes and seconds; the current year, the month; the day of the month and the day of the week; the dominical letter, epact, and thence, the day of all feast days, both fixed and movable; the solar cycle; the golden number; the Roman indiction; the dominant planet of any year and its sign; the phases of the moon and its mean age; and all the motions of the sun and the moon as to longitude, latitude, eccentricity, etc., are immediately seen, so accurately that [not only] the true new full moons and the true quadrature, etc., of the sun and the moon appear, but also, all solar and lunar eclipses—both visible and invisible; as in heaven, so on the clock, they are conspicuously celebrated in their true times, and those of the past and those of the future, with their circumstances of time and duration, magnitude, etc., can be quickly determined. All this was devised and brought to light by the author... (trans. Bedini, 39).

[31] Ibid., 56.

[32] Ibid., 31.

[33] Cf. J. G. A. Pocock's paper on aspects of continuity and contingency (being and becoming) in a context quite different from that of this paper: "Modes of Political and Historical Time in Early Eighteenth-Century England," *Studies in Eighteenth-Century Culture*, ed. R. C. Rosbottom (Madison: University of Wisconsin Press, 1976) 87–102.

[34] Haber, *The Age of the World*, 108–109.

[35] For an evaluation of La Mettrie's principles and work, see Aram Vartanian, *La Mettrie's L'Homme Machine, a Study in the Origins of an Idea* (Princeton: Princeton University Press, 1960).

[36] David Hume, *Dialogues Concerning Natural Religion* (1779), in *The English Philosophers from Bacon to Mill*, ed. E. A. Burtt (New York: The Modern Library, 1939), 726.

[37] For a reflection of the Darwinian revolution in literature see Samuel L. Macey's chapter, "Augustan Clockwork and Romantic Organicism," in his *Clocks and the Cosmos*, 167–81.

[38] Bentley Glass, "Heredity and Variation in the Eighteenth-Century Concept of the Species," in *Forerunners of Darwin*, 144–72.

[39] A. O. Lovejoy, "Kant and Evolution," in *Forerunners of Darwin*, 173–206.

[40] Adam Smith, *The Wealth of Nations*, Chapter 2.

[41] Gotthold Ephraim Lessing, *The Education of the Human Race*, trans. F. W. Robertson (London: Anthroposophical Publishing Company, 1927) verses 85, 91, 100.

[42] Immanuel Kant, *The Idea of a Universal History*, in *On History*, ed. L. W. Beck (New York: Bobbs Merrill, 1963) 18.

[43] Translated in *Kant's Cosmogony* by W. Hastie (Westport, CT.: Greenwood Press, 1968).

[44] G. J. Whitrow, "Kant and the Extragalactic Nebulae."

[45] The various quotations in this section are from *Kant's Cosmogony*, 47–48 and 123–33.

[46] IV.40–42.

[47] See *Mountain Gloom and Mountain Glory: The Development of the Aesthetic of the Infinite* (Ithaca: Cornell University Press, 1959).

[48] "Die Alpen," quoted in Kant's *Cosmogony*, 134.

[49] See Gregor Sebba's superb essay on "Time and the Modern Self: Descartes, Rousseau, Beckett," in *The Study of Time I*, ed. J. T. Fraser, F. C. Haber, and G. H. Müller (New York and Heidelberg: Springer Verlag, 1972) 452–69. Cf. M. L. Perkins, "Destiny, Sentiment, and Time in the *Confessions* of Jean Jacques Rousseau," in *Studies in Voltaire and the Eighteenth Century*, 67 (Genève: Institut et Musée Voltaire, 1969).

[50] For a full discussion of this issue which within the confines of this paper must remain obscure though clearly relevant, see J. T. Fraser, *Time as Conflict* (Basel: Birkhäuser Verlag, 1978).

[51] *Discourse sur l'Origin et les Fondements de l'Inégalité parmi les Hommes*. In *Oeuvres Complètes* (Paris: Furne, 1848) I: 526–78. This is an ideal edition for those who appreciate the idyllic engravings that nineteenth-century readers identified with the romanticism of Rousseau.

[52] See, e.g., J. W. Smeed, *Faust in Literature* (New York: Oxford University Press, 1975).

[53] For a recent analysis see Eviatar Zerubavel, "The French Republican Calendar: a Case Study in the Sociology of Time." Paper presented at the 71st Annual Meeting of the American Sociological Association, New York, 1976.

[54] The Posthumous History of the Enlightenment," in *Varieties of Literary Experience*, ed. Stanley Burnshaw (New York: New York University Press, 1962) 387–405.

13. That Awesome Gift

[1] For the principles and development of the hierarchical theory of time, first formulated in 1970, see "Perspectives upon Time and Conflict" in J. T. Fraser, *Time, Conflict, and Human Values*, Urbana: University of Illinois Press, 1999, pp. 21–43. The theory jettisons the identification of time entirely with the human experience of time. Such an identification—conventional and seldom as much as noted—has been imposing upon the study of time the same type of limitations as did, in its epoch, the Ptolemaic, earth-centered model of the universe upon an understanding of the cosmos. The theory also takes advantage of the striking similarities between the structure and properties of mathematics on the one hand and, on the other hand, the structures and processes of nature at large. See, J. T. Fraser, "Time and Mathematics" in Paul Harris and Michael Crawford, eds. *Time and Uncertainty—The Study of Time XI.* In preparation

[2] The debate is set forth in smoothly flowing translation in *Erasmus-Luther—Discourse on Free Will*, tr., and ed. Ernest F. Winter, New York: Ungar, 1961.

[3] *Lucretius on the Nature of the Universe*, Baltimore: Penguin Books, 1962, p. 92.

[4] Lawrence Fagg, *The Becoming of Time*, Toronto: The Scholars Press, 1995, p. 27.

[5] G. J. Whitrow, *The Natural Philosophy of Time*, 2nd ed. Oxford: Clarendon Press, 1980. p. 358. Also, "At the speed of light, a clock would come to a standstill." Wolfgang Rindler, "Time from Newton to Einstein to Friedman," *KronoScope*, No. 1–2 (2001), p. 68.

[6] For a pragmatic, operational definition of "reality," as that concept is used here, see J. T. Fraser, "The extended umwelt principle: Uexküll and the Nature of Time," *Semiotica*, v. 134–1/4 (2001), pp. 263–273.

[7] Wolfgang. Rindler, "Visual horizons in world-models," *Monthly Notices of the Royal Astronomical Society* 116, 662 (1956). For summary surveys see also Whitrow, op. cit. 307–13,

and Edward R. Harrison, *Cosmology—The Science of the Universe*, Cambridge: Cambridge University Press, 1981, pp. 369–87.

[8] Harrison, op. cit. p. 355.

[9] John D. Barrow, *The Book of Nothing*, New York: Pantheon, 2000, p. 230.

[10] The $10^{1,000,000}$ figure is from Gerald M. Edelman and Giulio Tononi, *A Universe of Consciousness: How Matter Becomes Imagination*, New York: Basic Books, 2000. p. 38. "Neural complexity" is discussed on pp. 130–4 and passim. Estimates much higher than the $10^{1,000,000}$ may be found in "The human brain: boundary to biological complexity," in J. T. Fraser, *Time, the Familiar Stranger*, Amherst: The University of Massachusetts Press, 1987, pp. 149–160. The "most complicated object" is cited from Gerald M. Edelman, *Bright Air, Brilliant Fire*, New York: Basic Books, 1991, p. 17.

[11] For the reasoning and assumptions involved, see, "Complexity and its Measure" in J. T. Fraser, *Time, Conflict, and Human Values*, Chicago: University of Illinois Press, 1999, pp. 235–42.

[12] On the early history of the human brain as the boundary of complexity and on the relation of this complexity to imagination, see J. T. Fraser, *Of Time, Passion, and Knowledge*. Princeton: Princeton University Press, 1975 and 1990, pp. 257–261.

[13] Harrison, op. cit. 360.

[14] In the hierarchical theory of time the term "timelessness" is reserved to signify the description of certain psychological experiences. See the section "The Ecstasies of Timelessness," in J. T. Fraser, *Time, the Familiar Stranger*, op. cit. pp. 296–302.

[15] This law of nature is usually identified as the Principle of Uncertainty. The name suggests that as our knowledge expands, uncertainty may be replaced by certainty. It speaks of an epistemic condition. But the law reflects an ontological condition. For that reason I prefer the name "indeterminacy" to identify the principle. This word is also the correct translation of the original German *Unbestimmtheit*.

[16] *Classical Dynamics and its Quantum Analogues*, New York: Springer Verlag, 1979, p. 1.

[17] Einstein believed that there was something essential and very important about the "now" but that whatever it was, it remained outside the realm of science. By "science" he meant physics. See, P. A. Schilpp, ed. *The Philosophy of Rudolf Carnap*, La Salle, Ill,: Open Court, 1963, p. 37.

The "sense of Now and the so called 'passage of time' derive their meaning from aspects of reality that cannot be expressed in terms referring to spacetime, which are the terms that physicists ordinarily use." David Park, "Brain Time and Mind Time," in *Time, Mind, and Behavior*, ed. J. A. Michon and J. L. Jackson, New York: Springer Verlag, 1985, p. 62.

"All the successful equations of physics are symmetrical in time.... The future and the past seem physically to be on a completely equal footing." Roger Penrose, *The Emperor's New Mind: Concerning Computers, Minds, and the Laws of Physics*. New York: Oxford University Press, 1989, p. 302.

[18] If every creature defines—creates and maintains—its own organic present, with respect to which future and past may acquire meaning, then how does one explain the fact that the "now" of all living things in the room, all organisms on the globe, all astronauts in the solar system and all galaxies in the universe are simultaneous?

Answer: Instants of "now" are not obviously and necessarily shared instants.

If every living organism divides the nonpresent categories of time into futurity and pastness, mentally and/or biologically, then how do we explain the fact that time's flow is the self same passage for all fleas on a dog, all dogs on the block, all creatures on earth and all astronauts in the solar system and all galaxies in the universe?

Answer: Time is not the self-same passage for all places.

See: J. T. Fraser, *Time, Conflict, and Human Values*. Chicago: University of Illinois Press, 1999, pp. 66–7. For a detailed discussion of "The determination of time at a distance" see Whitrow, op. cit., pp. 230–245.

[19] Whitrow, op. cit. p. 328.

[20] Ernst Mayr, *Toward a New Philosophy of Biology*, Cambridge, Mass.: Harvard University Press, 1988, p. 45.

[21] The "morphological distinction [between the sexes] is bound to find expression in differences of psychical development. 'Anatomy is destiny', to vary a saying of Napoleon." In "Dissolution of the Oedipus Complex." In *The Standard Edition of the Complete Psychological Works of Sigmund Freud*. James Strachey, tr. and ed. London: The Hogarth Press, 1966–74, v. 19, p. 178. I have not been able to identify the saying of Napoleon, referred to by Freud.

[22] Translated by Joseph Needham in his *History of Embryology*, Cambridge: Cambridge University Press, 1959. p. 93.

[23] S.G.F. Brandon, "A New Awareness of Time and History," paper read at the Edward F. Gallahue Conference on World Religions held in 1966 at the Princeton Theological Seminary. In: Edward J. Jurji, ed. *Religious Pluralism and World Community—Interfaith and Intercultural Communications*. Leiden: Brill, 1969. pp. 225–236. Quote from p. 228.

[24] *The City of God*, tr. Marcus Dodd, NY: Modern Library, 1950 p. 350.

[25] James Ussher, *Annales Veteris et Novi Testamenti a Prima Mundi Origine Deducti....* Bremae, 1686, p. 1.

[26] Whitrow, op. cit. p. 375.

[27] Alexander Argyros in S. L. Macey, ed. *Encyclopedia of Time*, New York: Garland, 1994. *s.v.* "Fraser, J. T."

[28] See Chapter 4 in this book.

[29] Paul A. Harris, in a review of *Time, Conflict, and Human Values*, in *KronoScope* v. 2:1 (2002) p. 120.

[30] W. Norris Clarke, "Is Natural Theology Still Possible Today?" in R. J. Russell, W. R. Stoeger and G. E. Coyne, eds. *Physics, Philosophy, and Theology: A Common Quest for Understanding*. Vatican City State: Vatican Observatory, 1988, pp. 103–23, quote from p. 211.

[31] *The Saviors of God—Spiritual Exercises*, New York: Simon and Schuster 1960, p. 106.

14. Opiates that Civilize

[1] Marx, Karl, *Early Texts*. Translated and edited by David McLellan. New York: Barnes & Noble, 1971. Both quotes are on p. 116.

[2] The S. G. F. Brandon, "The Deification of Time," in J. T. Fraser et al, eds. *The Study of Time*, New York: Springer Verlag, 1972, pp. 370–382. The quote is from p. 371.

[3] S. G. F. Brandon, *History, Time and Deity—A Historical and Comparative Study of the Conception of Time in Religious Thought and Practice*. Manchester: Manchester University Press, 1965.

[4] Robert Herrick (1591–1674), "To the Virgins, to make Much of Time."

[5] Opening lecture of a series sponsored by the Penn Humanities Forum at the University of Pennsylvania, Philadelphia, Pennsylvania, September 25, 2001.

[6] First Letter to the Corinthians, 14:10–11.

[7] For a detailed discussion of the problems of integrated studies, see *The Voices of Time*, 2nd ed., Amherst: University of Massachusetts Press, 1981, pp. xxv–xlviii.

[8] For the propositions and a detailed, critical summary of the theory see "Perspectives upon Time and Conflict" in J. T. Fraser, *Time, Conflict, and Human Values*, Urbana: University of Illinois Press, 1999, pp. 21–43.

[9] For reasons why the human brain should be considered as a unique system of the natural world, section, "The Human Boundaries of Nature" in the essay, "Human Freedom," *KronoScope*, v. 2–2 (2002), pp. 227–8.

[10] For a schematic representation of this structuring and of the corresponding division of organic and mental labor, see Merlin Donald, *Origins of the Modern Mind*, Cambridge: Harvard University Press, 1991, Figure 5–1, p. 143.

[11] A crucial notion here is the idea of reality. On the changing boundaries of human reality, see J. T. Fraser, "The extended umwelt principle: Uexküll and the nature of time," in *Semiotica*, 134–1/4 (2001) pp. 263–73.

[12] First letter to the Corinthians 13:11.

[13] "Human Temporality in a Nowless Universe," *Time and Society*, 1(2) 1992, pp. 159–73. Also, "Time and the Origin of Life," in J. T. Fraser and M. P. Soulsby, eds. *Dimensions of Time and Life, (The Study of Time VIII)* Madison: International Universities Press, 1996, pp. 3–17.

[14] W. H. Auden, "Progress?" in *Thank You Fog—Last Poems*. New York: Random House, 1974, p. 18.

[15] "Fern Hill," in *The Poems of Dylan Thomas*, New York: New Directions, 1971, p. 195.

[16] William Shakespeare, *As you like it*, ii–vii–20.

[17] "What the stone cannot say about time: reflections on time and number" in J. T. Fraser, *Time, the Familiar Stranger*, Amherst: University of Massachusetts Press, 1987, pp. 282–9.

[18] In Sir Isaac Newton's Mathematical Principles of Natural Philosophy and his System of the *World*, tr. A. Motte, ed. F. Cajori. Berkeley: University of California Press, 1966, p. xiii.

[19] William Blake, *Jerusalem*. Princeton: Princeton University Press, [1991] p. 133.

15. How to Perpetuate Conflicts

* The Founder's Lecture, July 10, 1989.

[1] *Of Time, Passion, and Knowledge* (Fraser, 1990) remains the main reference. The dialectical nature of the theory was elaborated in *Time as Conflict* (1978a). The theory's implications for physics were the subject of *The Genesis and Evolution of Time* (1982). A good first entry is *Time, the Familiar Stranger* (1987).

[2] The idea of different temporalities is closely tied to the question of what is to be meant by their reality and by reality in general (see the definitions and explications given, with slightly different foci, in Fraser [1982, pp. 19–23; 1987, pp. 108–110; 1990, pp. 75–76]).

[3] The details are rather technical, but the rounded out numbers that can stand as measures of complexity on a logarithmic scale are as follows: 0 for the atemporal, 2 for the prototemporal, 6–7 for the eotemporal, 30–40 for the biotemporal, 10^9 for the nootemporal (see Fraser [1982, pp. 154–156, 167]).

[4] On this see Fraser ([1987, pp. 136–142]; index entries under "Biogenesis" [1990]; the chapter on "Biogenesis and Organic Evolution" [1982, pp. 144–175]).

[5] The principle of unresolvable conflicts may be extended down into the domain of undirected temporalities, where it takes the form of opposition between entropy decreasing and increasing processes. Tracing these processes through physical cosmology to their cosmic origins suggests that what arose from the primeval chaos was not order but conflict between orderability and disorderability. This new perspective upon the standard model of cosmic creation—when traced forward in time—permits the recognition of continuity between inorganic and organic evolution and leads to an interpretation of biogenesis in the temporal domain (see Fraser [1978a] index entries under "Conflicts/Resolutions/Fiascos" and Fraser [1990] under "Conflicts").

[6] "Umwelt" is a German word first used by the theoretical biologist Jakob von Uexküll to signify a concept that has been variously translated as "self-world," "phenomenal world," and "species-specific universe." Because of the awkwardness of all these translations, the word became naturalized in English. The most satisfactory definition of "umwelt" is given in English and English (1958, s.v. "umwelt") as "the circumscribed portion of the environment which is meaningful and effective for a given animal species...." Note that the "environment" of which the animal's umwelt is a portion, is our own human unwelt (see Fraser [1988, pp. 108–110]).

[7] The strong human desire for identifying all things permanent is beautifully illustrated by a view held by Albert Einstein. He believed that the absence of deterministic causation in the quantum world, expressed in the probabilistic laws of particle behavior, is not a statement about the true behavior of particles but a sign of our ignorance of their true laws which, he felt, should permit precise predictability. Contemporary view firmly holds that the probabilistic nature of predictability in the quantum world is correct.

[8] Recognizing the ontological priority of motion to rest removes the reasons that have made Zeno's flying arrow paradoxical. Namely, contrary to all prior formulations that I know of, the question one must ask is not how to construct motion from elements of rest but rather, how to construct rest from elements of motion—and that is an easy task (see Fraser [1987, pp. 27, 254–255]).

[9] The central theme of Penrose (1989), the inadequacy of contemporary mathematical physics for the handling of mental functions, resembles but is not identical with what is being said here.

[10] The phrase, "the ascent of man" is a reference to Bronowski's popular television series and book by that title (1973). The phrase reflects an Enlightenment view which, in its current use, is associated with the belief that in the scientific method humanity has finally found a way for conquering all ills that plague mankind.

[11] The term is due to the French man of letters, Romain Rolland. It is described by Freud, who paraphrased a part of a letter he received from the author. It is "a peculiar feeling, which he himself [Romain Rolland] is never without, which he finds confirmed by many others... which he would like to call a sensation of 'eternity,' a feeling of something limitless, unbounded—as it were 'oceanic'" (Freud, 1930, p. 64).

[12] For a superb essay on the recognition of these moods in the literary arts see Soulsby (chapter 11, this volume). For a related analysis from the point of view of Derridian deconstruction, see Argyros (1991).

[13] If convincing reasons could be marshaled, one could measure anger by the distance that a person could throw a standard hand ball when in that state of mind. But the measurement would still be a comparison of like with like: distance with distance.

16. The True

[1] *New York Times*, Oct 29, 2005. Page A12.

[2] The Gospel of John 8:32.

[3] Donald Gordon, "'Veritas Filia Temporis' : Hadrianus Junius and Geoffrey Whitney," *Journal of the Warburg and Courtland Institutes*, vol. 3, no. 3–4 (1940) pp. 228–240. Quotes from pp. 228 and 229.

[4] For a detailed discussion of this dynamic role of the search for truth see "The Many Kinds of Truths: Guidelines for Beliefs" in J. T. Fraser, *Time, Conflict, and Human Values*. Urbana and Chicago: University of Illinois Press, 1999, pp. 44–73.

17. Music Do I Hear?

[1] Plotinus, *Ennead II*. 9.16.

[2] *Hesiod, the Homeric Hymns and Homeridae*, (Loeb Classical Library No. 57). Cambridge: Harvard University Press, 1954, p. 336.

[3] "Phänomenologische und psychiatrische Untersuchung der Homerischen Träume," *Acta Psychoteraeutica, Psychosomatica et Orthopaedagogica*, v. 3 (1955), pp. 205–19.

[4] For a summary discussion see D. D. Kirtley, *The Psychology of Blindness*, Chicago: Nelson-hall, 1975, pp. 179–80.

[5] Tragedy, that most time-conscious of all dramatic forms, is a post-Homeric refinement of Homer. See Jacqueline de Romilly, *Time in Greek Tragedy*. Ithaca: Cornell University Press, 1968.

[6] "The Maker" in *The Aleph and Other Stories*. Edited and translated by Thomas di Giovanni, New York: Dutton, 1970, pp. 155–7.

[7] Jorge Luis Borges, *Dreamtigers*. Austin: University of Texas Press, 1964, p. 93.

[8] See "Being, Becoming and Existential Tension" in J. T. Fraser, *Of Time, Passion, and Knowledge*, Princeton: Princeton University Press, 1990, pp. 43–6 also "Becoming before Being" J. T. Fraser, *Time, Conflict, and Human Values*. Chicago: University of Illinois Press, 1999, pp. 61–4.

[9] It is a 7.2 cm tall figurine carved from green serpentine. She has her right hand on her hip, her upper body is twisted so that her left breast is seen in profile. Galgenberg is the name of the Austrian village where it was found in 1988.

[10] Robert Browning, "The Pied Piper of Hamelin," Stanza XII.

[11] The relevant literature is extensive, though scattered. An entry may be had through www.pitt.edu/~dash/hameln.html

[12] On this, see "Dance, then, wherever you may be…" in J. T. Fraser, *Time, Conflict and Human Values*, Chicago: University of Illinois Press, 1999, pp. 133–6.

[13] The quote is from the summary of "Time and the Origin of Life," of Chapter 9.

18. A Different Wonder

[1] *The History—Herodotus*. Translated by David Greene. Chicago: University of Chicago Press, 1987, p. 186. Cheops was so wicked, added Herodotus, that when he needed more money for the Pyramid he sent his daughter into prostitution. But she "got the idea of leaving some memorial of her own; and so she asked each man that came at her to make her a present of one stone in the works, and from these stones, they say, a pyramid was built…in front of the great pyramid. Each side of it measures one-hundred and fifty feet." Ibid. 187.

[2] These thumbnail sketches of the Seven Wonders of the World are based on the respective articles of *Encyclopedia Britannica Online*.

[3] These thoughts are extensions of the opening lines of the chapter, "Ask the Stone to Say," in J. T. Fraser, *Time, the Familiar Stranger*. Amherst: University of Massachusetts Press, 1987, p. 222.

[4] Founder's Lecture. Fifth Conference of the International Society for the Study of Time, July 5, 1983.

[5] An example of criminal optimism is a forty-two-page supplement to the May 9, 1983, issue of *U.S. News & World Report* entitled, "What the Next 50 Years Will Bring." Many of its glorious predictions are plainly contradicted by other reports in the same issue, recording contemporary facts. The reader cannot help wondering how we are supposed to get from here and now, to there and then.

[6] Joseph Needham, "The Roles of Europe and China in the Evolution of Oecumenical Science," *Advancement of Science* 24 (1967–68): 11.

[7] J. T. Fraser, *The Genesis and Evolution of Time* (Amherst: University of Massachusetts Press, 1982); "Toward an Integrated Understanding of Time," in *The Voices of Time*, ed. J. T. Fraser, 2d ed. (Amherst: University of Massachusetts Press, 1981), pp. xxv–xlix; *Of Time, Passion and Knowledge* (New York: Braziller, 1975).

[8] Ludwig Feuerbach, *Principles of the Philosophy of Future*, trans. M. H. Vogel (Indianapolis: Bobbs-Merril, 1966), par. 59, p. 71.

[9] Joseph Needham, "The Roles of Europe and China in the Evolution of Oecumenical Science," p. 1.

[10] N. Sivin, "Why the Scientific Revolution Did Not Take Place in China—or Didn't It?" *Chinese Science* 5 (1982): 60.

[11] Benjamin Farrington, *Greek Science* (Baltimore: Penguin Press, 1969), pp. 48–49.

[12] Joseph Needham, "Integrative Levels: A Revaluation of the Idea of Progress," in *Time: The Refreshing River* (London: Allen and Unwin, 1944), p. 237.

[13] Augustine, *On Free Choice of the Will*, trans. A. S. Benjamin and L. H. Hacksfatt (Indianapolis: Bobbs-Merril, 1964), bk. 2, sec. 11: "How are the rules of number and wisdom related." The quote is from verse 129.

[14] Johannes Kepler, *Gesammelte Werke*, ed. Walter von Dyck and Max Caspar (München: C. H. Beck'sche Verlagsbuchhandlung, 1938–75), 8: 30 n. 8, and 6: 223.

[15] Albert Einstein, "Autobiographical Notes," in *Albert Einstein, Philosopher Scientist*, ed. P. Schilpp (New York: Tudor, 1949), p. 7.

[16] C. W. Misner, K. P. Thorne, and J. A. Wheeler, *Gravitation* (San Francisco: Freeman, 1973), p. 779.

[17] J. A. Wheeler, "Frontiers of Time," in *Proceedings of the International School of Physics, "Enrico Fermi," Course* 72 (Amsterdam: North Holland, 1979), p. 438.

[18] On the relationship between individual and collective personalities and preferred ways of knowledge, see the chapter on "Epistemology and the True," in Fraser, *Of Time, Passion, and Knowledge*, pp. 321–60, and its extensive references.

[19] G. J. Whitrow, *The Natural Philosophy of Time*, 2d ed. (Oxford: Clarendon Press, 1980), p. 1.

[20] Kepler, *Gessammelte Werke*, "Briefe," 15: 141–42. Free translation.

[21] Max Weber, *The Protestant Ethic and the Spirit of Capitalism*, trans. Talcott Parsons (New York: Scribner, 1904), p. 249 n. 145.

[22] Sir William Petty, *The Petty Papers* (London: Constable and Co., 1927), 1: 193–98.

[23] See Joseph Needham, *Science and Civilisation in China*, vol. 2, *History of Scientific Thought* (Cambridge: Cambridge University Press, 1956, "The *Fa Chia* (Legalists)," pp. 204–15.

[24] Ibid., p. 210.

[25] Joseph Needham, "Science and China's Influence on the World," in *The Legacy of China*, ed. Raymond Dawson (Oxford: Oxford University Press, 1964), p. 308. For a penetrating analysis of that intense and peculiar anxiety that came to inform post-Reformation Europe, to favor quantity substituted for quality, and to create the social temperament that was to build science and industry, see William J. Bouwsma's "Anxiety and the Formation of Early Modern Culture," in *After the Reformation*, ed. B. C. Malament (Philadelphia: University of Pennsylvania Press, 1980), pp. 215–46.

[26] The reference here is to the generalized umwelt principle of the hierarchical theory of time. See *The Genesis and Evolution of Time*, esp. sec. 2.1. See also John Michon, sec. 2.3, in this volume.

[27] Needham, *Science and Civilisation in China*, "Human Law and the Laws of Nature," 2: 582.

[28] For an entry to this debate, see Joseph Needham's essay, "Time and Knowledge in China and the West," in *The Voices of Time*, pp. 92–135.

[29] Ibid., p. 104, repeated in his message to this conference, included above.

[30] Benjamin Gal-Or, *Cosmology, Physics, Philosophy* (New York: Springer-Verlag, 1981), p. 41 (Gal-Or's italics).

[31] Needham, "The *Fa Chia* (Legalists)," p. 211.

[32] Eugene Wigner, *Symmetries and Reflections* (Bloomington: Indiana University Press, 1967), p. 222.

[33] See *The Genesis and Evolution of Time*, pp. 156–63; *Of Time, Passion and Knowledge*, pp. 294–99; *Time as Conflict* (Boston and Basel: Birkhäuser, 1978), index entries under "number"; and "Out of Plato's Cave: The Natural History of Time," *Kenyon Review*, n. s., 2 (1980): 143–62.

[34] Max Caspar and Walter von Dyck, *Kepler in Seinem Briefe* (Berlin: Oldenburg, 1930), 2: 308.

[35] In the hierarchical theory of time, the sociotemporal umwelt is not identical with "society," that is, a group of interacting people. It comprises, instead, the objects and processes generated by the symbolic transforms of human experience: the artifacts of industry and science (as structures and as functions), the creations of the arts and letters, and the myriad other forms that human communication can take.

[36] In Goethe's *Faust*, Wagner is the name of the protagonist's assistant, a character of mediocrity in spirit and mind.

[37] Tong B. Tang, "Will China Be Tainted by Western Science?" *Nature* 280 (1979): 100. By the same author, "China: Where the New and the Historical Materialism Interact"; "Research Centres, Libraries, Journals Flourish in New China"; and "How the Popularization of Science Narrows the Polarization of the People," ibid. 283 (1980): 423–24; 516–17; 616–17. See also the unsigned editorial, "Planting a Tall Tree," ibid. 301 (1983): 280–84; and Joseph Needham, "Science Reborn in China," ibid. 274 (1978): 832–34.

[38] From an address by Belisario Betancourt, President of the Republic of Colombia, to the United Nations, October 5, 1983.

20. Turmoil at the Anthill Threshold

* Founder's Lecture, July 6, 1998.

[1] The quote is from *Hamlet*, v–ii 395–7.

[2] *The Tempest*, IV–i–152–154 .

[3] Jorge Luis Borges, "The Library of Babel," in his *Labyrinths*, ed. Donald A. Yates and J. E. Irby (New York: New Directions, 1962). Quotes from pp. 11, 52 and 55.

[4] An entry to issues of time ecology may be had through Martin Held and Kartheiz A. Geissler, eds., *Ökologie der Zeit* (Stuttgart: Hirzel, 1993).

[5] For a detailed, critical discussion of the current crises in human values, see J. T. Fraser, *Time, Conflict, and Human Values*, Urbana and Chicago: University of Illinois Press, 1999.

[6] Paul Kennedy, *Preparing for the Twenty-first Century* (New York: Random House, 1993), 344 and 350.

[7] Winsor, Frederick, and Marian Parry, *The Space Child's Mother Goose* (New York: Simon and Schuster, 1963), 37. Reproduced with kind permission of Margaret W. Stubbs.

[8] Borges, "Library of Babel," 55.

[9] J. T. Fraser, "The Interdisciplinary Study of Time" in *Interdisciplinary Perspectives of Time*, [1966], Annals of the New York Academy of Sciences, vol. 138, art. 2. (1967), 822–847. Quotes from pp. 822, 843, and 844.

21. Whose Past is Our Prologue?

[1] "The Tempest," iv–i 110–112.

[2] "The Tempest," ii–i 259–261.

[3] Founder's Lecture, July 26, 2004, Conference of the International Society for the Study of Time, Clare College, Cambridge, England.

[4] "The Communist Internationale." The words are those of Eugene Pottier, a Parisian transport worker who, in 1871, wrote of "c'est la lutte final."

[5] "Midsummer-Night's Dream," iii–ii–115.

[6] Those youthful features, explained Michelangelo, were the result of her "never having entertained the slightest immodest thought which might have troubled her body." R. S. Liebert, *Michelangelo*. New Haven: Yale Universoity Press, 1983. pp. 67–8. An entry to that dialogue may be had through the comments on this Pietà in Umberto Baldini, *The Sculpture of Michelangelo*, New York: Rizzoli, 1982, pp. 39–40.

[7] J. T. Fraser, *Time, Conflict, and Human Values*. Chicago: University of Illinois Press, 1999, pp. 151–2. Also in the introduction to *Il tempo: una presenza sconosciuta*, Milano: Feltrinelli (1991) p. 7.

[8] Matt. xi, 28.

[9] "Give me your energetic, your rich / Your privileged yearning to be free / The executives of your teaming shore / Send these, the achievers to us: / We lift our lamp beside the golden door." Poem titled Invitation to New York Corporations Thinking of Moving Their Headquarters to Fairfield County. Author is E. J. Brennan, publisher, *Fairfield County Magazine*. 1975.

[10] T. S. Eliot, "Burnt Norton." In *Four Quartets*, New York: Harcourt, Brace & World, 1971, p. 14.

[11] S. G. F. Brandon, *Man and his Destiny in the Great Religions*. Manchester: Man chester University Press, 1962, p. 384. For a summary and continuation of this idea see his essay, "Time and the Destiny of Man," in J. T. Fraser. ed. *The Voices of Time*, 2nd ed. Amherst: University of Massachusetts Press, 1981, pp. 140–60.

[12] J. T. Fraser, ed. *The Voices of Time*, (1966) 2nd ed. Amherst: University of Massachusetts Press, p. xvii, 1981.

[13] This is No. 6 of his Holy Sonnets.

[14] J. T. Fraser, ed. *The Voices of Time* (1966) 2nd ed. Amherst: University of Massachusetts Press, 1981.

[15] 1 Corinthians 14: 10–11.

[16] New York Academy of Sciences, *Annals*. V. 138, Art. 2. "Interdisciplinary Perspectives of Time," pp. 822–48. Quote is from p. 845.

[17] G. J. Whitrow, "Foreword" in J. T. Fraser, F. C. Haber and G. H. Müller, eds. *The Study of Time*, Heidelberg, New York: Springer Verlag, 1972 p. v.

[18] Sam Walter Foss, "The House by the Side of the Road," Hazel Felleman, ed. The Best Loved Poems of the American People, Garden City: Garden City Publishing, 1936, p. 105.

[18] Joseph Needham, (1937) Reprinted in his, *Time, the Refreshing River*, London: George Allen and Unwin, 1943, pp. 233–72.

[19] "Complexity and its measure," (constructed with the help of algorithmic information theory) in J. T. Fraser, *Time, Conflict, and Human Values*, Chicago, University of Illinois Press, 1999, pp. 235–42.

[20] By "languages" is meant a coherent family of signs and symbols necessary to describe the structures and interpret the functions of the stable integrative levels of nature.

[21] J. T. Fraser, "Mathematics and Time," *KronoScope* 3–2 (2003) pp. 153–168.

[22] For its mature form, see "Perspectives on Time and Conflict," *ibid.*, pp. 21–43.

[23] Michael Heller, "The Origins of Time" in J. T. Fraser, N. Lawrence and D. Park, eds. *The Study of Time IV*, New York: Springer Verlag, 1981, pp. 90–93.

[24] P. C. W. Davies, *The Physics of Time Asymmetry*, U. Calif. Press, 1974, p. 21. The absence of time's flow and that of a present are corollaries. Namely, the flow of time must appeal to distinctions between future and past, and future and past have meaning only with reference to a present. If there are no distinctions between future and past, directed time can have no meaning. David Park even asked, "Should Physicists say that the Past Really Happened?" In J. T. Fraser, ed. *The Study of Time VI*. Madison, CT: International Universities Press, 1989, pp. 125–42.

[25] P. C. W. Davies, *The Physics of Time Asymmetry*, U. Calif. Press, 1974, p. 26.

[26] P. A. Schilpp, ed. *The Philosophy of Rudolf Carnap* (La Salle: Open Court, 1963) pp. 37–8.

[27] David Layzer, *Cosmogenesis—The Growth of Order in the Universe*. Oxford: Oxford University Press, 1990, p. 144.

[28] On this, see J. T. Fraser, "From Chaos to Conflict," J. T. Fraser, M. P. Soulsby and A. Argyros, *eds. Time, order, Chaos (The Study of Time IX)*, Madison: International Universities Press, 1998, pp. 3–17.

[29] G. J. Whitrow, *The Natural Philosophy of Time*, 2nd ed. Oxford: Clarendon Press, 1980, p. 253. See also "Conventionality of Simultaneity," in *Stanford Encyclopedia of Philosophy, http://plato,stanford.edu/entrieds/spacetime-comvensimul/*

[30] "And so from hour to hour we ripe and ripe / And then from hour to hour we rot and rot...." ("As You Like It," 2-7-26.).

[31] J. T. Fraser, "Temporal Levels and Reality Testing," *International Journal of Psycho-Analysis* (1981) v. 62, pp. 3–26.

[32] On the Kantian idea of time, see Charles M. Sherover's *The Human Experience of Time*, 2nd ed., Evanston, IL, Northwestern University Press, 2000, p. 109 ff.

[33] These lines are from Alfred Tennyson's "Ulysses."

[34] Walt Whitman, "Songs of Myself," No. 15 in *Complete Poetry and Selected Letters*, Emory Holloway, ed. London: The Nonesuch Press, 1938, p. 39.

[35] Alfred North Whitehead, *Adventures of Ideas*, New York: Macmillan, 1933, p. 41.

[36] Samuel P. Huntington, *The Clash of Civilizations and the Remaking of World Order*. New York: Simon and Schuster, 1996.

[37] "The past is not a frozen landscape that may be discovered and described once and for all, but a chart of landmarks and paths which is continuously redrawn in terms of new aspirations, values and understanding." In J. T. Fraser, *Of Time, Passion, and Knowledge*. 2nd ed. Princeton: Princeton University Press, 1990, p. xv. See also Robert Robertson, "The New Global History: History in a Global Age," in Scott Lash et al. eds. *Time and Value*, Oxford: Blackwell, 1988, pp. 210–27. J. Prager's essay, "Collective Memory, Psychology of" in N. J. Smelser and Paul B. Baltes eds. *International Encyclopedia of the Social & Behavioral Sciences*. Oxford: Pergamon, 2001, pp. 2223–2227.

[38] My inquiry has an ancestry in Thomas Ungvàri's "Time and the Modern Self," in J. T. Fraser, F.C. Haber and G.H. Müller, eds. *The Study of Time I* New York: Springer Verlag, 1972, pp. 470-478.

[39] Claudia Clausius, "Tragedy as Forgotten Memory in Wole Soyinka's Drama," paper at this conference.

[40] For a detailed discussion see "The Tragic," in J. T. Fraser, *Time, Conflict, and Human Values*. Chicago: University of Illinois Press, 199, pp. 157–62.

[41] Hamlet 1-5-188.

[42] "Hamlet" in Boris Pasternak, *Doctor Zhivago*, New York: Pantheon, 1958, p. 523.

[43] One of them sings the "Carmagnole," a folk dance, danced around the Guillotine. See J. T. Fraser, *Of Time, Passion, and Knowledge*, Princeton: Princeton University Press, (2nd ed.) 1990, pp. 392–3.

[44] *Selected Plays of Eugene O'Neill*, New York: Random House, 1940. p. 758.

[45] S. G. F. Brandon, "Time and the Destiny of Man," in J.T. Fraser, *The Voices of Time*, Amherst: University of Massachusetts Press, 1981, pp. 140–57. Quote from p. 154.

[46] Paul A. Harris, "www.timeandglobalization.com/narrative" in *Time and Society*, 9–2/3 (2000) pp. 319–29. The conference was jointly organized by ISST and the French Association for the Advancement of Science.

[47] In the chronology of 7980 years, used by astronomers, the minute the first airplane hit, was Julian Day 2,452,163.864,583,335. This cycle was devised in the 16th century by the Dutch philologist and chronologist Josephus Scaliger, to encompass all human history known in his time.

[48] Anne Schulenberger Lévy, "At the Limits of the Utopian Festival." *KronoScope* 4-1 (2004) pp. 75–91. Quote from p. 75.

[49] "Julius Caesar," III–I–111.

[50] "London." in *The Portable Blake*, New York: Viking, 1968, p. 112.

[51] Walter Russell Brain, "Time is the Essence..." *Lancet*, May 28, 1966.

[52] Walter Russell Brain, "Some Reflections on Brain and Mind." *Brain*, 86 (1963) p. 392.

[53] Robert Hillyer, *Collected Poems*, New York, 1961, p. 129.

22. Expanding the Universe

[1] Job 38:4–7.

[2] The statistics and the quotes are from "We Are the Web," article in *Wired* magazine http://wired-vig.wired.com/wired/archive/13.08/tech_pr.html

[3] On the dynamics of emergence see "A Nested Hierarchy of Causations, Language, Temporalities and Conflicts" in J. T. Fraser, *Time, Conflict, and Human Values*, Chicago: University of Illinois Press, 1999 pp. 26–43, especially p. 32 on the metastability of mesoforms.

[4] See the section, "From the Comintern to the Internet" in J. T. Fraser, *Time, Conflict, and Human Values*. Chicago: University of Illinois Press, 1999, pp. 204–9.

[5] "UN. Warns That Trafficking in Human Beings Is Growing." *The New York Times*, June 25, 2000 pp. 1 and 10. See also "Human Trafficking and Slavery," *The CQ Researcher*, March 26, 2004, p. 275.

[6] "Slavery in the 21st century." *Britannica Book of the Year, 2001*. Encyclopedia Britannica Premium Service, 23 March 2006. http://www.britannica.com/eb/article-9344443 Slavery is defined as "forced labor for little or no pay under the threat of violence."

[7] Cited from the Population Council of UNICEF. http://www.newint.org/issue337/facts.htm

[8] Rupert Brooke, *The Complete Poems*. London: Sidgwick and Jackson, 1942, p. 132.

[9] Frederick Winsor, *The Space Child's Mother Goose*, New York: Simon and Schuster, 1958, No. 37.

INDEX

www.ingramcontent.com/pod-product-compliance
Lightning Source LLC
LaVergne TN
LVHW080848300926
849162LV00003B/369